Stochastic Mechanics
Random Media
Signal Processing
and Image Synthesis
Mathematical Economics and Finance
Stochastic Optimization
Stochastic Control
Stochastic Models in Life Sciences

Stochastic Modelling and Applied Probability

(Formerly:
Applications of Mathematics)

8

A.N. Shiryaev

Optimal Stopping Rules

Translated by A.B. Aries

Reprint of the 1978 Edition
with a new Preface

Author

Albert N. Shiryaev

Steklov Mathematical Institute
Gubkina 8
119991 Moscow
Russia
albertsh@mi.ras.ru

Managing Editors

B. Rozovskii

Division of Applied Mathematics
Brown University
182 George Str.
Providence, RI, USA
rozovsky@dam.brown.edu

G. Grimmett

Centre for Mathematical Sciences
Wilberforce Road
Cambridge CB3 0WB
UK
g.r.grimmett@statslab.cam.ac.uk

Mathematics Subject Classification (2000): 60G40, 62L10, 62L15

Library of Congress Control Number: 2007934268

Soft cover reprint of the 1978 edition, with a new preface by the author
Translated from the Russian 2nd edition STATISTICHESKY POSLEDOVATELNY ANALYZ (OPTIMALNYJE PRAVILA OSTANOVKI) by Nauka, Moscow 1976

ISSN 0172-4568
ISBN 978-3-540-74010-0 Springer-Verlag Berlin Heidelberg New York

ISBN: 0-387-90256-2 hard cover Springer-Verlag, New York Inc.

Springer is a part of Springer Science+Business Media

springeronline.com

Printed on acid-free paper 41/3180/VTEX - 5 4 3 2 1 0

Preface to the Present Reprint

The Russian editions of this book were published in 1969 and 1976. The corresponding English editions appeared in 1973 and 1978 (by AMS and Springer respectively).

Although three decades have passed since first publication, the content remains up-to-date and interesting for many researchers as is shown by the many references to it in current publications. As mentioned in the original Preface, the "ground floor" of Optimal Stopping Theory was constructed by A. Wald in his *sequential analysis* in connection with the testing of statistical hypotheses by non-traditional (sequential) methods.

It was later discovered that these methods have, in idea, a close connection to the general theory of *stochastic optimization* for random processes.

The area of application of the Optimal Stopping Theory is very broad. It is sufficient at this point to emphasise that its methods are well tailored to the study of American (-type) options (in mathematics of finance and financial engineering), where a buyer has the freedom to exercise an option at any stopping time.

In this book, the general theory of the construction of optimal stopping policies is developed for the case of Markov processes in discrete and continuous time. Note that in the recent monograph "Optimal stopping and free-boundary problems" of G. Peskir and A. Shiryaev (Birkhäuser 2006), the reader may find substantial further material on the general theory of optimal stopping rules, with many new applications and new techniques for solving optimal stopping problems. In particular, material on the solution of free-boundary (Stefan) problems with detailed analysis of the "continuous fit" and "smooth fit" conditions that play a role in the additional conditions needed for solving these problems, is provided.

Moscow, July 2007 Albert N. Shiryaev

Preface

1

Along with conventional problems of statistics and probability, the investigation of problems occurring in what is now referred to as *stochastic theory of optimal control* also started in the 1940s and 1950s. One of the most advanced aspects of this theory is the theory of optimal stopping rules, the development of which was considerably stimulated by A. Wald, whose *Sequential Analysis*[1] was published in 1947.

In contrast to the classical methods of mathematical statistics, according to which the number of observations is fixed in advance, the methods of sequential analysis are characterized by the fact that the time at which the observations are terminated (*stopping time*) is random and is defined by the observer based on the data observed. A. Wald showed the advantage of sequential methods in the problem of testing (from independent observations) two simple hypotheses. He proved that such methods yield on the average a smaller number of observations than any other method using fixed sample size (and the same probabilities of wrong decisions). Furthermore, Wald described a specific sequential procedure based on his sequential probability ratio criterion which proved to be optimal in the class of all sequential methods.

By the *sequential method*, as applied to the problem of testing two simple hypotheses, we mean a rule according to which the time at which the observations are terminated is prescribed as well as the terminal decision as to which of the two hypotheses is true. It turns out that the problem of optimal terminal decision presents no particular difficulties and that the problem of finding the best sequential procedure can be reduced

[1]The Russian translation became available in 1960.

to that of finding the optimal stopping time for a Markov sequence constructed in a specific fashion (Sections 4.1, 4.2).

The necessity to use sequential methods did not seem very compelling in the problem of testing two simple hypotheses. However, the two problems given below require by their very nature, a sequential observation procedure, and associated optimal stopping times.

One such problem is the following optimal selection problem.

We are given n objects ordered in accordance with some common characteristic. We assume that the objects arrive in a random sequence. We wish to determine which object is the best one by pairwise comparison.

The problem is to optimize the selection scheme so as to maximize the probability of choosing the best object. (We assume that we have no access to the objects rejected.) We show in Section 2.3 that this problem can be also reduced to that of finding the optimal stopping time for a Markov chain.

The other problem (the so-called disruption problem: Sections 4.3, 4.4) is the following.

Let θ be a random variable taking on the values $0, 1, \ldots$, and let the observations $\xi_1, \xi_2 \ldots$ be such that for $\theta = n$ the variables $\xi_1, \xi_2, \ldots, \xi_{n-1}$ are independent and uniformly distributed with a distribution function $F_0(x)$, and $\xi_n, \xi_{n+1}, \ldots$ are also independent and uniformly distributed with a distribution function $F_1(x) \neq F_0(x)$. (Thus, the probability characteristics change in the observable process at time θ.) The problem is how to decide by observing the variables $\xi_1, \xi_2, \ldots$ at which instant of time one should give the "alarm signal" indicating the occurrence of discontinuity or disruption (in probabilistic terms). But this should be done as to (on the one hand) avoid a "false alarm," and (on the other hand) so that the interval between the "alarm signal" and the discontinuity occurrence (when the "alarm signal" is given correctly) is minimal. By analogy with the previous problems, the solution of this problem can be also reduced to finding the optimal stopping time for some Markov random sequence.

2

The present book deals with the general theory of optimal stopping rules for Markov processes with discrete and continuous time which enables us to solve, in particular, the problems mentioned above.

The general scheme of the book is the following.

Let $X = (x_n, \mathcal{F}_n, P_x)$, $n = 0, \ldots$, be a Markov chain[2] with state space $(E, \mathcal{B})$. Here x_n is the state of the chain at time n, the σ-algebra $\mathcal{F}_n$ is interpreted as the totality of events observed before time n inclusively, and P_x is the probability distribution corresponding to the initial state x. Let us assume that if we stop the observations at time n we shall have the gain

[2]The basic probabilistic concepts are given in Chapter 1.

$g(x_n)$. Then the average gain corresponding to the initial state x is the mathematical expectation $M_x g(x_n)$.

Next, let τ be a random variable taking on the values $0, 1, \ldots$ and such that the event $\{\tau = n\} \in \mathfrak{F}_n$ for each n.

We shall interpret τ as the instant of time at which the observations are terminated. Then the condition $\{\tau = n\} \in \mathfrak{F}_n$ implies that the solution of the problem whether the observations should be terminated at time n depends only on the events observed until and including the time n.

We shall consider the gain $M_x g(x_\tau)$ corresponding to the stopping time τ and the initial state x (assuming that the mathematical expectation $M_x g(x_\tau)$ is defined).

Set

$$s(x) = \sup_\tau M_x g(x_\tau).$$

The function $s(x)$ is said to be a payoff and the time τ_ε such that $s(x) \leqslant M_x g(x_{\tau_\varepsilon}) + \varepsilon$ for all $x \in E$ is said to be an ε-optimal time. The main questions discussed in this book are: What is the structure of the function $s(x)$?; How can this function be determined?; When do the ε-optimal and optimal (*i.e.*, 0-optimal) times coincide?; What is their structure?

Chapter 2 deals with the investigation of these questions for various classes of the functions $g(x)$ and various classes of the times τ (taking, in particular, the value $+\infty$, as well) for the case of discrete time.

Here is a typical result of this chapter. Let us assume that the function $g(x)$ is bounded, $|g(x)| \leqslant C \leqslant \infty$, $x \in E$. Then we can show that the payoff $s(x)$ is the smallest excessive majorant of the function $g(x)$, *i.e.*, the smallest function $f(x)$ satisfying the conditions:

$$g(x) \leqslant f(x), \qquad Tf(x) \leqslant f(x),$$

where $Tf(x) = M_x g(x_1)$.

The time

$$\tau_\varepsilon = \inf\{n \geqslant 0 \colon s(x_n) \leqslant g(x_n) + \varepsilon\}$$

is ε-optimal for any $\varepsilon > 0$, and the payoff $s(x)$ satisfies the equation

$$s(x) = \max\{g(x), Ts(x)\}.$$

Chapter 3 deals with the theory of optimal stopping rules for Markov processes (with continuous time). Most results obtained in this chapter are similar, at least formally, to the pertinent results related to the case of discrete time. We should, however, note that rather advanced tools of the theory of martingales and Markov processes with continuous time have been used in this chapter.

Chapter 1 is of an auxiliary nature. Here, the main concepts of probability theory and pertinent material from the theory of martingales and Markov processes are given and properties of Markov times and stopping times are detailed. Chapter 4 deals with the applications of the results of

Chapters 2 and 3 to the solution of the problem of sequential testing of two simple hypotheses and the problem of disruption for discrete and continuous time.

3

The structure of the present edition is similar to that of 1969; nevertheless, there is substantial difference in content. Chapter 3, which deals with the case of continuous time, has been changed significantly to take into account new and recent results.

Chapter 2 contains some new results as well. Also, simpler proofs are given for some lemmas and theorems.

Finally, we note that the references given consist mainly of textbooks and monographs. References to sources of new results as well as supplementary material can be found in the Notes at the end of each chapter. Each chapter has its own numeration of lemmas, theorems, and formulas; in referring to the lemmas and theorems within each chapter the chapter number is omitted.[3]

To conclude the Preface I wish to express my gratitude to A. N. Kolmogorov for introducing me to the study of sequential analysis and for his valuable advice. I am also grateful to B. I. Grigelionis for many useful discussions pertaining to sequential analysis. I am indebted to G. Yu. Engelbert and A. Engelbert for many helpful comments and suggestions in preparing this edition for publication. I would like also to thank N. N. Moisejev who initiated the writing of this book.

Moscow
March 1977

A. N. SHIRYAYEV

[3]*Editor's Note*: The author's numbering scheme is illustrated by the following examples. References made *in* Chapter 2 *to* Chapter 2 might take the form Theorem 15, Lemma 15, Section 15, Subsection 15.5 (*i.e.*, the fifth subsection of Section 15). References made *in* Chapter 2 *to* Chapter 3 might take the form Theorem 3.15, Lemma 3.15, Section 3.15, Subsection 3.15.5. However, formula numbers begin with the chapter number, whether the reference is to a formula in the same or to another chapter, so that (2.15) signifies a reference (in Chapter 2 *or* in Chapter 3) to the fifteenth formula of Chapter 2. Finally, figures are numbered sequentially from start to finish of the entire work, whereas: theorems, lemmas, formulas, and footnotes are numbered sequentially by chapter; definitions are numbered sequentially by section; and remarks are numbered sequentially by subsection.

Contents

Random processes: Markov times 1

1.1 Background material from the theory of probability

1.1.1

Let $(\Omega, \mathscr{F})$ be a measure space, i.e., a set Ω of points ω with a distinguished system $\mathscr{F}$ of its subsets forming a σ-algebra.

According to Kolmogorov's axiomatics the basis for all probability arguments is a probability space $(\Omega, \mathscr{F}, P)$ where $(\Omega, \mathscr{F})$ is a measure space and P is probability measure (probability) defined on sets from $\mathscr{F}$ and having the following properties:

$$P(A) \geq 0, A \in \mathscr{F} \qquad \text{(nonnegativity);}$$

$$P(\Omega) = 1 \qquad \text{(normability);}$$

$$P\left(\bigcup_{i=1}^{\infty} A_i\right) = \sum_{i=1}^{\infty} P(A_i) \qquad \text{(countable or } \sigma\text{-additivity);}$$

here $A_i \in \mathscr{F}$, $A_i \cap A_j = \varnothing$, $i \neq j$, where $\varnothing$ is the empty set.

The class of sets $\mathscr{F}^P$ is said to be the *completion of* $\mathscr{F}$ *with respect to measure* P if $\mathscr{F}^P$ contains the sets $A \subseteq \Omega$ when, for A_1 and $A_2 \in \mathscr{F}$, $A_1 \subseteq A \subseteq A_2$ and $P(A_2 - A_1) = 0$. The system of sets $\mathscr{F}^P$ is a σ-algebra, and the measure P extends uniquely to $\mathscr{F}^P$. A probability space $(\Omega, \mathscr{F}, P)$ is said to be *complete* if $\mathscr{F}^P$ coincides with $\mathscr{F}$.

Let $(\Omega, \mathscr{F})$ be a measure space and let $\overline{\mathscr{F}} = \bigcap_P \mathscr{F}^P$ where the intersection is taken over all probability measures P on $(\Omega, \mathscr{F})$. The system $\overline{\mathscr{F}}$ is a σ-algebra whose sets are said to be *absolutely measurable* sets in the space $(\Omega, \mathscr{F})$.

Let $(\Omega, \mathscr{F})$ and $(E, \mathscr{B})$ be two measure spaces. The function $\xi = \xi(\omega)$ defined on Ω and taking on values in E is said to be *$\mathscr{F}/\mathscr{B}$-measurable* if the set $\{\omega : \xi(\omega) \in B\} \in \mathscr{F}$ for each $B \in \mathscr{B}$. In the theory of probability such functions are known as *random elements with values in E*. If $E = R$ and $\mathscr{B}$ is the σ-algebra of Borel subsets of R, the $\mathscr{F}/\mathscr{B}$-measurable functions $\xi = \xi(\omega)$ are said to be *random variables*. (The $\mathscr{F}/\mathscr{B}$-measurable functions are frequently referred to as *$\mathscr{F}$-measurable functions*).

1.1.2

If $\xi = \xi(\omega)$ is a nonnegative random variable, its *mathematical expectation* (denoted $M\xi$) will be, by definition, a *Lebesgue integral* $\int_\Omega \xi(\omega)P(d\omega)$.

The expectation $M\xi$ or the Lebesgue integral[1] of an arbitrary random variable $\xi = \xi(\omega)$ can be defined only in the case where one of the expectations $M\xi^+$ or $M\xi^-$ is finite (here $\xi^+ = \max(\xi, 0)$, $\xi^- = -\min(\xi, 0)$), and then is equal to $M\xi^+ - M\xi^-$.

The random variable ξ is said to be *integrable* if

$$M|\xi| = M\xi^+ + M\xi^- < \infty.$$

The Lebesgue integral $\int_A \xi(\omega)P(d\omega)$ over a set $A \in \mathscr{F}$, denoted by $M(\xi; A)$ as well, is, by definition, $\int_\Omega \xi(\omega)I_A(\omega)P(d\omega)$ where $I_A = I_A(\omega)$ is the *indicator* (characteristic function) of the set A:

$$I_A(\omega) = \begin{cases} 1, & \omega \in A, \\ 0, & \omega \notin A. \end{cases}$$

Thus, $M(\xi; A) = M(\xi I_A)$ and $M(\xi; \Omega) = M\xi$.

If $\mathscr{G}$ is a sub-σ-algebra of $\mathscr{F}$, and ξ is a random variable for which $M\xi$ exists (i.e., $M\xi^+ < \infty$ or $M\xi^- < \infty$), $M(\xi|\mathscr{G})$ denotes the *conditional expectation ξ with respect to $\mathscr{G}$*, i.e., any $\mathscr{G}$-measurable random variable $\eta = \eta(\omega)$ for which, for any $A \in \mathscr{G}$,

$$\int_A \xi(\omega)P(d\omega) = \int_A \eta(\omega)P(d\omega). \tag{1.1}$$

By virtue of the Radon–Nikodym theorem such a random variable $\eta(\omega)$ exists and can be defined from (1.1) uniquely on sets of P-measure zero.

In the case where $\xi(\omega) = I_A(\omega)$ is the indicator of a set A, the conditional expectation $M(I_A|\mathscr{G})$ is denoted by $P(A|\mathscr{G})$ and is said to be the *conditional probability of an event A with respect to $\mathscr{G}$*.

A sequence of random variables ξ_n, $n = 1, 2, \ldots,$ is said to be a *sequence convergent in probability to the random variable ξ* (in this case the notation $\xi_n \xrightarrow{P} \xi$ or $\xi = P\text{-}\lim \xi_n$ is used) if for each $\varepsilon > 0$

$$\lim_{n\to\infty} P\{|\xi_n - \xi| > \varepsilon\} = 0.$$

[1] The Lebesgue integral is frequently denoted by $\int_\Omega \xi(\omega)dP(\omega)$, $\int_\Omega \xi\, dP$, or $\int \xi\, dP$.

The sequence of random variables ξ_n, $n = 1, 2, \ldots$, is said to be *convergent with probability one*, or *convergent almost surely*, to the random variable ξ (the notation is $\xi_n \to \xi$ or $\xi_n \to \xi$ (P-a.s.)) if the set $\{\omega : \xi_n(\omega) \nrightarrow \xi(\omega)\}$ has P-measure zero. It is also said that $\xi_n \to \xi$ *on the set* A if $P(A \cap \{\xi_n \nrightarrow \xi\}) = 0$. In this case the notation is $\xi_n \to \xi$ (A; (P-a.s.)).

If $\xi_n \to \xi$ (P-a.s.) and $\xi_n \leq \xi_{n+1}$ (P-a.s.) we shall write $\xi_n \uparrow \xi$ or $\xi_n \uparrow \xi$ (P-a.s.). The convergence $\xi_n \downarrow \xi$ can be defined similarly.

1.1.3

We shall give the main theorems on passage to the limit under the sign of a Lebesgue integral (expectation).

Theorem 1 (Monotone convergence). *If* $\xi_n \uparrow \xi$ (P-a.s.) *and* $M\xi_1^- < \infty$, *then*

$$M\xi_n \uparrow M\xi. \tag{1.2}$$

If $\xi_n \downarrow \xi$ (P-a.s.) *and* $M\xi_1^+ < \infty$, *then*

$$M\xi_n \downarrow M\xi. \tag{1.3}$$

Theorem 2 (Fatou's lemma). *If* $\xi_n \geq \eta$, $n = 1, 2, \ldots$, *and* $M\eta > -\infty$, *then*[2]

$$M \underline{\lim}\, \xi_n \leq \underline{\lim}\, M\xi_n. \tag{1.4}$$

If $\xi_n \leq \eta$, $n = 1, 2, \ldots$, *and* $M\eta < \infty$, *then*

$$\overline{\lim}\, M\xi_n \leq M\, \overline{\lim}\, \xi_n \tag{1.5}$$

Theorem 3 (Lebesgue's theorem on *dominated* convergence). *Let* $\xi_n \xrightarrow{P} \xi$ *and let there exist an integrable random variable* η *such that* $|\xi_n| \leq \eta$, $n = 1, 2, \ldots$. *Then* $M|\xi| < \infty$ *and*

$$M|\xi_n - \xi| \to 0, \qquad n \to \infty. \tag{1.6}$$

Remark 1. Theorems 1–3 still hold if the expectations are replaced by conditional expectations in the formulations of these theorems.

The following is the generalization of Lebesgue's theorem on dominated convergence.

Theorem 4. *Let* $\xi_n \xrightarrow{P} \xi$, $M|\xi_n| < \infty$, $n = 1, 2, \ldots$. *Then* $M|\xi| < \infty$ *and*

$$M|\xi_n - \xi| \to 0, \qquad n \to \infty, \tag{1.7}$$

if and only if the random variables ξ_n, $n = 1, 2, \ldots$, *are uniformly integrable.*[3]

[2] We denote by $\underline{\lim}\, \xi_n$ (or $\liminf \xi_n$, the lower limit of the sequence ξ_n, $n = 1, 2, \ldots$, i.e., $\sup_n \inf_{m \geq n} \xi_m$. Similarly, the upper limit $\overline{\lim}\, \xi_n$ (or $\limsup \xi_n$) is $\inf_n \sup_{m \geq n} \xi_m$.

[3] The family of random variables $\{\xi_\alpha, \alpha \in \mathfrak{A}\}$ is said to be *uniformly integrable* if

$$\lim_{x \to \infty} \sup_{\alpha \in \mathfrak{A}} \int_{\{|\xi_\alpha| > x\}} |\xi_\alpha| \, dP = 0.$$

Remark 2 (Generalized Fatou's lemma). Let $\xi_n \geq \eta_n$, $n = 1, 2, \ldots$, where the variables η_n, $n = 1, 2, \ldots$, are uniformly integrable and $\eta_n \xrightarrow{P} \eta_\infty$. Then (compare with (1.4))

$$M \,\underline{\lim}\, \xi_n \leq \underline{\lim}\, M\xi_n.$$

To prove this we shall note that by virtue of (1.4)

$$\begin{aligned} M \,\underline{\lim}\, \xi_n - M\eta_\infty &= M(\underline{\lim}\, \xi_n - \eta_\infty) \\ &= M \,\underline{\lim}(\xi_n - \eta_n) \leq \underline{\lim}(M\xi_n - M\eta_n) \leq \underline{\lim}\, M\xi_n - \lim M\eta_n. \end{aligned}$$

But by virtue of (1.7) $\lim M_{\eta_n} = M_{\eta_\infty}$, hence $M \,\underline{\lim}\, \xi_n \leq \underline{\lim}\, M\xi_n$.

1.1.4

Let $T = [0, \infty)$, $\bar{T} = T \cup \{\infty\}$, $N = \{0, 1, \ldots\}$, $\bar{N} = N \cup \{\infty\}$. The family of $\mathscr{F}/\mathscr{B}$-measurable functions (random elements) $X = \{\xi_t(\omega)\}$, $t \in T\,(t \in N)$, is said to be a *random process with continuous (discrete) time with values in E*. The random process with discrete time is said to be a *random sequence* as well.

For fixed $\omega \in \Omega$ the function of time $\xi_t(\omega)$, $t \in T$ (or $t \in N$) is said to be a *trajectory corresponding to the elementary outcome* ω.

Each random process $X = \{\xi_t(\omega)\}$, $t \in Z$ (where $Z = T$ in the case of continuous time and $Z = N$ in the case of discrete time), is naturally associated with σ-algebras $\mathscr{F}_t^\xi = \sigma\{\omega : \xi_s, s \leq t\}$, the smallest σ-algebras containing algebras $\mathscr{A}_t^\xi$ generated by sets $\{\omega : \xi_s \in \Gamma\}$, $s \leq t$, $\Gamma \in \mathscr{B}$.

The random process $X = \{\xi_t(\omega)\}$, $t \in T$, is said to be a *measurable process* if for any $\Gamma \in \mathscr{B}$ the set

$$\{(\omega, t) : \xi_t(\omega) \in \Gamma\} \in \mathscr{F} \times \mathscr{B}(T),$$

where $\mathscr{B}(T)$ is the σ-algebra of Borel sets on $T = [0, \infty)$.

The measurable random process $X = \{\xi_t(\omega)\}$, $t \in T$, is said to be a *process adapted to the family of σ-algebras* $F = \{\mathscr{F}_t\}$, $t \in T$, if for each $t \in T$ and $\Gamma \in \mathscr{B}$

$$\{\omega : \xi_t(\omega) \in \Gamma\} \in \mathscr{F}_t.$$

For short we shall denote these processes by $X = (\xi_t, \mathscr{F}_t)$, $t \in T$, or simply $X = (\xi_t, \mathscr{F}_t)$.

The random process $X = (\xi_t, \mathscr{F}_t)$, $t \in T$, is said to be a *progressively measurable process* if for each $t \in T$ and $\Gamma \in \mathscr{B}$

$$\{(\omega, s) : \xi_s(\omega) \in \Gamma, s \leq t\} \in \mathscr{F}_t \times \mathscr{B}([0, t]),$$

where $\mathscr{B}([0, t])$ is the σ-algebras of Borel sets on $[0, t]$.

Each progressively measurable process is measurable and adapted. The converse also holds in a precise sense: if the real[4] process $X = \{\xi_t(\omega)\}$, $t \in T$, is measurable and adapted to $F = \{\mathscr{F}_t\}$, $t \in T$, it permits a progressively

[4] This is a process with values in R or $\bar{R}$.

measurable modification[5] ([72], chap. 4, p. 42). Each real adapted process with right (left) continuous trajectories is a progressively measurable process ([72], chap. 4, p. 43).

1.2 Markov times

1.2.1

In the present section we shall define and discuss the properties of Markov times, which play a decisive role in the theory of optimal stopping rules. The discussion will deal only with the case of continuous time. The definitions and results can be carried almost automatically to the case of discrete time—in which case as a rule they become simpler.

Let $(\Omega, \mathscr{F})$ be a measure space, let $T = [0, \infty)$, and let $F = \{\mathscr{F}_t\}$, $t \in T$, be a nondecreasing sequence of sub-σ-algebras, i.e., $\mathscr{F}_s \subseteq \mathscr{F}_t \subseteq \mathscr{F}$ for $s \leq t$.

Definition 1. The random variable $\tau = \tau(\omega)$ with values in $\bar{T} = [0, \infty]$ is said to be a *Markov time with respect to the system* $F = \{\mathscr{F}_t\}$, $t \in T$,[6] if for each $t \in T$

$$\{\omega : \tau(\omega) \leq t\} \in \mathscr{F}_t.$$

Markov times can be interpreted as random variables independent of the "future."

Definition 2. The Markov time $\tau = \tau(\omega)$ defined in a probability space $(\Omega, \mathscr{F}, P)$ is said to be a *stopping time* or a *finite Markov time* if

$$P\{\tau(\omega) < \infty\} = 1.$$

With each Markov time $\tau = \tau(\omega)$ we may associate the aggregate $\mathscr{F}_\tau$ of the sets $A \in \mathscr{F}$ for which $A \cap \{\omega : \tau(\omega) \leq t\} \in \mathscr{F}_t$ for all $t \in T$. It is easy to verify that $\mathscr{F}_\tau$ is a σ-algebra.

An obvious interpretation of the σ-algebra $\mathscr{F}_\tau$ is the following. By $\mathscr{F}_t$ we shall understand the totality of events related to some physical process and observed before time t. (For example, let $\mathscr{F}_t = \sigma\{\omega : \xi_s, s \leq t\}$ be a σ-algebra generated by values ξ_s, $s \leq t$, of an observable process $X = \{\xi_t\}$, $t \in T$). Then $\mathscr{F}_\tau$ is the totality of events to be observed over the random time τ.

[5] The process $\tilde{X} = \{\tilde{\xi}_t(\omega)\}$, $t \in T$, is said to be a modification of the process $X = \{\xi_t(\omega)\}$, $t \in T$, if, for each $t \in T$, $P(\omega : \tilde{\xi}_t(\omega) \neq \xi_t(\omega)) = 0$.

[6] The words "with respect to the system $F = \{\mathscr{F}_t\}$, $t \in T$" will be omitted when ambiguity is impossible.

Definition 3. The system of σ-algebras $F = \{\mathscr{F}_t\}$, $t \in T$, is said to be a *right continuous* system[7] if for each $t \in T$

$$\mathscr{F}_t = \mathscr{F}_{t+},$$

where $\mathscr{F}_{t+} = \bigcap_{s>t} \mathscr{F}_s$.

Lemma 1. *Let τ be a Markov time. Then the events $\{\tau < t\}$ and $\{\tau = t\}$ belong to $\mathscr{F}_t$ for each $t \in T$.*

Proof follows immediately from the fact that

$$\{\tau < t\} = \bigcup_{k=1}^{\infty} \left\{\tau \le t - \frac{1}{k}\right\} \quad \text{and} \quad \left\{\tau \le t - \frac{1}{k}\right\} \in \mathscr{F}_{t-1/k} \subseteq \mathscr{F}_t.$$

Lemma 2. *Let a family $F = \{\mathscr{F}_t\}$, $t \in T$, be right continuous and let $\tau = \tau(\omega)$ be a random variable with values in $\bar{T} = [0, \infty]$ such that $\{\tau < t\} \in \mathscr{F}_t$ for all $t \in T$. Then τ is a Markov time, i.e., $\{\tau \le t\} \in \mathscr{F}_t$, $t \in T$.*

PROOF. If $\{\tau < t\} \in \mathscr{F}_t$, then $\{\tau \le t\} \in \mathscr{F}_{t+\varepsilon}$ for each $\varepsilon > 0$. Consequently,

$$\{\tau \le t\} \in \bigcap_{\varepsilon > 0} \mathscr{F}_{t+\varepsilon} = \mathscr{F}_{t+} = \mathscr{F}_t.$$

The above lemma implies that in the case of right continuous families $F = \{\mathscr{F}_t\}$, $t \in T$, we need only to prove that $\{\tau < t\} \in \mathscr{F}_t$, $t \in T$, in order to verify whether the random variable τ is a Markov time.

In general the condition "$\{\tau < t\} \in \mathscr{F}_t$, $t \in T$" is weaker than the condition "$\{\tau \le t) \in \mathscr{F}_t$, $t \in T$." To convince ourselves that this is the fact we shall put $\Omega = T$. Let $\mathscr{F}$ be a σ-algebra of Lebesgue sets on T,

$$x_t(\omega) = \begin{cases} 0, & t \le \omega, \\ 1, & t > \omega, \end{cases}$$

and let $\mathscr{F}_t = \sigma\{\omega : x_s(\omega),\ s \le t\}$. Then the random variable $\tau(\omega) = \inf\{t \ge 0 : x_t(\omega) = 1\}$ satisfies the condition $\{t < t\} \in \mathscr{F}_t$, whereas $\{\tau \le t\} \notin \mathscr{F}_t$, $t \in T$. □

Remark. Let $t \in N = \{0, 1, \ldots\}$ and let $\tau = \tau(\omega)$ be a random variable with values in $\bar{N} = \{0, 1, \ldots, \infty\}$. Then the condition "$\{\tau \le n\} \in \mathscr{F}_n$, $n \in N$" is equivalent to the condition "$\{\tau < n\} \in \mathscr{F}_n$, $n \in N$."

Lemma 3. *If τ and σ are Markov times, then $\tau \wedge \sigma \equiv \min(\tau, \sigma)$, $\tau \vee \sigma = \max(\tau, \sigma)$, and $\tau + \sigma$ are also Markov times.*

[7] This definition is no longer meaningful in the case of discrete time $t \in N = \{0, 1, \ldots\}$.

Proof follows from the relations[8]

$$\{\tau \wedge \sigma \le t\} = \{\tau \le t\} \cup \{\sigma \le t\} \in \mathscr{F}_t,$$

$$\{\tau \vee \sigma \le t\} = \{\tau \le t\} \cap (\sigma \le t\} \in \mathscr{F}_t,$$

$$\begin{aligned}\{\tau + \sigma > t\} &= \{\tau = 0, \tau + \sigma > t\} + \{0 < \tau < t, \tau + \sigma > t\} \\ &\quad + \{\tau \ge t, \tau + \sigma > t\} \\ &= \{\tau = 0\} \cap \{\sigma > t\} + \{0 < \tau < t, \tau + \sigma > t\} \\ &\quad + \{\tau > t\} \cap \{\sigma = 0\} + \{\tau \ge t\} \cap \{\sigma > 0\},\end{aligned}$$

where

$$\{\tau = 0\} \cap \{\sigma > t\} \in \mathscr{F}_t, \qquad \{\tau > t\} \cap \{\sigma = 0\} \in \mathscr{F}_t,$$

and

$$\{0 < \tau < t, \tau + \sigma > t\} = \bigcap_{r \in (0,t)} (\{r < \tau < t) \cap \{\sigma > t - r\}) \in \mathscr{F}_t, \tag{1.8}$$

$$\{\tau \ge t\} \cap \{\sigma > 0\} = \{\tau \ge t) \cap \overline{\{\sigma = 0\}} \in \mathscr{F}_t.$$

(In (1.8) the summation is carried out over all rational numbers r in the interval $(0, t)$.)

Lemma 4. *Let τ_n, $n = 1, 2, \ldots$ be a sequence of Markov times. Then $\sup_n \tau_n$ is also a Markov time. If, in addition, the family $F = \{\mathscr{F}_t\}$, $t \in T$, is right continuous, then $\inf \tau_n$, $\overline{\lim}\, \tau_n$ and $\underline{\lim}\, \tau_n$ will be also Markov times.*

Proof is based on the fact that

$$\left\{\sup_n \tau_n \le t\right\} = \bigcap_n \{\tau_n \le t\} \in \mathscr{F}_t,$$

$$\left\{\inf_n \tau_n < t\right\} = \bigcup_n \{\tau_n < t\} \in \mathscr{F}_t,$$

$$\overline{\lim}\, \tau_n = \inf_m \sup_{n \ge m} \tau_n,$$

$$\underline{\lim}\, \tau_n = \sup_m \inf_{n \ge m} \tau_n.$$

Lemma 5. *Any Markov time τ (with respect to $F = \{\mathscr{F}_t\}$, $t \in T$) is an $\mathscr{F}_\tau$-measurable random variable. If τ and σ are two Markov times and $\tau(\omega) \le \sigma(\omega)$, $\omega \in \Omega$, then $\mathscr{F}_\tau \subseteq \mathscr{F}_\sigma$.*

PROOF. To prove the first assertion we need to prove that the event $\{\tau \le s\} \in \mathscr{F}_\tau$ for any $s \ge 0$. We have

$$\{\tau \le s\} \cap \{\tau \le t\} = \{\tau \le t \wedge s\} \in \mathscr{F}_{t \wedge s} \subseteq \mathscr{F}_t,$$

i.e., τ is an $\mathscr{F}_\tau$-measurable random variable.

[8] If $A \cap B = \varnothing$, then instead of $A \cup B$ we write $A + B$.

Let an event $A \in \mathscr{F}_\tau$. Then

$$A \cap \{\sigma \leq t\} = (A \cap \{\tau \leq t\}) \cap \{\sigma \leq t\} \in \mathscr{F}_t, t \in T,$$

and consequently, $A \in \mathscr{F}_\sigma$. □

Lemma 6. *Let τ_n, $n = 1, 2, \ldots$, be a sequence of Markov times with respect to a right continuous family of σ-algebras $F = \{\mathscr{F}_t\}$, $t \in T$, and let $\tau = \inf_n \tau_n$. Then $\mathscr{F}_\tau = \bigcap_n \mathscr{F}_{\tau_n}$.*

PROOF. By virtue of Lemma 4, τ is a Markov time. Hence, by virtue of Lemma 5, $\mathscr{F}_\tau \subseteq \bigcap_n \mathscr{F}_{\tau_n}$. On the other hand, if $A \in \bigcap_n \mathscr{F}_{\tau_n}$, then

$$A \cap \{\tau < t\} = A \cap \left(\bigcup_n \{\tau_n < t\}\right) = \bigcup_n (A \cap \{\tau_n < t\}) \in \mathscr{F}_t,$$

which implies (because the family F is right continuous) that $A \cap \{\tau \leq t\} \in \mathscr{F}_t$, and, therefore, $A \in \mathscr{F}_\tau$. □

Lemma 7. *Let τ and σ be two Markov times. Then each of the events $\{\tau < \sigma\}$, $\{\tau > \sigma\}$, $\{\tau \leq \sigma\}$, $\{\tau \geq \sigma\}$, $\{\tau = \sigma\}$ belongs to $\mathscr{F}_\tau$ and $\mathscr{F}_\sigma$.*

PROOF. For each $t \in T$

$$\{\tau < \sigma\} \cap \{\sigma \leq t\} = \bigcap_{0 \leq r < t} (\{\tau < r\} \cap \{r < \sigma \leq t\}),$$

where the r are rational numbers. Consequently, $\{\tau < \sigma\} \in \mathscr{F}_\sigma$. On the other hand,

$$\{\tau < \sigma\} \cap \{\tau \leq t\} = \bigcap_{0 \leq r < t} (\{\tau \leq r\} \cap \{r < \sigma\}) \cup (\{\tau \leq t) \cap \{t < \sigma\}) \in \mathscr{F}_t,$$

i.e., $\{\tau < \sigma\} \in \mathscr{F}_\tau$.

It can be proved similarly that $\{\sigma < \tau\} \in \mathscr{F}_\tau$, $\{\sigma < \tau\} \in \mathscr{F}_\sigma$. Consequently, $\{\tau \leq \sigma\}$, $\{\sigma \leq \tau\}$, and $\{\tau = \sigma\}$ belong to both $\mathscr{F}_\tau$ and $\mathscr{F}_\sigma$. □

Lemma 8. *Let $X = (\xi_t, \mathscr{F}_t)$, $t \in T$, be a progressively measurable random process with values in a measurable space $(E, \mathscr{B})$. Let τ be a finite Markov time (with respect to the system $F = \{\mathscr{F}_t\}$, $t \in T$) with values in $[0, \infty)$. Then the function $\xi_{\tau(\omega)}(\omega)$ is $\mathscr{F}_\tau/\mathscr{B}$-measurable.*

PROOF. Let $B \in \mathscr{B}$, $t \in T$. We need to show that

$$\{\xi_{\tau(\omega)}(\omega) \in B\} \cap \{\tau(\omega) \leq t\} \in \mathscr{F}_t.$$

Let $\sigma = \min(\tau, t)$. Then

$$\begin{aligned}\{\xi_\tau \in B\} \cap \{\tau \leq t\} &= \{\xi_\tau \in B\} \cap [\{\tau < t\} \cup \{\tau = t\}] \\ &= [\{\xi_\sigma \in B\} \cap \{\sigma < t\}] \cup [\{\xi_t \in B\} \cap \{\tau = t\}].\end{aligned}$$

It is clear that $\{\xi_t \in B\} \cap \{\tau = t\} \in \mathscr{F}_t$. Hence, if we show that ξ_σ is an $\mathscr{F}_t/\mathscr{B}$-measurable function, then $\{\xi_t \in B\} \cap \{\sigma < t\} \in \mathscr{F}_t$.

To this end we note that the mapping $\omega \to (\omega, \sigma(\omega))$ is a measurable mapping of $(\Omega, \mathscr{F}_t)$ onto $(\Omega \times [0, t], \mathscr{F}_t \times \mathscr{B}([0, t]))$, and the mapping $(\omega, s) \to \xi_s(\omega)$ of the space $(\Omega \times [0, t], \mathscr{F}_t \times \mathscr{B}([0, t]))$ onto $(E, \mathscr{B})$ is measurable by virtue of the fact that the process X is progressively measurable. Consequently, the mapping $\xi_\sigma(\omega)$ of the space $(\Omega, \mathscr{F}_t)$ onto $(E, \mathscr{B})$ is also measurable since it is the composition of two measurable mappings. □

1.2.2

We shall give examples involving Markov times.

Let $X = \{\xi_t\}$, $t \in T$, be a real process, $\mathscr{F}_t = \mathscr{F}_t^\xi \equiv \sigma\{\omega : \xi_s, s \le t\}$, let A be a Borel set of the real line, and also let

$$\sigma_A = \inf\{t \ge 0 : \xi_t \in A\}, \tag{1.9}$$

$$\tau_A = \inf\{t > 0 : \xi_t \in A\} \tag{1.10}$$

be times of (the first and the first after $+0$) attainment of a set A. We shall agree to put $\sigma_A = \infty$ and $\tau_A = \infty$ if the sets $\{\cdot\}$ in (1.9) and (1.10) are empty. We shall note that the times σ_A and τ_A do not coincide only in the case where $\xi_0(\omega) \in A$ and that there exists $\varepsilon > 0$ such that $\xi_t(\omega) \notin A$ for all $t \in (0, \varepsilon)$.

It can be easily verified that the times σ_A and τ_A possess the following properties—

$$A \subseteq B \Rightarrow \sigma_A \ge \sigma_B, \qquad \tau_A \ge \tau_B,$$

$$\sigma_{A \cup B} = \min(\sigma_A, \sigma_B), \qquad \tau_{A \cup B} = \min(\tau_A, \tau_B),$$

$$\sigma_{A \cap B} \ge \max(\sigma_A, \sigma_B), \qquad \tau_{A \cap B} \ge \max(\tau_A, \tau_B),$$

$$\text{if } A = \bigcup_n A_n, \text{ then } \sigma_A = \inf_n \sigma_{A_n}, \tau_A = \inf_n \tau_{A_n}.$$

Lemma 9. *Let a real process $X = (\xi_t, \mathscr{F}_t)$, $t \in T$, have right continuous trajectories, $\mathscr{F}_{t+} = \mathscr{F}_t$ and let C be an open set (in $\bar{R}$). Then the times σ_C and τ_C are Markov times (with respect to the system $F = \{\mathscr{F}_t\}$, $t \in T$).*

PROOF. Let $D = \bar{R} - C$. Then, by virtue of the fact that the set D is closed and the trajectories of the process X are right continuous we have

$$\{\sigma_C \ge t\} = \{\omega : \xi_s(\omega) \in D, s < t\} = \bigcap_{0 \le r < t} \{\omega : \xi_r(\omega) \in D\} \in \mathscr{F}_t,$$

where the r are rational numbers. Consequently, $\{\sigma_C < t\} \in \mathscr{F}_t$ and by virtue of Lemma 2, σ_C is a Markov time.

We can prove similarly that τ_C is a Markov time. □

The following more general result, of which Lemma 9 is a particular case, will be used frequently from now on.

Theorem 5. *Let $X = (\xi_t, \mathscr{F}_t)$, $t \in T$, be a progressively measurable random process with values in a measurable space $(E, \mathscr{B})$. Let $\mathscr{F}_t = \mathscr{F}_t^{\mathrm{P}}, \mathscr{F}_{t+} = \mathscr{F}_t$, $t \in T$. Then for any absolutely measurable set $\Gamma \in \bar{\mathscr{B}}$ the times*

$$\sigma_\Gamma^s = \inf\{t \geq s : \xi_t \in \Gamma\}, \qquad \tau_\Gamma^s = \inf\{\tau > s : \xi_t \in \Gamma\},$$

where $s \geq 0$, are Markov.

For a proof see: [72], Chap. 4, pp. 46–49; [30], §2, suppl.

1.1.3

Let $(\Omega, \mathscr{F})$ be a measurable space and let $X = \{\xi_t\}$, $t \in T$, be a measurable random process with values in a measurable space $(E, \mathscr{B})$. We shall assume that the primary space Ω is sufficiently "rich" in the sense that for each $t \geq 0$ and $\omega \in \Omega$ there will be $\omega' \in \Omega$ such that[9]

$$\xi_s(\omega') = \xi_{s \wedge t}(\omega) \tag{1.11}$$

for all $s \geq 0$.

Let $\mathscr{F}_t^\xi = \sigma\{\omega : \xi_s, s \leq t\}$, $\mathscr{F}_\infty^\xi = \sigma(\bigcup_{t \in T} \mathscr{F}_t^\xi)$ and let $\tau = \tau(\omega)$ be a Markov time (with respect to the system $F^\xi = \{\mathscr{F}_t^\xi\}$, $t \in T$) with values in $[0, \infty)$.

With each of such Markov times naturally associate a σ-algebra $\mathscr{F}_\tau^\xi$ consisting of the sets $A \in \mathscr{F}_\infty^\xi$ for which $A \cap \{\tau \leq t\} \in \mathscr{F}_t^\xi$ for all $t \geq 0$.

Denote by $\mathscr{G}_\tau^\xi$ the σ-algebra generated by sets $\{\omega : \xi_{t \wedge \tau(\omega)}(\omega) \in B\}$, $t \geq 0$, $B \in \mathscr{B}$.

Theorem 6. *Under the assumption given in* (1.11) *the σ-algebras $\mathscr{F}_\tau^\xi$ and $\mathscr{G}_\tau^\xi$ coincide:*

$$\mathscr{F}_\tau^\xi = \mathscr{G}_\tau^\xi. \tag{1.12}$$

Proof follows immediately from Lemmas 12 and 13, which follow.

It follows from this theorem that the σ-algebra $\mathscr{F}_\tau^\xi$ is generated by random variables $\xi_{\tau \wedge t}$, $t \geq 0$, i.e., $\mathscr{F}_\tau^\xi$ is the smallest σ-algebra with respect to which we shall measure the "stopped" process $\{\xi_{\tau \wedge t}\}$, $t \geq 0$.

The definitions and auxiliary facts which follow (in, particular, Lemmas 12 and 13) are of independent interest on their own.

1.1.4

Definition 4. We shall say that the points ω, ω' of a space Ω are *t-equivalent* ($\omega \overset{t}{\sim} \omega'$) if for all $s \leq t$

$$\xi_s(\omega) = \xi_s(\omega'). \tag{1.13}$$

[9] *Editor's Note*: This assumption is made in the sections that follow.

Let a set $A \subseteq \Omega$. Denote by $[A]_t$ the set of all $\omega \in \Omega$ for which there is $\omega' \in A$ such that $\omega' \overset{t}{\sim} \omega$. We shall denote by $\tilde{\mathscr{F}}_t^\xi$ all the sets $A \in \mathscr{F}_\infty^\xi$ for which $[A]_t = A$. It can be easily verified that $\tilde{\mathscr{F}}_t^\xi$ forms a σ-algebra.

Lemma 10. *The σ-algebras $\mathscr{F}_t^\xi$ and $\tilde{\mathscr{F}}_t^\xi$ coincide:*

$$\mathscr{F}_t^\xi = \tilde{\mathscr{F}}_t^\xi. \tag{1.14}$$

PROOF. Let $\alpha_t = \alpha_t(\omega)$ denote the mapping of Ω into Ω such that

$$\xi_s(\alpha_t(\omega)) = \xi_{s \wedge t}(\omega), \qquad s \geq 0. \tag{1.15}$$

(By virtue of the assumption given by (1.11) there exists such a mapping.) It can be easily verified that the mapping α_t is a measurable mapping of $(\Omega, \mathscr{F}_t^\xi)$ into $(\Omega, \mathscr{F}_\infty^\xi)$ for each $t \in T$.

Let us show that $\tilde{\mathscr{F}}_t^\xi \subseteq \mathscr{F}_t^\xi$. To this end it suffices to verify that all sets $A \in \mathscr{F}_\infty^\xi$ such that $A = [A]_t$ also belong to $\mathscr{F}_t^\xi$.

Thus, let $A \in \mathscr{F}_\infty^\xi$, $A = [A]_t$. Then $\alpha_t^{-1}A \in \mathscr{F}_t^\xi$. By virtue of the relation $\alpha_t(\omega) \overset{t}{\sim} \omega$ the set $A \subseteq \alpha_t^{-1}A$. Suppose that $\omega \in \alpha_t^{-1}A$. Then $\alpha_t(\omega) \in A$ and it follows from the relations $\alpha_1(\omega) \overset{t}{\sim} \omega$ and $[A]_t = A$ that $\omega \in A$. Therefore, $A = \alpha_t^{-1}A$. But $\alpha_t^{-1}A \in \mathscr{F}_t^\xi$, and hence $A \in \mathscr{F}_t^\xi$.

We shall establish next that $\mathscr{F}_t^\xi \subseteq \tilde{\mathscr{F}}_t^\xi$. For all $s \leq t$, $\Gamma \in \mathscr{B}$, $\{\xi_s \in \Gamma\} \in \mathscr{F}_\infty^\xi$, and, furthermore, $[\{\xi_s \in \Gamma\}]_t = \{\xi_s \in \Gamma\}$. Consequently, for all $s \leq t$ and $\Gamma \in \mathscr{B}$, $\{\xi_s \in \Gamma\} \in \tilde{\mathscr{F}}_t^\xi$ and therefore $\mathscr{F}_t^\xi \subseteq \tilde{\mathscr{F}}_t^\xi$. □

Lemma 11. *If the set $A \in \mathscr{F}_t^\xi$, then $[A]_t = A$.*

PROOF. Denote by $\mathscr{D}_t$ the aggregate of all $A \in \mathscr{F}_t^\xi$ for which $[A]_t = A$, and show that $\mathscr{D}_t = \mathscr{F}_t^\xi$.

(a) If the set $A = \{\omega : \xi_{s_1} \in \Gamma_1, \ldots, \xi_{s_n} \in \Gamma_n\}$ where $\Gamma_i \in \mathscr{B}$, $s_i \leq t$, $i = 1, \ldots, n$, then obviously $A \in \mathscr{D}_t$.

(b) It can be easily verified that

$$\left[\bigcup_n A_n\right]_t = \bigcup_n [A_n]_t,$$

$$\left[\bigcap_n A_n\right]_t \subseteq \bigcap_n [A_n]_t.$$

(c) Let $\mathscr{U}_t$ be an algebra of sets being the union of the finite number of nonoverlapping sets of the form $\{\xi_{s_1} \subseteq \Gamma_1, \ldots, \xi_{s_n} \in \Gamma_n\}$, $\Gamma_i \in \mathscr{B}$, $s_i \leq t, i = 1, \ldots, n, n \geq 1$. Then it follows from (a) and (b) that $\mathscr{D}_t \supseteq \mathscr{U}_t$.

(d) We shall prove next that $\mathscr{D}_t$ is a monotone class of sets. Let $A_1 \subseteq A_2 \subseteq \cdots$ be a sequence of sets from $\mathscr{D}_t$. We shall show that $\bigcup_n A_n \in \mathscr{D}_t$. Since $A_n \in \mathscr{D}_t$, then by virtue of (b) $[\bigcup_n A_n]_t = \bigcup_n [A_n]_t = \bigcup_n A_n$, i.e., $\bigcup_n A_n \in \mathscr{D}_t$.

Let, next, $A_1 \supseteq A_2 \supseteq \cdots$ where $A_n \in \mathscr{D}_t$. Then by virtue of (b) $[\bigcap_n A_n]_t \subseteq \bigcap_n [A_n]_t = \bigcap_n A_n \in \mathscr{D}_t$. On the other hand, it is obvious that $[\bigcap_n A_n]_t \supseteq \bigcap_n A_n$ and, therefore, $[\bigcap_n A_n]_t = \bigcap_n A_n$, i.e., $\bigcap_n A_n \in \mathscr{D}_t$.

It follows from (c) and (d) that $\mathscr{D}_t \supseteq \sigma(\mathscr{U}_t)$. But $\sigma(\mathscr{U}_t) = \mathscr{F}_t^\xi$ and hence $\mathscr{D}_t \subseteq \mathscr{F}_t^\xi$ which fact together with the obvious inclusion $\mathscr{D}_t \supseteq \mathscr{F}_t^\xi$ proves the equality $\mathscr{D}_t = \mathscr{F}_t^\xi$.

The following important result will be formulated as:

Theorem 7. *A necessary and sufficient condition for the function $\tau = \tau(\omega)$ with values in $T = [0, \infty)$ to be a Markov time (with respect to $F^\xi = \{\mathscr{F}_t^\xi\}$, $t \in T$), is that this function be $\mathscr{F}_\infty^\xi$-measurable and that for all $t \in T$*

$$[\{\tau = t\}]_t = \{\tau = t\}. \tag{1.16}$$

PROOF.

Necessity: If τ is a Markov time, obviously this time is $\mathscr{F}_\infty^\xi$-measurable. Further, if τ is a Markov time, the sets $\{\tau \le t\}$ and $\{\tau < t\}$ belong to $\mathscr{F}_t^\xi$. Consequently, $\{\tau = t\} \in \mathscr{F}_t^\xi$. Then by Lemma 11 $[\{\tau = t\}]_t = \{\tau = t\}$.

Sufficiency: The function τ is $\mathscr{F}_\infty^\xi$-measurable. Hence, if we show that $[\{\tau \le t\}]_t = \{\tau \le t\}$, the definition of the σ-algebra $\tilde{\mathscr{F}}_t^\xi$ will imply that the set $\{\tau \le t\} \in \tilde{\mathscr{F}}_t^\xi$. But then by Lemma 10 the set $\{\tau \le t\} \in \mathscr{F}_t^\xi$.

Thus, let $\tau(\omega) \le t$ for $\omega \in \Omega$. Suppose $\omega' \overset{t}{\sim} \omega$. Denote $s = \tau(\omega)$. Then, obviously, $\omega' \overset{s}{\sim} \omega$, and by hypothesis of the theorem $[\{\tau = s\}]_s = \{\tau = s\}$. Consequently, $\tau(\omega') = s$, and, therefore, $\tau(\omega') \le t$. Therefore, $[\{\tau \le t\}]_t = \{\tau \le t\}$, which fact, as indicated above, implies that $\{\tau \le t\} \in \mathscr{F}_t^\xi$.

Corollary. *The function $\tau = \tau(\omega)$ with values from $T = [0, \infty)$ is a Markov time if and only if it is $\mathscr{F}_\infty^\xi$-measurable and the set $\{\tau = t\} \in \mathscr{F}_t^\xi$ for any $t \ge 0$.*

PROOF. The necessity is obvious. Let the function $\tau = \tau(\omega)$ be an $\mathscr{F}_\infty^\xi$-measurable function and let $\{\tau = t\} \in \mathscr{F}_t^\xi$. Then by virtue of Lemma 11, $[\{\tau = t\}]_t = \{\tau = t\}$, which fact by Theorem 7 indicates that τ is a Markov time.

1.1.5

Definition 5. Let τ be a Markov time (with respect to $F^\xi = \{\mathscr{F}_t^\xi\}$, $t \ge 0$) with values in $T = [0, \infty)$. We shall say that the points ω, ω' of a space Ω are *τ-equivalent* ($\omega \overset{\tau}{\sim} \omega'$) if for all $s \le \tau(\omega)$

$$\xi_s(\omega) = \xi_s(\omega'). \tag{1.17}$$

We shall show that the relation $\overset{\tau}{\sim}$ is an equivalence relation.

First, it is clear that $\omega \overset{\tau}{\sim} \omega$. Let $\omega \overset{\tau}{\sim} \omega'$. Denote $u = \tau(\omega)$. Then $\omega \overset{u}{\sim} \omega'$, and, by virtue of Theorem 7, $\tau(\omega') = u$. Consequently, $\omega' \overset{\tau}{\sim} \omega$.

Finally, let $\omega \overset{\tau}{\sim} \omega'$, $\omega' \overset{\tau}{\sim} \omega''$. We need to show that $\omega \overset{\tau}{\sim} \omega''$.

As in the proof of the implication $\omega \overset{\tau}{\sim} \omega' \Rightarrow \omega' \overset{\tau}{\sim} \omega$, we can show here that $\tau(\omega) = \tau(\omega')$ and $\tau(\omega') = \tau(\omega'')$. Denote $u = \tau(\omega) = \tau(\omega') = \tau(\omega'')$. Then, since $\omega \overset{\tau}{\sim} \omega'$ and $\omega' \overset{\tau}{\sim} \omega''$,

$$\xi_s(\omega) = \xi_s(\omega'), \qquad \xi'_s(\omega') = \xi_s(\omega''), \qquad s \leq u = \tau(\omega),$$

and, consequently, $\xi_s(\omega) = \xi_s(\omega'')$, $s \leq \tau(\omega)$, and, therefore, $\omega \overset{\tau}{\sim} \omega''$.

Let a set $A \subseteq \Omega$ and let τ be a Markov time with values in $T = [0, \infty)$. Denote by $[A]_\tau$ the aggregate of all those $\omega \in \Omega$ for which there is a point $\omega' \in A$ with the property $\omega' \overset{\tau}{\sim} \omega$. Denote by $\tilde{\mathscr{F}}_\tau^\xi$ the smallest σ-algebra in the space Ω containing all sets $A \in \mathscr{F}_\infty^\xi$ for which $[A]_\tau = A$.

Lemma 12. *The σ-algebras $\tilde{\mathscr{F}}_\tau^\xi$ and $\mathscr{G}_\tau^\xi$ coincide:*

$$\tilde{\mathscr{F}}_\tau^\xi = \mathscr{G}_\tau^\xi. \tag{1.18}$$

Proof. Let $\alpha_\tau = \alpha_\tau(\omega)$ be a mapping of Ω onto Ω such that

$$\xi_s(\alpha_\tau(\omega)) = \xi_{s \wedge \tau}(\omega), \qquad s \geq 0. \tag{1.19}$$

(According to the assumption given by (1.11) there exists such a mapping.) It can be easily seen that this mapping α_τ is a measurable mapping of $(\Omega, \mathscr{G}_\tau^\xi)$ into $(\Omega, \mathscr{F}_\infty^\xi)$.

We shall establish the inclusion $\tilde{\mathscr{F}}_\tau^\xi \subseteq \mathscr{G}_\tau^\xi$. We need only to show that the set $A \in \mathscr{F}_\infty^\xi$ with property $[A]_\tau = A$ belongs to a σ-algebra $\mathscr{G}_\tau^\xi$. Since $A \in \mathscr{F}_\infty^\xi$, it follows that $\alpha_\tau^{-1} A \in \mathscr{G}_\tau^\xi$.

Let us show that $\alpha_\tau^{-1} A = A$ (then $A \in \mathscr{G}_\tau^\xi$). In fact, let $\omega \in A$. Then $\omega \overset{\tau}{\sim} \alpha_\tau(\omega)$, i.e.,

$$\xi_s(\alpha_\tau(\omega)) = \xi_{s \wedge \tau}(\omega) = \xi_s(\omega), \qquad s \geq \tau(\omega).$$

But $[A]_\tau = A$, hence $\alpha_\tau(\omega) \in A$, and, therefore, $\alpha_\tau^{-1} A \supseteq A$.

Let now $\omega \in \alpha_\tau^{-1} A$. Then $\alpha_\tau(\omega) \in A$ and by virtue of the equivalence $\omega \overset{\tau}{\sim} \alpha_\tau(\omega)$ and the equality $[A]_\tau = A$ we see that $\omega \in A$.

Therefore, $A = \alpha_\tau^{-1} A$ which fact implies that $\tilde{\mathscr{F}}_\tau^\xi \subseteq \mathscr{G}_\tau^\xi$.

To prove the inverse inclusion $\mathscr{G}_\tau^\xi \subseteq \tilde{\mathscr{F}}_\tau^\xi$ we need only to show that the set $\{\xi_{t \wedge \tau} \in \Gamma\} \in \tilde{\mathscr{F}}_\tau^\xi$ for all $t \geq 0$, $\Gamma \in \mathscr{B}$.

The mapping $\omega \to (\tau(\omega) \wedge t, \omega)$ is a measurable mapping of $(\Omega, \mathscr{F}_\infty^\xi)$ into $([0, \infty) \times \Omega, \mathscr{B}([0, \infty)) \times \mathscr{F}_\infty^\xi)$. Since the process $X = \{\xi_t\}$, $t \in T$, is a measurable mapping of $([0, \infty) \times \Omega, \mathscr{B}([0, \infty)) \times \mathscr{F}_\infty^\xi)$ into $(E, \mathscr{B})$, the function $\xi_{t \wedge \tau}(\omega)$ is a measurable mapping of $(\Omega, \mathscr{F}_\infty^\xi)$ into $(E, \mathscr{B})$. Consequently,

$$\{\omega : \xi_{t \wedge \tau}(\omega) \in \Gamma\} \in \mathscr{F}_\infty^\xi,$$

which fact, together with the obvious relation

$$[\{\xi_{t \wedge \tau} \in \Gamma\}]_\tau = \{\xi_{t \wedge \tau} \in \Gamma\},$$

shows that (by the definition of $\tilde{\mathscr{F}}_\tau^\xi$) the set $\{\xi_{t \wedge \tau} \in \Gamma\} \in \tilde{\mathscr{F}}_\tau^\xi$. □

Lemma 13. *The σ-algebras $\mathscr{F}_\tau^\xi$ and $\tilde{\mathscr{F}}_\tau^\xi$ coincide*:

$$\mathscr{F}_\tau^\xi = \tilde{\mathscr{F}}_\tau^\xi. \tag{1.20}$$

PROOF. Let $A \in \mathscr{F}_\tau^\xi$. Obviously, $\mathscr{F}_\tau^\xi \subseteq \mathscr{F}_\infty^\xi$. Hence $A \in \mathscr{F}_\infty^\xi$. To prove that the set $A \in \tilde{\mathscr{F}}_\tau^\xi$, it suffices to show that $[A]_\tau = A$.

Let $\omega \in A$ and let $\omega' \overset{\tau}{\sim} \omega$. Set $s = \tau(\omega)$. Then, obviously, $\omega' \overset{s}{\sim} \omega$ and, furthermore,

$$\omega \in \{\tau = s\} \cap A.$$

But $A \cap \{\tau = s\} \in \mathscr{F}_s^\xi$, and by Lemma 11

$$[A \cap \{\tau = s\}]_s = A \cap \{\tau = s\}.$$

Consequently, since $\omega' \overset{s}{\sim} \omega$ and $\omega' \in A \cap \{\tau = s\}$, it follows that the point $\omega' \in A$. Therefore, $[A]_\tau = A$ and $A \in \tilde{\mathscr{F}}_\tau^\xi$, and, in consequence, $\mathscr{F}_\tau^\xi \subseteq \tilde{\mathscr{F}}_\tau^\xi$.

To prove the inverse inclusion $\mathscr{F}_\tau^\xi \supseteq \tilde{\mathscr{F}}_\tau^\xi$, we shall suppose that $A \in \mathscr{F}_\infty^\xi$ and $[A]_\tau = A$.

We shall show that then for all $t \geq 0$, $A \cap \{\tau \leq t\} \in \mathscr{F}_t^\xi$. By virtue of the fact that $A \in \mathscr{F}_\infty^\xi$ and $\{\tau \leq t\} \in \mathscr{F}_t^\xi \subseteq \mathscr{F}_\infty^\xi$, the set $A \cap \{\tau \leq t\} \in \mathscr{F}_\infty^\xi$. Let $\omega \in A \cap \{\tau \leq t\}$ and let $\omega' \in \Omega$ be such that $\omega \overset{t}{\sim} \omega'$. Then, obviously, $\omega \overset{\tau}{\sim} \omega'$ and, therefore, $\omega' \in [A]_\tau = A$.

Further, let $s = \tau(\omega)$. Since $\{\tau \leq t\} \leq \mathscr{F}_t^\xi$ and $\omega \overset{t}{\sim} \omega'$, then $\omega \overset{s}{\sim} \omega'$. Hence $\tau(\omega') = s$ (Theorem 7), and, therefore, $\tau(\omega') \leq t$.

Thus, we have proved that if $\omega \in A \cap \{\tau \leq t\}$ and $\omega \overset{t}{\sim} \omega'$, then $\omega' \in A \cap \{\tau \leq t\}$, i.e.,

$$[A \cap \{\tau \leq t\}]_t = A \cap \{\tau \leq t\},$$

and by the definition of the σ-algebras $\tilde{\mathscr{F}}_t^\xi$,

$$A \cap \{\tau \leq t\} \in \tilde{\mathscr{F}}_t^\xi.$$

Taking Lemma 10 into account we obtain thereby that $A \cap \{\tau \leq t\} \in \mathscr{F}_t^\xi$ and, t being arbitrary, $A \in \mathscr{F}_\tau^\xi$, which was to be proved. □

Remark 1. A necessary and sufficient condition for the set $A \subseteq \Omega$ to belong to $\mathscr{F}_\tau^\xi$ is that $A \in \mathscr{F}_\infty^\xi$ and $[A]_\tau = A$.

In fact, let $A \in \mathscr{F}_\tau^\xi$. Then, as in proving the inclusion $\tilde{\mathscr{F}}_\tau^\xi \subseteq \mathscr{F}_\tau^\xi$ in Lemma 13, we can prove that $A \in \mathscr{F}_\infty^\xi$ and $[A]_\tau = A$.

Conversely, let $A \in \mathscr{F}_\infty^\xi$ and $[A]_\tau = A$. Then $A \in \tilde{\mathscr{F}}_\tau^\xi$, and by virtue of Lemma 13, $A \in \mathscr{F}_\tau^\xi$.

Remark 2. A necessary and sufficient condition for the set $A \subseteq \Omega$ to belong to $\mathscr{F}_\tau^\xi$ is that $A \in \mathscr{F}_\infty^\xi$ and $A \cap \{\tau = t\} \in \mathscr{F}_t^\xi$ for all $t \geq 0$.

The necessity is obvious. To prove sufficiency we shall assume that $\omega \in A$, $\omega \overset{\tau}{\sim} \omega'$. Set $t = \tau(\omega)$. Then $\omega \overset{t}{\sim} \omega'$. By virtue of Lemma 2,

$[A \cap \{\tau = t\}]_t = A \cap \{\tau = t\}$. Consequently, $\omega' \in A \cap \{\tau = t\}$, and, therefore, $\omega' \in A$.

Thus, $[A]_\tau = A$. Hence, $A \in \mathscr{F}_\tau^\xi$ by virtue of the definition of the σ-algebra $\mathscr{F}_\tau^\xi$ and Lemma 13.

1.1.6

As noted above, Lemmas 12 and 13 imply immediately Theorem 6, which states that the σ-algebras $\mathscr{F}_\tau^\xi$ and $\mathscr{G}_\tau^\xi$ coincide. The result which follows is a useful corollary of this theorem (compare with Lemma 8).

Theorem 8. *Let* $X = \{\xi_t\}$, $t \in T = [0, \infty]$, *be a measurable random process and let the* (1.11) *be satisfied. Then for any Markov time* τ *(with respect to* $F^\xi = \{\mathscr{F}_t^\xi\}$, $t \in T$, *with values in* $T = [0, \infty)$) *the function* ξ_τ *is* $\mathscr{F}_\tau^\xi/\mathscr{B}$-*measurable.*

PROOF. It follows from the definition of σ-algebra $\mathscr{G}_\tau^\xi$ that the function $\xi_{\tau \wedge n}$ is $\mathscr{G}_\tau^\xi$-measurable and by virtue of Theorem 6 it is $\mathscr{F}_\tau^\xi$-measurable for all $n \geq 0$. Clearly, the function $\xi_\tau = \lim_{n\to\infty} \xi_{\tau \wedge n}$ is also $\mathscr{F}_\tau^\xi$-measurable. □

1.3 Martingales and semimartingales

1.3.1

The development of the theory of optimal stopping rules (for both Markov and non-Markov processes) is essentially based on the results related to the theory of martingales and semimartingales. Hence we shall dwell next on the main definitions and some results to be used in the chapters which follow.

Let $(\Omega, \mathscr{F}, P)$ be a probability space and let $F = \{\mathscr{F}_t\}$, $t \in Z$, be a non-decreasing family of σ-algebras $\mathscr{F}_t \subseteq \mathscr{F}$.

Definition 1. The random process $X = (\xi_t, \mathscr{F}_t)$, $t \in Z$, is said to be a *martingale* if

$$M|\xi_t| < \infty, \qquad t \in Z, \tag{1.21}$$

$$M(\xi_t | \mathscr{F}_s) = \xi_s \qquad (P\text{-a.s.}), \qquad t \geq s. \tag{1.22}$$

Definition 2. The random process $X = (\xi_t, \mathscr{F}_t)$, $t \in Z$, is said to be a *supermartingale* or a *submartingale* if $M|\xi_t| < \infty$, $t \in Z$, and

$$M(\xi_t | \mathscr{F}_s) \leq \xi_s \qquad (P\text{-a.s.}), \qquad t \geq s, \tag{1.23}$$

$$M(\xi_t | \mathscr{F}_s) \geq \xi_s \qquad (P\text{-a.s.}), \qquad t \geq s, \tag{1.24}$$

respectively.

Supermartingales and submartingales together are referred to as *semimartingales*.

Obviously, if $X = (\xi_t, \mathscr{F}_t)$ is a supermartingale, the process $Y = (-\xi_t, \mathscr{F}_t)$ is a submartingale. Hence, it suffices to consider only, say, supermartingales in investigating semimartingale properties.

Definition 3. In Definitions 1 and 2 instead of the condition $M|\xi_t| < \infty$, $t \in Z$, the condition

$$M\xi_t^+ < \infty \quad \text{or} \quad M\xi_t^- < \infty, \qquad t \in Z \tag{1.25}$$

may be used (assuming the existence of the expectations $M\xi_t$ and the expectations $M(\xi_t | \mathscr{F}_s)$); the process $X = (\xi_t, \mathscr{F}_t)$, $t \in Z$, satisfying (1.22) ((1.23) or (1.24)) is said to be a *generalized martingale (supermartingale or submartingale).*

1.3.2

In the case of continuous time $t \in T$ we shall assume from now on that the trajectories of the semimartingales considered are continuous (P-a.s.) at the right and that the σ-algebras $\mathscr{F}_t$ are augmented by sets from $\mathscr{F}$ of P-measure zero.

Theorem 9.

1°. *Let $X = (\xi_t, \mathscr{F}_t)$, $t \in Z$, be a supermartingale satisfying the condition*

$$\sup_{t \in Z} M\xi_t^- < \infty. \tag{1.26}$$

Then with probability one, $\lim_{t\to\infty} \xi_t$ $(= \xi_\infty)$ exists and is finite, and $M\xi_\infty^- < \infty$.

2°. *If, in addition, the random variables $\{\xi_t, t \in Z\}$ are uniformly integrable, the system $\bar{X} = (\xi_t, \mathscr{F}_t)$, $t \in \bar{Z} = Z \cup \{\infty\}$, with $\xi_\infty = \lim_{t\to\infty} \xi_t$ and $\mathscr{F}_\infty = \sigma(\bigcup_{t \in Z} \mathscr{F}_t)$, also forms a supermartingale.*

3°. *Let $X = (\xi_t, \mathscr{F}_t)$, $t \in Z$, be a generalized supermartingale. Then for almost all ω such that*

$$\inf_{s \in Z} \sup_{t \in Z} M(\xi_t^- | \mathscr{F}_s) < \infty,$$

$\lim_{t\to\infty} \xi_t$ $(= \xi_\infty)$ exists and is finite or equal to $+\infty$.

We can deduce from this theorem an essential result related to properties of conditional mathematical expectations.

Theorem 10.

1°. *Let $\eta = \eta(\omega)$ be an integrable random variable, $M|\eta| < \infty$. Then the martingale $X = (\xi_t, \mathscr{F}_t)$, $t \in Z$, with $\xi_t = M(\eta | \mathscr{F}_t)$ is such that with probability one the limit $\lim_{t\to\infty} M(\eta | \mathscr{F}_t)$ exists and*

$$\lim_{t\to\infty} M(\eta | \mathscr{F}_t) = M(\eta | \mathscr{F}_\infty) \qquad (P\text{-a.s.}), \tag{1.27}$$

where $\mathscr{F}_\infty = \sigma(\bigcup_{t \in Z} \mathscr{F}_t)$.

2°. *If the random variable* $\eta = \eta(\omega)$ *is such that* $M\eta^+ < \infty$, *then with probability one the limit* $\lim_{t\to\infty} M(\eta|\mathscr{F}_t)$ *exists and*

$$\lim_{t\to\infty} M(\eta|\mathscr{F}_t) \le M(\eta\ \mathscr{F}_\infty) \qquad (P\text{-a.s.}). \tag{1.28}$$

The following result plays an essential role in the theory of optimal stopping rules.

Theorem 11. *Let* $X = (\xi_t, \mathscr{F}_t)$, $t \in Z$, *be a supermartingale such that* (P-a.s.)

$$\xi_t \ge M(\eta|\mathscr{F}_t), \qquad t \in Z, \tag{1.29}$$

where $\eta = \eta(\omega)$ *is an integrable random variable. Then for any two Markov times* τ *and* σ (*with respect to* $F = \{\mathscr{F}_t\}$, $t \in Z$) *such that* $P(\tau \ge \sigma) = 1$, *the random variables* ξ_τ *and* ξ_σ *are integrable and*

$$\xi_\sigma \ge M(\xi_\tau|\mathscr{F}_\sigma) \qquad (P\text{-a.s.}). \tag{1.30}$$

If, in particular, $X = (\xi_t, \mathscr{F}_t)$, $t \in Z$, *is a martingale and the variables* $\{\xi_t, t \in Z\}$ *are uniformly integrable, then*

$$\xi_\sigma = M(\xi_\tau|\mathscr{F}_\sigma) \qquad (P\text{-a.s.}). \tag{1.31}$$

Remark. In (1.30) and (1.31)

$$\xi_\tau = \begin{cases} \xi_{\tau(\omega)}(\omega) & \text{if } \tau(\omega) < \infty, \\ \xi_\infty(\omega) = \lim\limits_{t\to\infty} \xi_t(\omega) & \text{if } \tau(\omega) = \infty, \end{cases}$$

where $\lim_{t\to\infty} \xi_t(\omega)$ exists by virtue of Theorem 9.

It follows from (1.31) that for a uniformly integrable martingale $X = (\xi_t, \mathscr{F}_t)$, $t \in Z$, for any Markov time τ

$$M\xi_\tau = M\xi_0. \tag{1.32}$$

The theorem which follows includes the conditions for Equation (1.32) to be satisfied for a *fixed* Markov time satisfying the condition $P(\tau < \infty) = 1$ without assuming that the family $\{\xi_t, t \in Z\}$ is uniformly integrable.

Theorem 12. *Let* $X = (\xi_t, \mathscr{F}_t)$, $t \in Z$, *be a martingale and let* $\tau = \tau(\omega)$ *be a finite Markov time* ($P(\tau < \infty) = 1$). *Then it suffices for Equation* (1.32) *to be satisfied that*

$$M|\xi_\tau| < \infty, \tag{1.33}$$

$$\lim_{t\to\infty} \int_{\{\omega:\tau(\omega) > t\}} \xi_t(\omega) P(d\omega) = 0. \tag{1.34}$$

For proofs of Theorems 9–12 see [72], [97], [45], [68].

1.4 Markov processes

1.4.1

We shall give the main definitions and properties of Markov processes (for discrete as well as continuous time) that we shall need for problems of optimal stopping.

Let $(\Omega, \mathscr{F})$ be a measure space of elementary outcomes, let $F = \{\mathscr{F}_t\}$, $t \in Z$, be a nondecreasing family of sub-σ-algebras $\mathscr{F}$, and let $(E, \mathscr{B})$ be a state space.[10]

Further, let $(x_t, \mathscr{F}_t)$, $t \in Z$, be a random process with values in $(E, \mathscr{B})$ and let probability measure P_x be given on a σ-algebra $\mathscr{F}$ for each $x \in E$.

Definition 1. The system $X = (x_t, \mathscr{F}_t, P_x)$, $t \in Z$, $x \in E$, is said to be a *(homogeneous, nonterminating) Markov process with values in a state space* $(E, \mathscr{B})$ if the following conditions are satisfied:

(1) For each $A \in \mathscr{F}$, $P_x(A)$ is a $\mathscr{B}$-measurable function for x;
(2) For all $x \in E$, $B \in \mathscr{B}$, $s, t \in Z$,

$$P_x(x_{t+s} \in B \mid \mathscr{F}_t) = P_{x_t}(x_s \in B) \qquad (P_x\text{-a.s.}); \tag{1.35}$$

(3) $P_x(x_0 = x) = 1$, $x \in E$;
(4) For each $t \in Z$ and $\omega \in \Omega$ there will be a unique $\omega' \in \Omega$ such that

$$x_s(\omega') = x_{s+t}(\omega) \tag{1.36}$$

for all $s \in Z$.

Condition (1.35) implies the Markov principle of "the future" independent of "the past" for the fixed "present." Condition (1.36) implies that the initial space of elementary outcomes Ω has to be sufficiently "rich" and that the set of trajectories $\{x_t(\omega)\}$, $t \in Z$, has a property of homogeneity.

If $Z = N = \{0, 1, \dots\}$, then $X = (x_t, \mathscr{F}_t, P_x)$ is said to be a *Markov process with discrete time, a Markov random sequence, or a Markov chain.*

Write $\mathscr{F}_t^* = \sigma(\omega : x_3, s \leq t\}$. It is easy to see that along with $X = (x_t, \mathscr{F}_t, P_x)$ the process $X^* = (x_t, \mathscr{F}_t^*, P_x)$ will be also a Markov process.

Definition 2. The progressively measurable Markov process $X = (x_t, \mathscr{F}_t, P_x)$, $t \in Z$, $x \in E$, is said to be a *strong Markov process* if for any Markov time τ (with respect to the system $F = \{\mathscr{F}_t\}$, $t \in Z$) (2) of Definition 1 is strengthened as follows:

(2)* For all $x \in E$, $B \in \mathscr{B}$, $s, t \in Z$,

$$P_x(x_{\tau+s} \in B \mid \mathscr{F}_\tau) = P_{x_\tau}(x_s \in B) \qquad (\{\tau < \infty\}; \qquad (P_x\text{-a.s.})). \tag{1.37}$$

[10] The measure space $(E, \mathscr{B})$ is said to be a *state space* if all one-point sets $\{x\} \in \mathscr{B}$.

It is a known fact that the Markov process with discrete time is always a strong Markov process [45]. In the case of continuous time, this fact is, generally speaking, not true any longer.

Definition 3. The progressively measurable Markov process $X = (x_t, \mathscr{F}_t, P_x)$, $t \in T$, is said to be a *left quasicontinuous* process if for any Markov time σ, x_σ is $\mathscr{F}_\sigma/\mathscr{B}$-measurable and for any nondecreasing sequence of Markov times $\tau_n \uparrow \tau$

$$x_{\tau_n} \to x_\tau \qquad (\{\tau < \infty\}; \qquad (P_x\text{-a.s.})) \tag{1.38}$$

for all $x \in E$.

Write $P(t, x, \Gamma) = P_x(x_t \in \Gamma)$, $x \in E$, $\Gamma \in \mathscr{B}$, $t \in Z$. The function $P(t, x, \Gamma)$ is said to be a *transition function of the Markov process* X. Its properties immediately follow from Definition 1:

(1) $P(t, x, \cdot)$ is a measure on $(E, \mathscr{B})$ for all $x \in E$, $t \in Z$;
(2) $P(t, \cdot, \Gamma)$ is a $\mathscr{B}$-measurable function of x for each $t \in Z$ and $\Gamma \in \mathscr{B}$;
(3) (Kolmogorov–Chapman equation)

$$P(t + s, x, \Gamma) = \int_E P(t, y, \Gamma)P(s, x, dy); \tag{1.39}$$

(4) $P(0, x, \Gamma) = I_\Gamma(x)$.

For the case of discrete time $t \in Z$ the transition function $P(t, x, \Gamma)$ can be completely defined by the transition function $P(x, \Gamma) \equiv P(1, x, \Gamma)$ for one step.

1.4.2

In the case of continuous time the above definition of a Markov process becomes too general to yield a fruitful theory. The present section deals with the crucial concept of a standard Markov process; problems of optimal stopping will be investigated for such processes in Chapter 3.

Definition 4. The strong Markov, right continuous,[11] left quasicontinuous process $X = (x_t, \mathscr{F}_t, P_x)$ $t \in T$, with values in a state space $(E, \mathscr{B})$ is said to be *standard* if:

(A) $\mathscr{F}_t = \mathscr{F}_{t+}$, $t \in T$;
(B) E is a semicompact (i.e., a locally compact Haussdorf space with a countable basis); $\mathscr{B}$ is the σ-algebra of Borel sets in E (i.e., $\mathscr{B} = \sigma(\mathscr{C})$ where $\mathscr{C}$ is a system of open sets in the topological space E).

[11] This is that for each $\omega \in \Omega$ the trajectories $\{x_t(\omega), t \geq 0\}$ are right-continuous functions with respect to $t \geq 0$.

In many problems related to random process theory the σ-algebras $\mathscr{B}$, $\mathscr{F}$, $\mathscr{F}_t$ become too narrow and one has to introduce into consideration the enlarged σ-algebras $\hat{\mathscr{B}}$, $\bar{\mathscr{B}}$, $\bar{\mathscr{F}}$, $\bar{\mathscr{F}}_t$ obtained as follows:

Let μ be a probability measure on $(E, \mathscr{B})$, let $\mathscr{B}^{\mu}$ be the completion of the σ-algebra $\mathscr{B}$ with respect to measure μ, and let $P^{\mu}(A) = \int_E P_x(A)\mu(dx)$, $A \in \mathscr{F}$. Denote (see Section 1) by $\bar{\mathscr{B}} = \bigcap_{\mu} \mathscr{B}^{\mu}$, $\bar{\mathscr{F}} = \bigcap_{\mu} \mathscr{F}^{P^{\mu}}$, $\bar{\mathscr{F}}_t = \bigcap_{\mu} \mathscr{F}_t^{P^{\mu}}$, and $\bar{P}_x$ the extension of measure P_x on $\bar{\mathscr{F}}$.

We shall say that the set $\Gamma \in \hat{\mathscr{B}}$ if for any probability measure μ there will be sets Γ_1 and Γ_2 from $\mathscr{B}$ such that $\Gamma_1 \subseteq \Gamma \subseteq \Gamma_2$ and $P^{\mu}\{I_{\Gamma_1}(x_t) = I_{\Gamma_2}(x_t), t \in T\} = 1$. The system $\hat{\mathscr{B}}$ forms a σ-algebra, $\mathscr{B} \subseteq \hat{\mathscr{B}} \subseteq \bar{\mathscr{B}}$; the sets from $\hat{\mathscr{B}}$ are said to be *almost Borel sets*; the $\hat{\mathscr{B}}$-measurable functions are also said to be *almost Borel functions*.

It is known ([31], Theorem 3.12) that if the process $X = (x_t, \mathscr{F}_t, P_x)$ is standard, the process $\bar{X} = (x_t, \bar{\mathscr{F}}_t, \bar{P}_x)$ is also standard. Hence we can immediately assume in investigating these processes that

$$\mathscr{B} = \bar{\mathscr{B}}, \qquad \mathscr{F} = \bar{\mathscr{F}}, \qquad \mathscr{F}_t = \bar{\mathscr{F}}_t, \qquad P_x = \bar{P}_x. \tag{1.40}$$

Let $B(E, \mathscr{B})$ be a space of $\mathscr{B}$-measurable bounded functions $f = f(x)$, $x \in E$, with a norm $\|f\| = \sup_x |f(x)|$. For each function $f \in B(E, \mathscr{B})$, let us define the transformation

$$T_t f(x) = \int_E f(y)P(t, x, dy), \qquad t \in Z, \tag{1.41}$$

mapping $B(E, \mathscr{B})$ into itself.

(1.41) defines a one-parameter family of linear operators $\{T_t\}$, $t \in Z$. By virtue of (1.39) this family forms a semigroup:

$$T_s \cdot T_t = T_{s+t}, \qquad s, t \in Z.$$

It can be easily seen that this semigroup is a contraction:

$$\|T_t f\| \leq \|f\|, \qquad t \in Z.$$

Let $C(E, \mathscr{B}) \subseteq B(E, \mathscr{B})$ be the space of bounded $\mathscr{B}$-measurable continuous functions given on a space $(E, \mathscr{B})$ satisfying (B) of Definition 4.

Definition 5. The semigroup of operators $\{T_t\}$, $t \in T$, is said to be a *Feller semigroup* (the corresponding transition function $P(t, x, \Gamma)$ is said to be a *Feller function* and the Markov process X a *Feller process*), if for each function $f \in C(E, \mathscr{B})$ the function $T_t f$ is continuous with respect to x for any $t \in T$.

1.4.3

Let $X = (x_t, \mathscr{F}_t, P_x)$, $t \in Z$, be a Markov process. By definition of (1.36), for each $t \in Z$ and $\omega \in \Omega$ there will be a unique element $\omega' \in \Omega$ such that $x_s(\omega') = x_{s+t}(\omega)$ for all $s \in Z$.

Denote by $\theta_t \omega$ the operator which associates the element ω' with the element ω for given $t \in Z$. Therefore,

$$x_s(\theta_t \omega) = x_{s+t}(\omega), \qquad s \in Z,$$

i.e., the operator θ_t "shifts" by t the trajectory $\{x_s(\omega), s \in Z\}$ "to the left."

The operator θ_t thus defined operates in a space of elementary outcomes Ω. It is also useful to define the shift operator which effects the random variables. In fact, if $\eta = \eta(\omega)$ is a random variable, we shall denote by $\theta_t \eta = \theta_t \eta(\omega)$ a random variable to be defined by the equality

$$\theta_t \eta(\omega) = \eta(\theta_t \omega).$$

(Thus, if $\eta(\omega) = x_u(\omega)$, then $\theta_t x_u(\omega) = x_u(\theta_t \omega) = x_{u+t}(\omega)$.)

Let $\tau = \tau(\omega)$ be a random variable with values in $T = [0, \infty)$ (in the case of continuous time) or in $N = \{0, 1, \ldots\}$ (in the case of discrete time). Let us define the operator θ_τ as an operator which associates the elementary event ω with an elementary event $\theta_{\tau(\omega)}\omega$.

In the case where the random variable τ takes on values in $\overline{T} = [0, \infty]$, the operator θ_τ can be defined only on a set $\Omega_\tau = \{\omega : \tau(\omega) < \infty\}$ and not on the whole space Ω.

Using the operator θ_τ introduced we can write the strictly Markov property given by (1.37) in the equivalent form: If $\eta = \eta(\omega)$ is an integrable $\mathscr{F}^x = \sigma\{\omega : x_s, s < \infty\}$-measurable random variable, then for any Markov time $\tau = \tau(\omega)$ (with respect to the system $\{\mathscr{F}_t\}$, $t \in Z$):

$$M_x(\theta_\tau \eta | \mathscr{F}_\tau) = M_{x_\tau} \eta \qquad (\{\tau < \infty\}; \quad (P_x\text{-a.s.})), x \in F. \tag{1.42}$$

It follows from (1.42) that if the random variable ξ is $\mathscr{F}_\tau$-measurable, if η is $\mathscr{F}^x$-measurable, and if $M_x|\xi| < \infty$ and $M_x|\xi \theta_\tau \eta| < \infty$ for $x \in E$, then

$$M_x(\xi \theta_\tau \eta) = M_x(\xi M_{x_\tau} \eta). \tag{1.43}$$

1.4.4

Let $(E, \mathscr{B})$ be a state space and let $B(E, \mathscr{B})$ be the Banach space of bounded measurable functions $f = f(x)$, $x \in E$, with $\|f\| = \sup_x |f(x)|$.

The infinitesimal operator $\mathscr{A}$ of the semigroup $\{T_t\}$, $t \in T$, is defined by the formula

$$\mathscr{A} f(x) = \lim_{t \downarrow 0} \frac{T_t f(x) - f(x)}{t}. \tag{1.44}$$

To define the operator $\mathscr{A}$ completely, we need to prescribe the domain of its definition $\mathscr{D}_{\mathscr{A}}$. We take $\mathscr{D}_{\mathscr{A}}$ to consist of all functions $f \in B(E, \mathscr{B})$ for which the limit in the right-hand side of (1.44) exists uniformly over $x \in E$.

If $\{T_t\}$, $t \in T$, is a semigroup corresponding to the Markov process $X = (x_t, \mathscr{F}_t, P_x)$, $\mathscr{A}$ is said to be an infinitesimal operator of the process X.

Let $X = (x_t, \mathscr{F}_t, P_x)$, $t \in T$, $x \in E$, be a Markov process with state space $(E, \mathscr{B})$ and let $\mathscr{C}$ be some topology ([31], suppl. to §5) in E. We shall fix a certain $\mathscr{B}$-measurable function $f = f(x)$ and a certain point $x \in E$. Denote by $\mathscr{U}(\mathscr{C})$ the aggregate of sets $V \in \mathscr{C} \cap \mathscr{B}$ for which time $\tau(U) = \inf\{t \geq 0 : x_t \in E - U\}$ is $\mathscr{F}$-measurable, $M_x|f(x_{\tau(U)})| < \infty$ ($f(x_\infty)$ is assumed to be zero) and the expression

$$\frac{M_x f(x_{\tau(U)}) - f(x)}{M_x \tau(U)} \tag{1.45}$$

makes sense (if $M_x \tau(U) = \infty$, this expression is assumed to be zero). Denote

$$\mathfrak{U} f(x) = \mathscr{C}\text{-}\lim_{U \downarrow x} \frac{M_x f(x_{\tau(U)}) - f(x)}{M_x \tau(U)}, \tag{1.46}$$

where $\mathscr{C}\text{-}\lim_{U \downarrow x}$ designates the limit with respect to a system of neighborhoods $\mathscr{U}(\mathscr{C})$ contracting to a point x (for details see [31], chap. 5, §3).

The set of all $\mathscr{B}$-measurable functions for which the limit (1.46) exists at the point x is denoted by $\mathscr{D}_{\mathfrak{U}}(x)$ and defines the operator $\mathfrak{U}$ (known as *the characteristic operator* of the process X in topology $\mathscr{C}$) at the point x. If $f \in \mathscr{D}_{\mathfrak{U}}(x)$ for all $x \in G$, we shall write $f \in \mathscr{D}_{\mathfrak{U}}(G)$. If $G = E$, we shall write $f \in \mathscr{D}_{\mathfrak{U}}$.

The characteristic operator is an expansion of the infinitesimal operator (and also of the so-called weak infinitesimal operator; for details, see [31], chap. 5) for a wide class of Markov processes.

1.4.5

Along with the concept of a Markov process we shall deal with the related concept of a Markov random function and that of a Markov family of random functions. We shall give definitions which follow.

Let $(\Omega, \mathscr{F}, P)$ be a probability space, let $F = \{\mathscr{F}_t\}$, $t \in Z$, be a nondecreasing family of sub-σ-algebras $\mathscr{F}$, and let $(x_t, \mathscr{F}_t)$, $t \in Z$, be a random process with values in a measure space $(E, \mathscr{B})$.

Definition 6. The system $X = (x_t, \mathscr{F}_t, P)$, $t \in Z$, is said to be a (*homogeneous, nonterminating*) *Markov random function* if (1.36) is satisfied and if for all $s, t \in Z$, $B \in \mathscr{B}$,

$$P(x_{t+s} \in B | \mathscr{F}_t) = P(x_{t+s} \in B | x_t) \qquad (P\text{-a.s.}). \tag{1.47}$$

We shall assume next that each $x \in E$ is associated with a Markov random function $X^x = (x_t^x, \mathscr{F}_t^x, P^x)$, $t \in Z$, with values in a state space $(E, \mathscr{B})$ (x_t^x are $\mathscr{F}_t^x$-measurable for each $t \in Z$, $\mathscr{F}_t^x \subseteq \mathscr{F}_t$).

Definition 7. The system $X^x = (x_t^x, \mathscr{F}_t^x, P^x)$, $t \in Z$, $x \in E$, is said to be a *Markov family of random functions* if for each fixed $x \in E$, $X^x = (x_t^x, \mathscr{F}_t^x, P^x)$, $t \in Z$, forms a Markov random function and:

(1) $P(t, x, \Gamma) = P^x(x_t^x \in \Gamma)$ is a $\mathscr{B}$-measurable function of $x (\Gamma \in \mathscr{B})$;
(2) $P(0, x, E - \{x\}) = 0$;
(3) $P^x(x_{t+s}^x \in \Gamma | x_t^x) = P(s, x_t^x, \Gamma)$ (P^x-a.s.) for any $s, t \in Z$ and $\Gamma \in \mathscr{B}$.

The concepts of a Markov process and of a Markov family of random functions are closely related. In fact, each Markov process $X = (x_t, \mathscr{F}_t, P_x)$, $t \in Z$, can be treated as a Markov family of random functions with $P^x = P_x$, $x_t^x \equiv x_t$, $\mathscr{F}_t^x \equiv \mathscr{F}_t$. The converse is also true: If $X^x = (x_t^x, \mathscr{F}_t^x, P^x)$, $t \in Z$, $x \in E$, is a Markov family of random functions, one can construct a Markov process with the same transition probability $P(t, x, \Gamma)$ at the expense of the pertinent completion of a space of elementary events. (For details, see [31], p. 119.)

1.4.6

The problems of optimal stopping will be considered from now on preferably for (homogeneous, nonterminating) Markov processes. The theory being discussed can be applied, however, almost without changes for both terminating and nonhomogeneous Markov processes, in connection with which we shall give a few definitions.

Let $(E, \mathscr{B})$ be a state space. Write $E_\Delta = E \cup \{\Delta\}$ where Δ is a (fictitious) point which does not belong to E, and let $\mathscr{B}_\Delta$ be a σ-algebra of subsets E_Δ generated by sets from $\mathscr{B}$. We shall note that the one-point set $\{\Delta\} \in \mathscr{B}_\Delta$ so that the space $(E_\Delta, \mathscr{B}_\Delta)$ is a state space.

We shall assume next that we are given: a measurable space of elementary outcomes $(\Omega, \mathscr{F})$; a family of σ-algebras $\{\mathscr{F}_t^s\}$, $0 \le s \le t$, $s, t \in \bar{Z}$ such that $\mathscr{F}_t^s \subseteq \mathscr{F}_v^u \subseteq \mathscr{F}$ if $0 \le u \le s \le t \le v$; and[12] probability measures $P_{s,x}$ on a σ-algebra $\mathscr{F}^s = \mathscr{F}_\infty^s$ for each pair $(s, x) \in Z \times E_\Delta$. We shall also consider as given a random process $(x_t, \mathscr{F}_t)$, $t \in Z$, with values in $(E_\Delta, \mathscr{B}_\Delta)$ having the property that if $x_{t_0}(\omega) = \Delta$, then $x_t(\omega) = \Delta$, $t \ge t_0$.

Definition 8. The system $X = (x_t, \mathscr{F}_t^s, P_{s,x})$, $s, t \in Z$, $x \in E_\Delta$, is said to be a *(nonhomogeneous, terminating) Markov process with state space* $(E, \mathscr{B})$ *with an adjoined point* Δ if the following conditions are satisfied:

(1) $P_{s,x}(A)$ is a $\mathscr{B}_\Delta$-measurable function of x for each $s \in Z$ and $A \in \mathscr{F}^s$;
(2) For all $x \in E_\Delta$, $B \in \mathscr{B}_\Delta$, $0 \le s \le t \le u$,

$$P_{s,x}(x_u \in B \mid \mathscr{F}_t^s) = P_{t,x_t}(x_u \in B) \qquad (P_{s,x}\text{-a.s.});$$

(3) $P_{s,x}(x_s = x) = 1;\qquad P_{s,x}(x_t \in E_\Delta) = 1,\qquad x \in E_\Delta,\ s \le t.$

The function $P(s, x; t, B) = P_{s,x}(x_t \in B)$ is said to be the *transition function of the Markov process X*.

The variable $\zeta(\omega) = \inf\{t : x_t(\omega) = \Delta\}$ is said to be the *life time* or *termination time* of the Markov process X.

It is a well-known fact that a nonhomogeneous process can be reduced to a homogeneous process at the expense of the primary state space and the space of elementary events, in doing which we replace the spaces Ω and E with spaces $\Omega' = Z \times \Omega$ and $E' = Z \times E$ and assume that $x_t'(\omega') = (t + z, x_{t+z}(\omega))$ for $\omega' = (z, \omega)$; for details see, for example, [30], chap. 4.

[12] Instead of $\mathscr{F}_t^0$ we shall write $\mathscr{F}_t$ and set $\mathscr{F}^s = \mathscr{F}_\infty^s$.

Notes to Chapter 1

1.1. The axiomatics of probability theory have been developed in Kolmogorov's fundamental book *Foundations of the Theory of Probability* [58]. The problems concerning measurability and progressive measurability of random processes are discussed in Dynkin [31] and Meyer [72].

1.2. Additional information concerning the properties of Markov times can be found in Blumenthal and Getoor [12], Gikhman and Skorokhod [45], Dynkin [31], Meyer [72], and Liptser and Shiryayev [69]. For the results and proofs of the assertions stated in Subsections 1.2.3–5 see also the paper by Courrège and Priouret [23].

1.3. The proofs of the theorems on martingales and semimartingales can be found in Gikhman and Skorokhod [45], Doob [28], Meyer [72], and Liptser and Shiryayev [69]. Generalized martingales and semimartingales were investigated in Snell's paper [97].

1.4. The basic definitions and assertions of the theory of Markov processes follow Blumenthal and Getoor [12], Gikhman and Skorokhod [45], and Dynkin [31].

Optimal stopping of Markov sequences

2

2.1 Statement of the problem of optimal stopping

2.1.1

Let $X = (x_n, \mathscr{F}_n, P_x)$, $n \in N = \{0, 1, \ldots\}$, be a homogeneous Markov chain with values in a state space $(E, \mathscr{B})$. Denote by B the aggregate of $\mathscr{B}$-measurable functions $g = g(x)$ taking on values in $(-\infty, \infty]$. Let $\tau = \tau(\omega)$ be a Markov time (with respect to the system $F = \{\mathscr{F}_n\}$, $n \in N$) such that for all $\omega \in \Omega$, $\tau(\omega) \leq N$, $N < \infty$.

We shall associate each Markov time indicated (denoting by $\mathfrak{M}(N)$ the aggregate of these Markov times) and the function $g \in B$ with the random variable[1]

$$g(x_\tau) = g(x_{\tau(\omega)}(\omega)),$$

which will be interpreted as the *gain* to be obtained in the state x_τ with the observation being stopped at time τ.

Assume that for a given $x \in E$ and for the Markov time τ considered the expectation $M_x g(x_\tau)$ (i.e., $M_x g^+(x_\tau)$ and $M_x g^-(x_\tau)$ are not equal to ∞ at the same time). Then it will be natural to interpret the value $M_x g(x_\tau)$ as the mean gain corresponding to a chosen time $\tau \in \mathfrak{M}(N)$ and an initial state x.

Let

$$s_N(x) = \sup M_x g(x_\tau), \tag{2.1}$$

[1] By definition we assume that $x_{\tau(\omega)}(\omega) = x_n(\omega)$ on the set $\{\omega : \tau(\omega) = n\}$; hence $g(x_\tau) = \sum_{n=0}^{N} g(x_n) I_{\{\tau = n\}}(\omega)$ where $I_{\{\tau = n\}}(\omega)$ is the indicator of the set $\{\tau = n\}$.

where the supremum is taken over the Markov times $\tau \in \mathfrak{M}(N)$ for which the expectations $M_x g(x_\tau)$ are defined for all $x \in E$.

The function $s_N(x)$ defined in (2.1) is said to be a *payoff*, and the time τ_N^* from the class $\mathfrak{M}(N)$ for which the expectations $M_x g(x_{\tau_N^*})$ are defined for all $x \in E$ and

$$M_x g(x_{\tau_N^*}) = s_N(x), \qquad x \in E$$

is said to be an *optimal stopping time* (in the class $\mathfrak{M}(N)$).

Our objective in Sections 2–3 will be to determine the structure of the payoff $s_N(x)$ and methods for finding it and the optimal stopping times τ (the existence of such optimal times will be proved below).

Denote by

$$\mathfrak{M}_g(N) = \{\tau \in \mathfrak{M}(N) : M_x g^-(x_\tau) < \infty, x \in E\}$$

the class of Markov times from $\mathfrak{M}(N)$ for which the expectations $M_x g(x_\tau)$ are defined for all $x \in E$ and are larger than $-\infty$. We note that the set $\mathfrak{M}_g(N)$ is nonempty; it contains the time $\tau \equiv 0$.

We shall show that it is sufficient to take the supremum in (2.1) only from the class $\mathfrak{M}_g(N)$, i.e.,

$$s_N(x) = \sup_{\tau \in \mathfrak{M}_g(N)} M_x g(x_\tau). \tag{2.2}$$

Let $\tau \in \mathfrak{M}(N)$ be such that the $M_x g(x_\tau)$ are defined for all $x \in E$. We shall fix a point $x_0 \in E$. To prove the assertion it suffices to show that there will be a time $\tau_0 \in \mathfrak{M}_g(N)$ such that

$$M_{x_0} g(x_{\tau_0}) \geq M_{x_0} g(x_\tau)$$

and at the same time for all $x \in E$

$$M_x g^-(x_{\tau_0}) < \infty.$$

If $M_{x_0} g^-(x_\tau) = \infty$, we can set $\tau_0 \equiv 0$. Then it is clear that $M_x g^-(x_{\tau_0}) = g^-(x) < \infty$ and at the same time $-\infty = M_{x_0} g(x_\tau) < g(x_0) = M_{x_0} g(x_{\tau_0})$.

If $M_{x_0} g^-(x_\tau) < \infty$, we set

$$\tau_0(\omega) = \begin{cases} \tau(\omega) & \text{if } x_0(\omega) = x_0 \\ 0 & \text{if } x_0(\omega) \neq x_0. \end{cases}$$

Since the one-point set $\{x_0\} \in \mathscr{B}$, the time τ_0 is Markov. It is also clear that $M_{x_0} g(x_\tau) = M_{x_0} g(x_{\tau_0})$ and $M_x g^-(x_\tau) < \infty$ for all $x \in E$, thus justifying (with the aid of (2.2)) the definition of the payoff $s_N(x)$.

Remark. To avoid stipulating that in defining the payoff $s_N(x)$ we take the supremum in (2.1) over the Markov times $\tau \in \mathfrak{M}(N)$ for which the expectations $M_x g(x_\tau)$ are defined, we shall agree to set $M_x g(x_\tau) = -\infty$ if $M_x g^+(x_\tau) = \infty$ and $M_x g^-(x_\tau) = \infty$ at the same time. We can then show, with the aid of the construction given above, that (2.2) also holds for the payoff $s_N(x)$. We will then also not have to stipulate that the mathematical expectations considered are defined.

It is clear that the payoff $s_N(x)$ is the maximum possible mean gain to be obtained when observation time is bounded by the number N. The problems of optimal stopping involving unbounded observation time are also of some interest. We shall introduce some necessary notations and concepts to make appropriate formulations.

Let $X = (x_n, \mathscr{F}_n, P_x)$, $n \in N$, be a (nonterminating, homogeneous) Markov chain with values in a state space $(E, \mathscr{B})$. Let $\overline{\mathfrak{M}}$ be a class of all Markov times (with respect to a system $F = \{\mathscr{F}_n\}$, $n \in N$) with values in $\bar{N} = N \cup \{+\infty\}$, and let $\mathfrak{M}$ be a class of all finite Markov times, i.e., times $\tau \in \overline{\mathfrak{M}}$ for which

$$P_x(\tau < \infty) = 1, \qquad x \in E.$$

(Times $\tau \in \mathfrak{M}$ are said to be stopping times.)

In addition to the class $\mathfrak{M}_g(N)$ introduced above we shall introduce the class $\mathfrak{M}_g = \{\tau \in \mathfrak{M} : M_x g^-(x_\tau) < \infty, x \in E\}$ and the function

$$s(x) = \sup_{\tau \in \mathfrak{M}_g} M_x g(x_\tau). \tag{2.3}$$

This function is also said to be a *payoff*. A stopping time $\tau_\varepsilon \in \mathfrak{M}_g (\varepsilon \geq 0)$ is said to be an *(ε, s)-optimal time* if for all $x \in E$

$$s(x) - \varepsilon \leq M_x g(x_{\tau_\varepsilon}).$$

A $(0, s)$-optimal stopping time is said to be simply an *optimal time*.

We shall note that as in the case $s_N(x)$, in this case also the payoff $s(x)$ does not change if we take the supremum in (2.3) not over the class $\mathfrak{M}_g$ but over a wider class of stopping times τ for which the expectations $M_x g(x_\tau)$ are defined for all $x \in E$ or even over the class $\mathfrak{M}$ (bearing in mind the remark made above).

The investigation of the problems of existence and structure of optimal stopping times shows that it is wise to introduce into consideration also the payoff

$$\bar{s}(x) = \sup_{\tau \in \overline{\mathfrak{M}}_g} M_x g(x_\tau), \tag{2.4}$$

where

$$\overline{\mathfrak{M}}_g = \{\tau \in \overline{\mathfrak{M}} : M_x g^-(x_\tau) < \infty, x \in E\},$$

and[2]

$$g(x_\infty) = \overline{\lim_{n \to \infty}}\, g(x_n). \tag{2.5}$$

As in the cases of $s_N(x)$ and $s(x)$, the payoff $\bar{s}(x)$ does not change here if in defining it we take the supremum not over the class $\overline{\mathfrak{M}}_g$ but over the class of

[2] The extension of the definition of the function $g(x_\tau)$ for $\tau = \infty$ by means of the expression in the right-hand side of (2.5) will be clear from the material to follow (see Remark 1 to Theorem 4 in Section 5).

the Markov times $\tau \in \overline{\mathfrak{M}}$ for which the $M_x g(x_\tau)$ are defined for all $x \in E$ or even over the class $\mathfrak{M}$ (bearing in mind the remark made above).

We shall say that the time $\tau_\varepsilon \in \overline{\mathfrak{M}}_g$ is an $(\varepsilon, \bar{s})$-optimal time if for all $x \in E$

$$\bar{s}(x) - \varepsilon \leq M_x g(x_{\tau_\varepsilon}).$$

It will be clear from the material which follows that each (ε, s)-optimal time is an $(\varepsilon, \bar{s})$-optimal time at the same time. Hence, for brevity the (ε, s)-optimal times are said to be *ε-optimal times* and the 0-optimal times are said to be *optimal stopping times*.

Each stopping time (and, in general, each Markov time) determines, as one says, a *stopping rule*. The principal problem to be discussed in the present chapter is the one of finding optimal (or near optimal) stopping rules. The theory of optimal stopping rules for Markov chains to be investigated here will be used later on for constructing the pertinent theory for Markov random processes involving continuous time (Chapter 3) as well as for investigating specific problems in statistics (Chapter 4).

2.2 Optimal stopping rules in the classes $\mathfrak{M}(n)$ and $\mathfrak{M}(m; n)$

2.2.1

The main result of the present section is contained in Theorem 1, in which the structure of the payoff $s_N(x)$ will be described and methods for finding it will be given; also an optimal stopping time will be found.

We shall say that the function $g = g(x)$ belongs to the class L if $g \in B$ and $M_x g^-(x_1) < \infty$, $x \in E$.

Let us now define the operator Q operating on functions $g \in L$ by

$$Qg(x) = \max\{g(x), Tg(x)\}, \tag{2.6}$$

where $T_g(x) = M_x g(x_1)$.

Also, let

$$s_n(x) = \sup_{\tau \in \mathfrak{M}_g(n)} M_x g(x_\tau), \qquad n = 0, 1, \ldots.$$

Theorem 1. *Let a function $g \in L$. Then:*

(1) $$s_n(x) = Q^n g(x), \qquad n = 0, 1, \ldots; \tag{2.7}$$

(2) $$s_n(x) = \max\{g(x), Ts_{n-1}(x)\}, \tag{2.8}$$

where $s_0(x) = g(x)$;

(3) *In the class $\mathfrak{M}_g(n)$ the time*

$$\tau_n^* = \min\{0 \leq m \leq n : s_{n-m}(x_m) = g(x_m)\} \tag{2.9}$$

is an optimal time:

$$M_x g(x_{\tau_n^*}) = s_n(x);$$

(4) *If* $M_x g^-(x_k) < \infty$, $k = 1, \ldots, n$, *then the time* τ_n^* *belongs to the class* $\mathfrak{M}(n)$ *and is an optimal time in the class* $\mathfrak{M}(n)$.

Before proving the theorem, we shall consider in detail the structure of operators Q^n and give two auxiliary propositions.

By definition, for $g \in L$,

$$Q^n g(x) = \max\{Q^{n-1}g(x), TQ^{n-1}g(x)\}, \qquad n = 1, 2, \ldots,$$

where $Q^0 g(x) = g(x)$. It is useful to note that the sequence of operators Q^n, $n = 0, 1, \ldots$, satisfies also (simpler) recursion relations:

$$Q^n g(x) = \max\{g(x), TQ^{n-1}g(x)\}, \qquad n = 1, \ldots. \tag{2.10}$$

In fact, since

$$T[\max g, Tg)] \geq Tg,$$

then

$$\begin{aligned} Q^2 g(x) &= \max\{Qg(x), TQg(x)\} \\ &= \max\{\max[g(x), Tg(x)], T[\max(g, Tg)](x)\} \\ &= \max\{g(x), T[\max(g, Tg)](x)\} = \max\{g(x), TQg(x)\}. \end{aligned}$$

Thus, we have proved (2.10) for $n = 2$. It can be proved in the general case, by induction, for example.

The two lemmas for $g \in L$ which follow will be used essentially for proving Theorem 1.

Lemma 1. *For any Markov time* $\tau \in \mathfrak{M}_g(n)$

$$M_x g(x_\tau) \leq Q^n g(x), \qquad x \in E, \tag{2.11}$$

and, therefore,

$$s_n(x) \leq Q^n g(x), \qquad x \in E. \tag{2.12}$$

Lemma 2. *For any* $n = 0, 1, \ldots$

$$Q^n g(x) = M_x g(x_{\sigma_n}), \qquad x \in E, \tag{2.13}$$

where

$$\sigma_n = \min\{0 \leq k \leq n : Q^{n-k}g(x_k) = g(x_k)\}. \tag{2.14}$$

PROOF OF LEMMA 1. For $n = 0$, (2.11) is obvious. Let $\tau \in \mathfrak{M}_g(n)$, $n > 0$. Set $A = \{\omega : \tau = n\}$. Obviously, the event[3]

$$A = \Omega - \sum_{i=0}^{n-1} \{\tau = i\} \in \mathscr{F}_{n-1}$$

[3] $I_A = I_A(\omega)$ is the indicator of the set A; $\sum_{i \geq 1} A_i$ indicates the union $\bigcup_{i \geq 1} A_i$ for sets A_1, $A_2, \ldots$ that are pairwise nonintersecting.

and, therefore,

$$\begin{aligned}
M_x g(x_\tau) &= M_x I_{\bar{A}} g(x_\tau) + M_x I_A g(x_\tau) \\
&= M_x I_{\bar{A}} g(x_{\tau \wedge (n-1)}) + M_x I_A g(x_n) \\
&= M_x I_{\bar{A}} g(x_{\tau \wedge (n-1)}) + M_x \{ I_A M_x [g(x_n) | \mathscr{F}_{n-1}] \} \\
&= M_x I_{\bar{A}} g(x_{\tau \wedge (n-1)}) + M_x I_A M_{x_{n-1}} g(x_1) \\
&= M_x I_{\bar{A}} g(x_{\tau \wedge (n-1)}) + M_x I_A M_{x_{\tau \wedge (n-1)}} g(x_1) \\
&\le M_x \max[g(x_{\tau \wedge (n-1)}), M_{x_{\tau \wedge (n-1)}} g(x_1)] \\
&= M_x Q g(x_{\tau \wedge (n-1)}).
\end{aligned}$$

Thus,

$$M_x g(x_\tau) \le M_x Q g(x_{\tau \wedge (n-1)}), \tag{2.15}$$

from which we have

$$\begin{aligned}
M_x g(x_\tau) &\le M_x Q g(x_{\tau \wedge (n-1)}) \\
&\le M_x Q^2 g(x_{\tau \wedge (n-2)}) \le \cdots \le M_x Q^n g(x_{\tau \wedge 0}) = Q^n g(x),
\end{aligned} \tag{2.16}$$

thus proving (2.11) and (2.12). □

Remark 1. Let $\mathfrak{M}(m; n)$ be a class of Markov times $\tau \in \mathfrak{M}(n)$ such that for all $\omega \in \Omega$

$$m \le \tau(\omega) \le n.$$

Then (under the assumption that $g \in L$) we can show by using the method for deducing inequalities (2.15) and (2.16) that, for $\tau \in \mathfrak{M}_g(m; n) = \{\tau \in \mathfrak{M}(m; n) : M_x g^-(x_\tau) < \infty, x \in E\}$,

$$M_x[g(x_\tau) | \mathscr{F}_m] \le Q^{n-m} g(x_m) \qquad (P_x\text{-a.s.}),\ x \in E. \tag{2.17}$$

Proof of Lemma 2. Induction will be used. For $n = 0$ the assertion of the lemma is obvious.

Let Equation (2.13) hold for some $n \ge 0$. We shall show that it can be satisfied for $n + 1$.

We fix a point $x \in E$. If $P_x\{\sigma_{n+1} = 0\} = 1$, then by virtue of (2.14)

$$P_x\{Q^{n+1} g(x_0) = g(x_0)\} = 1$$

and, therefore,

$$Q^{n+1} g(x) = g(x) = M_x g(x_{\sigma_{n+1}}).$$

Let $P_x\{\sigma_{n+1} = 0\} < 1$. Then, since $\{\sigma_{n+1} = 0\} \in \mathscr{F}_0$, according to the zero-one law ([31], p. 124) we have

$$P_x\{\sigma_{n+1} = 0\} = 0,$$

and hence $P_x\{\sigma_{n+1} \ge 1\} = 1$. We shall show that in this case

$$\sigma_{n+1} = 1 + \theta_1 \sigma_n \qquad (P_x\text{-a.s.}).$$

In fact, by virtue of the properties of the operators θ_n (Section 1.4)

$$\begin{aligned}\theta_1\sigma_n(\omega) &= \theta_1 \min\{0 \le k \le n : Q^{n-k}g(x_k(\omega)) = g(x_k(\omega))\}\\ &= \min\{0 \le k \le n : Q^{n-k}g(x_k(\theta_1\omega)) = g(x_k(\theta_1\omega))\}\\ &= \min\{0 \le k \le n : Q^{n-k}g(x_{k+1}(\omega)) = g(x_{k+1}(\omega))\}\\ &= \min\{0 \le k \le n : Q^{n+1-(k+1)}g(x_{k+1}(\omega)) = g(x_{k+1}(\omega))\}.\end{aligned}$$

Therefore, we have

$$\begin{aligned}1 + \theta_1\sigma_n(\omega) = {}& \min\{1 \le k+1 \le n+1 : Q^{n+1-(k+1)}g(x_{k+1}(\omega))\\ &= g(x_{k+1}(\omega))\}\\ &= \min\{1 \le l \le n+1 : Q^{n+1-l}g(x_l(\omega))\\ &= g(x_l(\omega))\} = \sigma_{n+1}(\omega),\end{aligned}$$

where the last equality follows from the definition of the time σ_{n+1} and the assumption $P_x\{\sigma_{n+1} \ge 1\} = 1$.

In the case in question (i.e., where $P_x\{\sigma_{n+1} \ge 1\} = 1$), $Q^{n+1}g(x) > g(x)$. From this and also by virtue of the inductive assumption and (2.10) we have

$$\begin{aligned}Q^{n+1}g(x) &= \max\{g(x), M_x Q^n g(x_1)\}\\ &= M_x M_{x_1} g(x_{\sigma_n})\\ &= M_x \theta_1 g(x_{\sigma_n}) = M_x g(x_{1+\theta_1\sigma_n}) = M_x g(x_{\sigma_{n+1}}),\end{aligned}$$

thus proving Lemma 2. □

2.2.2

PROOF OF THEOREM 1. It follows from (2.12) and (2.13) that

$$s_n(x) \le Q^n g(x) = M_x g(x_{\sigma_n}).$$

The time $\sigma_n \in \mathfrak{M}_g(n)$ and, obviously,

$$s_n(x) = M_x g(x_{\sigma_n}).$$

Hence, for any $n = 0, 1, \ldots,$

$$s_n(x) = Q^n g(x) = M_x g(x_{\sigma_n})$$

and, therefore, the time $\sigma_n(=\tau_n^*)$ is an optimal time in the class $\mathfrak{M}_g(n)$. The recursion relations (2.8) follow from (2.7) and (2.10). The latter assertion of the theorem follows from the fact that in the case considered the class $\mathfrak{M}_g(n)$ coincides with $\mathfrak{M}(n)$, thus completing the proof. □

2.2.3

Remark 2. For a fixed N we set

$$\Gamma_n^N = \{x : s_{N-n}(x) = g(x)\}, \qquad 0 \le n \le N.$$

Theorem 1 implies that the optimal stopping time τ_N^* can be described in terms of *stopping domains* Γ_n^N, $0 \le n \le N$, as follows:

$$\tau_N^* = \min\{0 \le n \le N : x_n \in \Gamma_n^N\}. \tag{2.18}$$

In other words, if $x_0 \in \Gamma_0^N$, then the optimal stopping rule prescribes an instant stop. If $x_0 \notin \Gamma_0^N$, we need to carry out an observation and to stop the observation (in the case $x_1 \in \Gamma_1^N$) or to carry out the next observation (in the case $x_1 \notin \Gamma_1^N$) etc., depending on the value of x_1 obtained. It is clear that the observation process is known to stop at time N since $\Gamma_N^N = E$.

Along with the stopping domains Γ_n^N, $0 \leq n \leq N$, we shall consider the *domains of continued observations* $C_n^N = E - \Gamma_n^N$, $0 \leq n \leq N$. It is clear that

$$\begin{aligned} C_N^N &= \varnothing \\ C_{N-1}^N &= \{x : s_1(x) > g(x)\} \\ &= \{x : Qg(x) > g(x)\} \\ &= (x : Tg(x) > g(x)\} = \{x : Lg(x) > 0\}, \end{aligned}$$

where the operator $L = T - I$ (I is the identity operator). Since $g(x) \leq s_1(x) \leq \cdots \leq s_N(x)$, the domains C_n^N, $0 \leq n \leq N$, satisfy the following chain of inclusions:

$$\varnothing = C_N^N \subseteq C_{N-1}^N \subseteq \cdots \subseteq C_0^N.$$

In particular,

$$C_0^N = \{x : s_N(x) > g(x)\} \supseteq \{x : Lg(x) > 0\},$$

which fact indicates that if "it is advantageous to make one observation" at a point x (i.e., $T_g(x) > g(x)$), then this point is known to belong to the domain of continued observation. (These arguments will be given in detail in Section 12 while finding so-called "truncation" criteria for optimal stopping rules.)

Remark 3. We shall denote by $\mathfrak{C}(N)$ the class of stopping times $\tau \in \mathfrak{M}(N)$ which are times of first entry into the sets (i.e., $\tau \in \mathfrak{C}(N)$ if

$$\tau = \min\{0 \leq n \leq N : x_n \in C_n^N\}$$

for some Borel sets C_n^N, $0 \leq n \leq N$). Then

$$s_N(x) = \sup_{\tau \in \mathfrak{C}(N)} M_x g(x_\tau).$$

In this sense the class $\mathfrak{C}(N)$ can be referred to as a *sufficient class of stopping times.*

Remark 4. It follows from Theorem 1 that to find the optimal stopping time $\tau_N^* \in \mathfrak{M}(N)$ we need to know the payoffs $s_n(x)$ for all $0 \leq n \leq N$. Therefore, to solve the problem of optimal stopping in the class $\mathfrak{M}(N)$ we need to solve analogous problems in the classes $\mathfrak{M}(1), \ldots, \mathfrak{M}(N-1)$. In this case the pertinent payoffs $s_1(x), \ldots, s_{N-1}(x)$ can be found with the help of iterations of the operator Q:

$$s_1(x) = Qg(x), \ldots, s_{N-1}(x) = Q^{N-1}g(x),$$

or, equivalently, with the help of the recursion relations

$$s_n(x) = \max\{g(x), T_{s_{n-1}}(x)\}.$$

Remark 5. We shall say that the random process $\xi = \{\xi_n, 0 \le n \le N\}$ majorizes a process $\eta = \{\eta_n, 0 \le n \le N\}$ if for each $0 \le n \le N$, $\xi_n \ge \eta_n$ (with probability 1). It follows from Theorem 1 that the process $(s_{N-n}(x_n), \mathscr{F}_n, P_x)$, $0 \le n \le N$, is the smallest supermartingale majorizing the process $\{g(x_n), 0 \le n \le N\}$ for each $x \in E$.

Remark 6. The condition $g \in L$ stipulated in the theorem guaranteed the existence of all expectations considered. If we consider problems of optimal stopping in the class $\mathfrak{M}(N)$, we can replace this condition with the condition $M_x g^+(x_n) < \infty$, $n = 0, 1, \ldots, N$, $x \in E$. The assertions of Lemmas 1 and 2 and Theorem 1 will hold in this case for all $n \le N$.

2.2.4

The present subsection deals with a generalization of the problem of optimal stopping investigated above in the classes $\mathfrak{M}(N)$. (This generalization proves, in particular, to be useful for reducing problems of optimal stopping for nonhomogeneous Markov chains to the case considered involving homogeneous Markov sequences. For details, see Section 12.)

In addition to the class $\mathfrak{M}(m; n)$ introduced above (see Remark 1) we shall consider the class $\mathfrak{M}_g(m; n) = \{\tau \in \mathfrak{M}(m; n) : M_x g^-(x_\tau) < \infty, x \in E\}$. It is clear that in the case $m = 0$ the class $\mathfrak{M}(0; n)$ ($\mathfrak{M}_g(0; n)$) coincides with the class $\mathfrak{M}(n)$ ($\mathfrak{M}_g(n)$).

Set

$$s_{m,n}(x) = \sup_{\tau \in \mathfrak{M}_g(m; n)} M_x g(x_\tau), \tag{2.19}$$

and let[4]

$$\gamma_{m,n}(x; \omega) = \operatorname*{ess\,sup}_{\tau \in \mathfrak{M}_g(m; n)} M_x[g(x_\tau) | \mathscr{F}_m]. \tag{2.20}$$

Theorem 2. *Let a function $g \in L$. Then for all $0 \le m \le n < \infty$:*

(1) $$s_{m,n}(x) = M_x s_{n-m}(x_m), \qquad x \in E; \tag{2.21}$$

(2) *For each $x \in E$,*

$$\gamma_{m,n}(x; \omega) = s_{n-m}(x_m) \qquad (P_x\text{-a.s.}); \tag{2.22}$$

(3) *The time*

$$\tau^*_{m,n} = \min\{m \le k \le n : s_{n-k}(x_k) = g(x_k)\} \tag{2.23}$$

[4] For $\{\xi_\alpha, \alpha \in \mathfrak{U}\}$, a family of random variables given on a probability space $(\Omega, \mathscr{F}, P)$, $\operatorname{ess\,sup}_{\alpha \in \mathfrak{U}} \xi_\alpha$ will be understood as a random variable η such that: $P(\eta \ge \xi_\alpha) = 1$ for each $\alpha \in \mathfrak{U}$; if $\tilde{\eta}$ is such that $P(\tilde{\eta} \ge \xi_\alpha) = 1$ for each $\alpha \in \mathfrak{U}$, then $P(\eta \le \tilde{\eta}) = 1$. For proof of the existence of this random variable, see [77], [22].

is an optimal time (in the class $\mathfrak{M}_g(m; n)$*) in the sense that* (P_x-a.s.)

$$M_x[g(x_{\tau^*_{m,n}})|\mathscr{F}_m] = \operatorname*{ess\,sup}_{\tau \in \mathfrak{M}_g(m,n)} M_x[g(x_\tau)|\mathscr{F}_m], \tag{2.24}$$

and

$$M_x g(x_{\tau^*_{m,n}}) = \sup_{\tau \in \mathfrak{M}_g(m;n)} M_x g(x_\tau), \tag{2.25}$$

where $x \in E$.

PROOF. By virtue of (2.17), for any $\tau \in \mathfrak{M}_g(m; n)$ we have

$$M_x[g(x_\tau)|\mathscr{F}_m] \le Q^{n-m}g(x_m) \qquad (P_x\text{-a.s.}),\ x \in E.$$

Since $Q^{n-m}g(y) = s_{n-m}(y)$, $y \in E$, we have

$$M_x[g(x_\tau)|\mathscr{F}_m] \le s_{n-m}(x_m).$$

We shall show that for the time $\tau^*_{m,n}$ introduced in (2.23)

$$M_x[g(x_{\tau^*_{m,n}})|\mathscr{F}_m] = Q^{n-m}g(x_m).$$

Let

$$\tau^0_{n-m} = \min\{0 \le k \le n - m : s_{n-m-k}(x_k) = g(x_k)\}.$$

Then, since

$$g(x_{\tau^*_{n,m}}) = g(\theta_m x_{\tau^0_{n-m}}) = \theta_m g(x_{\tau^0_{n-m}}),$$

by virtue of the Markovianness and Theorem 1 (P_x-a.s.)

$$\begin{aligned} M_x[g(x_{\tau^*_{m,n}})|\mathscr{F}_m] &= M_x[\theta_m g(x_{\tau^0_{n-m}})|\mathscr{F}_m] \\ &= M_{x_m} g(x_{\tau^0_{n-m}}) \\ &= s_{n-m}(x_m) = Q^{n-m}g(x_m), \end{aligned}$$

where $x \in E$.

Therefore, for any $\tau \in \mathfrak{M}_g(m; n)$ (P_x-a.s.)

$$M_x[g(x_\tau)|\mathscr{F}_m] \le M_x[g(x_{\tau^*_{m,n}})|\mathscr{F}_m] = s_{n-m}(x_m), \qquad x \in E, \tag{2.26}$$

and, therefore,

$$M_x g(x_\tau) \le M_x g(x_{\tau^*_{m,n}}) = M_x s_{n-m}(x_m). \tag{2.27}$$

Inequalities (2.26) and (2.27) imply immediately that all the assertions of Theorem 2 are true.

2.2.5

Remark 7. Theorem 2 substantiates the so-called "principle of backward induction" which is used in subsequent analysis (in dynamic programming) as the basis for finding optimal stopping rules (optimal strategies). The considerations leading us to the "principle of backward induction" (compare [9]) are as follows.

We shall consider the problem of optimal stopping for a function $g \in L$ in the class $\mathfrak{M}_g(N; N)$. It is clear that in this case the class $\mathfrak{M}_g(N; N)$ consists of one time $\tau = N$ and, therefore, the gain obtained (in the state x_N) is equal to $g(x_N)$. We shall consider next the class $\mathfrak{M}_g(N-1; N)$. In this case we can either stop observation (at time $N-1$), thus obtaining a gain $g(x_{N-1})$, or make another observation, thus obtaining a gain equal on the average to $M_x[g(x_N)|\mathscr{F}_{N-1}]$. Hence, intuitively, the fact seems to be justified that the stopping time

$$\tilde{\tau} = \begin{cases} N-1 & \text{if } g(x_{N-1}) \geq M_x[g(x_N)|\mathscr{F}_{N-1}], \\ N & \text{if } g(x_{N-1}) < M_x[g(x_N)|\mathscr{F}_{N-1}] \end{cases}$$

is an optimal time.

If we take into account that (P_x-a.s.)

$$M_x[g(x_N)|\mathscr{F}_{N-1}] = M_x[s_0(x_N)|\mathscr{F}_{N-1}] = M_{x_{N-1}}s_0(x_N),$$

$$s_1(x) = \max\{g(x), M_x s_0(x_1)\},$$

then the time $\tilde{\tau}$ can be written as follows:

$$\tilde{\tau} = \min\{N-1 \leq K \leq N;\ s_{N-1}(x_k) = g(x_k)\},$$

i.e., $\tilde{\tau} = \tau^*_{N-1,N}$ (see (2.23)).

Therefore, the intuitive method (implying that the gain resulted from "stopping" is compared with the gain resulted from "continuing") leads us to a stopping time which, as proved in Theorem 2, is, in fact, optimal (in the class $\mathfrak{M}_g(N-1; N)$). Similarly, we would consider optimality of $\tau^*_{N-2,N}$, $\tau^*_{N-3,N}, \ldots$ in the classes $\mathfrak{M}_g(N-2); N)$, $\mathfrak{M}_g(N-3; N), \ldots$.

Remark. If the condition $g \in L$ is replaced with the condition $M_x g^+(x_n) < \infty$, $n = 0, \ldots, N$, the theorem will still hold for all $m \leq n \leq N$ (compare Remark 6 to Theorem 1).

2.3 An optimal selection problem

2.3.1

As the first illustration of the methods given above for finding optimal stopping rules we shall give a solution of the optimal selection problem mentioned above, with the conditions which follow.

We have n objects numbered $1, 2, \ldots, n$, so that, say, an object numbered 1 is classified as "the best"..., and an object numbered n is classified as "the worst." It is assumed that the objects arrive at times $1, 2, \ldots, n$ in a random order (i.e., all $n!$ permutations are equally probable). It is clear from comparing any two of these objects which one is better, although their actual numbers still remain unknown. After having known each sequential object, we either reject this object (in this case we cannot have it back) or take it (in this case the choosing process is stopped). The problem consists in choosing the "best" object with maximal probability, i.e., the object numbered 1.

We shall show that the problem considered can be formulated as a problem of optimal stopping of a certain (relatively simple) Markov chain.

As the space of elementary events $\Omega = \{\omega\}$, $\omega = (\omega_1, \ldots, \omega_n)$, we shall consider the set of all permutations of numbers $1, 2, \ldots, n$, assuming in this case that

$$P(\omega) = \frac{1}{n!}.$$

For each $\omega = (\omega_1, \ldots, \omega_n)$ and $k = 1, 2, \ldots, n$ let a variable $y_k(\omega)$ be equal to the number of terms of $\omega_1, \ldots, \omega_k$, smaller or equal to ω_k, and let $\mathscr{F}_k = \sigma\{\omega : y_1, \ldots, y_k\}$. We note that the events in the σ-algebra $\mathscr{F}_k$ constitute the complete information available to the observer before time k as a result of comparing the objects observed at times $1, 2, \ldots, k$.

It is easy to verify that

$$P\{y_k(\omega) = i) = \frac{1}{k}, \qquad i = 1, 2, \ldots, k, \tag{2.28}$$

the random variables $y_1, \ldots, y_n$ being independent.

Let $\tau = \tau(\omega)$ be a stopping time (with respect to the system of σ-algebras $\{\mathscr{F}_k\}$, $k = 1, 2, \ldots, n$) taking on values $1, 2, \ldots, n$. The problem on hand of choosing the best object can be formulated as a problem of finding a time τ^* for which

$$P\{\omega_{\tau^*} = 1\} = \sup P\{\omega_\tau = 1\}, \tag{2.29}$$

where sup is taken over all the stopping times τ described.

To apply the theory discussed above of constructing optimal stopping rules we need to consider a homogeneous Markov chain $Z = (z_k, \mathscr{F}_k, P_z)$, $k = 1, \ldots, n$, with $z_k = (k, y_k)$, and for $z = (k, y)$ set

$$g(z) = \begin{cases} \dfrac{k}{n} & \text{if } y = 1, \\ 0 & \text{if } y > 1. \end{cases}$$

Simple calculations yield

$$P\{\omega_k = 1 \mid \mathscr{F}_k\} = P\{\omega_k = 1 \mid y_k\} = \begin{cases} \dfrac{k}{n} & \text{if } y_k = 1, \\ 0 & \text{if } y_k > 1. \end{cases}$$

Hence

$$\begin{aligned} P\{\omega_\tau = 1\} &= \sum_{k=1}^{n} \int_{\{\tau = k\}} I_{\{\omega_k = 1\}}(\omega) P(d\omega) \\ &= \sum_{k=1}^{n} \int_{\{\tau = k\}} P\{\omega_k = 1 \mid \mathscr{F}_k\} P(d\omega) = M_{(1,1)} g(z_\tau), \end{aligned} \tag{2.30}$$

from which fact it follows that finding the optimal time for choosing the best object can be reduced to finding a time τ^* for which

$$M_{(1,1)}g(z_{\tau^*}) = \sup M_{(1,1)}g(z_\tau), \tag{2.31}$$

where sup is taken over all Markov times τ (with respect to the system $\{\mathscr{F}_k\}$, $k = 1, \ldots, n$).

For $m = 1, \ldots, n$ (compare with (2.20)), write

$$\gamma_{m,n}(\omega) = \operatorname*{ess\,sup}_{\tau \in \mathfrak{M}(m;n)} M_{(1,1)}[g(z_\tau) | \mathscr{F}_m].$$

By virtue of Theorem 2 the variable $\gamma_{m,n}(\omega)$ is dependent of ω in $y_m(\omega)$. Hence there will be a function $s^m(y)$ such that

$$\gamma_{m,n}(\omega) = s^m(y_m(\omega)).$$

According to the same theorem, $\gamma_{m,n}(\omega) = s^m(y_m(\omega)) = s_{n-m}(y_m(\omega))$ where the functions $s_{n-m}(y)$ are subordinate to the recursion relations given by (2.8). Therefore, for the functions $s^m(y)$ $(m = 1, \ldots, n)$ we have

$$s^m(y) = \max\left\{g(m, y), \frac{1}{m+1} \sum_{i=1}^{m+1} s^{m+1}(i)\right\}, \tag{2.32}$$

where $m = 1, 2, \ldots, n$, and

$$s^n(y) = g(n, y) = \begin{cases} 1 & \text{if } y = 1, \\ 0 & \text{if } y > 1. \end{cases}$$

The system of equations given by (2.32) can be easily solved by means of "backward induction." In fact, let $m^* = m^*(n)$ be found from the inequalities

$$\frac{1}{n-1} + \frac{1}{n-2} + \cdots + \frac{1}{m^*} \leq 1 < \frac{1}{n-1} + \cdots + \frac{1}{m^*-1}.$$

(We shall note that $m^*(n) \sim n/e$ for large n.) Then for all $m \geq m^*$ from(2.32) we find that $(i > 1)$

$$s^{n-1}(1) = \frac{n-1}{n}, \qquad s^{n-1}(i) = \frac{n-1}{n} \cdot \frac{1}{n-1},$$

$$s^{n-2}(1) = \frac{n-2}{n}, \qquad s^{n-2}(i) = \frac{n-2}{n}\left[\frac{1}{n-1} + \frac{1}{n-2}\right],$$

$$\vdots$$

$$s^m(1) = \frac{m}{n}, \qquad s^m(i) = \frac{m}{n}\left[\frac{1}{n-1} + \cdots + \frac{1}{m}\right].$$

Further,

$$\begin{aligned} s^{m^*-1}(1) &= \max\left\{\frac{m^*-1}{n}, \frac{1}{m^*}\left[\frac{m^*}{n} + \frac{(m^*-1)m^*}{n}\left(\frac{1}{n-1} + \cdots + \frac{1}{m^*}\right)\right]\right\} \\ &= \frac{m^*-1}{n}\max\left\{1, \frac{1}{n-1} + \frac{1}{n-2} + \cdots + \frac{1}{m^*-1}\right\} \\ &= \frac{m^*-1}{n}\left[\frac{1}{n-1} + \cdots + \frac{1}{m^*-1}\right] \end{aligned}$$

and for $i > 1$

$$\begin{aligned} s^{m^*-1}(i) &= \frac{1}{m^*}\sum_{j=1}^{m^*} s^{m^*}(j) \\ &= \frac{1}{m^*}\left\{\frac{m^*}{n} + (m^*-1)\left[\frac{1}{n-1} + \cdots + \frac{1}{m^*}\right]\right\} = s^{m^*-1}(1). \end{aligned}$$

From the above and from (2.32) we deduce that

$$s^1(1) = s^2(1) = \cdots = s^{m^*-1}(1) = \frac{m^*-1}{n}\left[\frac{1}{n-1} + \cdots + \frac{1}{m^*-1}\right]. \tag{2.33}$$

We shall describe an optimal stopping rule. By virtue of (2.23) the time

$$\tau^*_{1,n} = \min\{1 \le m \le n : s^m(y_m) = g(m, y_m)\}$$

can be taken as an optimal time. Since $s^m(y) > 0$ for all $1 \le y \le m$, and $g(m, y) = 0$ for $y > 1$, then $\tau^*_{1,n}$ coincides with the minimal time m for which $y_m = 1$ and also $s^m(y_m) = m/n$. But, as seen from (2.33), for all $m < m^*$

$$s^m(1) > \frac{m}{n}.$$

Hence the time $\tau^* = \tau^*_{1,n}$ is the first time $m \ge m^*$ for which $y_m = 1$, i.e.,

$$\tau^* = \min\{m \ge m^* : y_m = 1\}.$$

This result can be formulated in a different way: An optimal selection rule implies that one observes and lets go an $(m^* - 1)$ object and continues observing till the time τ^*, at which time the best object from all the preceding objects makes its first appearance.

According to this rule the probability of choosing the best object is

$$\begin{aligned} P\{\omega_{\tau^*} = 1\} &= M_{(1,1)}g(z_{\tau^*}) = s^1(1) = s^{m^*-1}(1) \\ &= \frac{m^*-1}{n}\left[\frac{1}{n-1} + \cdots + \frac{1}{m^*-1}\right]. \end{aligned}$$

For large n, $m^* \sim n/e$. Hence

$$P\{\omega_{\tau^*} = 1\} \sim \frac{1}{e}. \tag{2.34}$$

Therefore, for sufficiently large n with approximate probability 0.368 ($1/e \approx 0.368$) it is possible to choose the best object, although it seems at first that this probability must tend to zero as the number of the objects observed increases. Since for large n, $m^* \sim n/e$, the optimal selection rule implies that one needs to let go approximately a third of the total number of objects before choosing the first best object.

2.4 Excessive functions and smallest excessive majorants

2.4.1

Excessive functions and smallest excessive majorants play an essential role[5] in investigating the structure and properties of the payoffs $s(x)$ and $\bar{s}(x)$ introduced in Section 1. We shall give here some results related to these functions and also methods for constructing them.

Let $X = (x_n, \mathscr{F}_n, P_x)$, $n \in N$, be a (homogeneous, nonterminating) Markov chain with values in a state space $(E, \mathscr{B})$.

Definition 1. A function $f \in B$ is said to be an excessive function (for a process X or with respect to an operator T), if for all $x \in E$ the expectations $Tf(x)$ $(= M_x f(x_1))$ are defined and

$$Tf(x) \le f(x). \tag{2.35}$$

Definition 2. An excessive function $f \in B$ is said to be an *excessive majorant of the function* $g \in B$ if $f(x) \ge g(x)$, $x \in E$. An excessive majorant $f(x)$ of a function $g(x)$ is said to be the *smallest excessive majorant of* $g(x)$ if $f(x)$ is less or equal to any excessive majorant of the function $g(x)$.

From now on we shall deal with various classes of functions from B. We shall give the most essential functions here.

We shall denote by $B(\mathrm{A}^-)$ and $B(\mathrm{A}^+)$ the functions f from B which satisfy the conditions

$$\mathrm{A}^- : M_x\left[\sup_n f^-(x_n)\right] < \infty, \qquad x \in E,$$

$$\mathrm{A}^+ : M_x\left[\sup_n f^+(x_n)\right] < \infty, \qquad x \in E,$$

[5] See, for example, Theorem 3.

respectively. We shall also write

$$B(A^-, A^+) = B(A^-) \cap B(A^+),$$
$$L(A^-) = L \cap B(A^-),\ L(A^+) = L \cap B(A^+),$$
$$L(A^-, A^+) = L(A^-) \cap L(A^+),$$

where (see Section 2) the class L is defined as the class of functions $f \in B$ for which $M_x g^-(x_1) < \infty$, $x \in E$.

We shall say that the function $f \in B$ belongs to the class $B(a^-)$ if the condition

$$a^- : M_x f^-(x_\infty) < \infty, \qquad x \in E,$$

is satisfied where $f(x_\infty) = \overline{\lim}_n\ f(x_n)$. It can easily be seen that

$$B(A^-) \subseteq B(a^-) \subseteq B$$

and

$$B(A^-) = L(A^-) \subseteq L \subseteq B.$$

2.4.2

We shall sketch here the main properties of excessive functions.

I. The function $f(x) \equiv \text{const.}$ is excessive.

II. If f and g are nonnegative excessive functions and a, b are nonnegative constants, the function $af + bg$ is excessive.

III. Let $\{f_n(x), n = 1, 2, \ldots\}$ be a nondecreasing sequence of excessive functions of the class L. Then the function $f(x) = \lim_n f_n(x)$ is excessive.

IV. Let f be an excessive function and let $M_x f^-(x_n) < \infty$, $n = 0, 1, \ldots$. Then the system $(f(x_n), \mathscr{F}_n, P_x)$ forms a generalized supermartingale

$$M_x[f(x_{n+1}) | \mathscr{F}_n] \leq f(x_n), \qquad n = 0, 1, \ldots. \tag{2.36}$$

V. If an excessive function f satisfies the condition A^-, then for any $m = 1, 2, \ldots$ the function $f_m(x) = T_m f(x)$ is also excessive and

$$T_m f(x) \leq T_n f(x), \qquad m \geq n. \tag{2.37}$$

VI. If the excessive functions f and g belong to the class L, the function $f \wedge g = \min(f, g)$ is also excessive.

VII. If the excessive function f satisfies the condition

$$\sup_n M_x f^-(x_n) < \infty, \tag{2.38}$$

then with P_x-probability 1 there exists a (finite or equal to $+\infty$) limit $\lim_{n\to\infty} f(x_n)$. In particular, (2.38) will be satisfied if there exists a random variable η with $M_x|\eta| < \infty$ such that

$$f(x_n) \geq M_x(\eta | \mathscr{F}_n) \qquad (P_x\text{-a.s.}),\ n = 0, 1, \ldots. \tag{2.39}$$

(Properties I–VI follow readily from the definition of the excessiveness; Property VII follows from IV and Theorem 1.9.)

2.4.3

The following lemma plays a fundamental role in investigating properties of payoffs $s(x)$ and $\bar{s}(x)$.

Lemma 3. *Let an excessive function $f \in L(A^-)$ $(=B(A^-))$. Then for any two Markov times τ and σ such that $P_x(\tau \geq \sigma) = 1, x \in E$, we have the inequalities*

$$M_x[f(x_\tau)|\mathscr{F}_\sigma] \leq f(x_\sigma) \qquad (P_x\text{-a.s.}),\ x \in E, \tag{2.40}$$

and, in particular,

$$M_x f(x_\tau) \leq M_x f(x_\sigma) \leq f(x), \qquad x \in E. \tag{2.41}$$

Proof. First we shall note that by virtue of Property VII the limit $\lim_{n\to\infty} f(x_n)$ exists and, according to (2.5), $f(x_\infty)$ is understood as $\lim_{n\to\infty} f(x_n)$ in (2.40) and (2.41). Further, if $f \in B(A^-, A^+)$, (2.40) follows immediately from Theorem 1.11, since the system $(f(x_n), \mathscr{F}_n, P_x)$, $n \leq \infty$, forms a supermartingale. To prove this fact in the general case we put $f^c(x) = f(x) \wedge c$. Then $f^c \in B(A^-, A^+)$ and

$$M_x[f^c(x_\tau)|\mathscr{F}_\sigma] \leq f^c(x_\sigma) \qquad (P_x\text{-a.s.}).$$

From this we get, by virtue of Fatou's lemma,

$$M_x\left[\lim_{c\to\infty} f^c(x_\tau)|\mathscr{F}_\sigma\right] \leq \lim_{c\to\infty} f^c(x_\sigma). \tag{2.42}$$

But

$$\begin{aligned}
\lim_{c\to\infty} f^c(x_\sigma) &= \lim_{c\to\infty} f^c(x_\sigma) I_{\{\sigma<\infty\}} + \lim_{c\to\infty} f^c(x_\infty) I_{\{\sigma=\infty\}} \\
&= f(x_\sigma) I_{\{\sigma<\infty\}} + \lim_{c\to\infty} \lim_{n} f^c(x_n) I_{\{\sigma=\infty\}} \\
&= f(x_\sigma) I_{\{\sigma<\infty\}} + \lim_{c\to\infty} \lim_{n} f^c(x_n) I_{\{\sigma=\infty\}} \\
&= f(x_\sigma) I_{\{\sigma<\infty\}} + \lim_{n} \lim_{c\to\infty} f^c(x_n) I_{\{\sigma=\infty\}} \\
&= f(x_\sigma) I_{\{\sigma<\infty\}} + \lim_{n} f(x_n) I_{\{\sigma=\infty\}} \\
&= f(x_\sigma) I_{\{\sigma<\infty\}} + f(x_\infty) I_{\{\sigma=\infty\}} = f(x_\sigma),
\end{aligned}$$

and, similarly, $\lim_{c\to\infty} f^c(x_\tau) = f(x_\tau)$ $(P_x\text{-a.s.})$. This, together with (2.42), leads us to the required inequality, (2.40), thus proving the lemma. □

Corollary. *Let a gain function $g = g(x)$ be excessive. Then, if $g \in L(A^-)$, then*

$$\bar{s}(x) \equiv s(x) \equiv g(x), \tag{2.43}$$

and the optimal stopping rule (in the problem $s(x) = \sup M_x g(x_\tau)$) implies that one need to stop immediately, since by virtue of (2.41) *for any Markov time $\tau \in \mathfrak{M}$ we have*

$$M_x g(x_\tau) \le M_x g(x_0) = g(x). \tag{2.44}$$

Lemma 4. *Let an excessive function $f \in L(A^-)$. Then the function*

$$f_A(x) = M_x f(x_{\sigma_A}) \tag{2.45}$$

with[6]

$$\sigma_A = \inf\{n \ge 0 : x_n \in A\}, \qquad A \in \mathscr{B},$$

is also excessive.

PROOF. Put $\sigma_A^1 = \inf\{n \ge 1 : x_n \in A\}$. Obviously, σ_A^1 is a Markov time and $\sigma_A^1 \ge \sigma_A$. Then

$$Tf_A(x) = M_x f_A(x_1) = M_x M_{x_1} f(x_{\sigma_A}).$$

By virtue of (1.41)

$$M_x M_{x_1} f(x_{\sigma_A}) = M_x \theta_1 f(x_{\sigma_A}) = M_x f(\theta_1 x_{\sigma_A}),$$

and it is a known fact ([31], p. 153, prop. 4.1D) that if

$$\sigma_A^s = \inf\{n \ge s : x_n \in A\} \qquad (\sigma_A^0 = \sigma_A)$$

then

$$\theta_t x_{\sigma_A^s} = x_{\sigma_A^{s+t}}.$$

Hence

$$\theta_1 x_{\sigma_A} = x_{\sigma_A^1}$$

and, therefore,

$$Tf_A(x) = M_x f(\theta_1 x_{\sigma_A}) = M_x f(x_{\sigma_A^1}) \le M_x f(x_{\sigma_A}) = f_A(x).$$

2.4.4

If the function $f(x)$ is an excessive majorant of the function $g(x)$, then, obviously,

$$f(x) \ge \max\{g(x), Tf(x)\}. \tag{2.46}$$

But if, in addition, $f(x)$ is the smallest excessive majorant of the function $g(x)$, we have equality in (2.46). Since this assertion is of essential importance we shall formulate it as follows.

[6] Recall that σ_A is assumed equal to $+\infty$ if the set $\{n \ge 0 : x_n \in A\} = \varnothing$.

Lemma 5. *Let a function $g \in L$ and let $v = v(x)$ be its smallest excessive majorant. Then*

$$v(x) = \max\{g(x), Tv(x)\}. \tag{2.47}$$

PROOF. Denote $v_1(x) = \max\{g(x), Tv(x)\}$. Then from (2.46) we have $g(x) \le v_1(x) \le v(x)$ and, since $g \in L$, $-\infty < Tg(x) \le Tv_1(x) \le \max\{g(x), Tv(x)\} = v_1(x)$. Consequently, $v_1(x)$ is an excessive function majorizing the function $g(x)$. But $v(x)$ is the smallest excessive majorant of the function $g(x)$, hence $v(x) \le v_1(x) = \max\{g(x), Tv(x)\}$. Therefore, $v(x) = v_1(x)$, thus proving (2.47). □

Remark 1. As can be seen from this lemma, the smallest excessive majorant $v(x)$ of the function $g \in L$ is a solution of the equation

$$f(x) = \max\{g(x), Tf(x)\}.$$

However, not every function that is a solution of the above equation is the smallest majorant of the function $g(x)$. In fact, if $|g(x)| \le C < \infty$, any function $f(x) \equiv K$ where the constant $K \ge C$ will be a solution of this equation.

2.4.5

Let a function $g \in L$. Does the function $g(x)$ have the smallest excessive majorants? A positive answer to the question is given in the lemma which follows, describing, in particular, a convenient practical method for finding the smallest excessive majorants.

Lemma 6. *Let a function $g \in L$ and let*

$$Qg(x) = \max\{g(x), Tg(x)\}. \tag{2.48}$$

Then the function $v(x) = \lim_{n\to\infty} Q^n g(x)$, where Q^n is the n*th power of the operator Q, is the smallest excessive majorant of the function $g(x)$.*

PROOF. We note first that $Q^{n+1}g(x) \ge Q^n g(x)$, hence the limit $v(x) = \lim_n Q^n g(x)$ exists. It is also clear that $v(x) \ge g(x)$. We shall verify that the function $v(x)$ is excessive.

Since $Q^n g(x) \ge TQ^{n-1}g(x)$, with $Q^{n-1}g(x) \ge -g^-(x)$ where $M_x g^-(x_1) < \infty$, by the Lebesgue theorem on monotone convergence

$$\begin{aligned} v(x) &= \lim_n Q^n g(x) \\ &\ge \lim TQ^{n-1}g(x) = T\left(\lim_n Q^{n-1}g\right)(x) = Tv(x). \end{aligned}$$

Therefore, $v(x)$ is the excessive majorant of the function $g(x)$.

Let $f(x)$ be also an excessive majorant of $g(x)$. Then $f(x) \geq Tf(x)$, $f(x) \geq g(x)$, and, therefore,

$$Qf(x) = \max\{f(x), Tf(x)\} = f(x),$$
$$f(x) = Q^n f(x) \geq Q^n g(x).$$

Therefore, $f(x) \geq v(x)$, i.e., $v(x)$ is the smallest excessive majorant of the function $g(x)$.

Remark 2. If we set[7]

$$\tilde{Q}g(x) = \sup\{g(x), Tg(x), T^2 g(x), \ldots\}, \tag{2.49}$$

we can show in similar fashion that $\tilde{v}(x) = \lim_n \tilde{Q}^n g(x)$ is also the smallest excessive majorant of the function g and, therefore, $\tilde{v}(x) = v(x)$.

Remark 3. Let a function $g \in L$ and let $v(x)$ be its smallest excessive majorant. Write

$$g^b(x) = \min(b, g(x)), \qquad b \geq 0,$$

and let $v^b(x)$ be its smallest excessive majorant. Then

$$v(x) = \lim_{b \to \infty} v^b(x) \tag{2.50}$$

and, furthermore,

$$v(x) = \lim_b \lim_n Q^n g^b(x) = \lim_n \lim_b Q^n g^b(x). \tag{2.51}$$

In fact, since $\lim_b Q^n g^b(x) = Q^n g(x)$, then we have, as has been proved,

$$v(x) = \lim_n Q^n g(x) = \lim_n \left(\lim_b Q^n g^b(x)\right).$$

Furthermore,

$$v^b(x) = \lim_n Q^n g^b(x) \leq \lim_n Q^n g(x) = v(x)$$

and, therefore, $\lim_b v^b(x) \leq v(x)$. But $\lim_b v^b(x)$ is an excessive function (by virtue of Property III) and $\lim_b v^b(x) \geq g(x)$. Hence

$$v(x) = \lim_b v^b(x) = \lim_b \lim_n Q^n g^b(x).$$

2.4.6

Lemma 7. *Let $g \in B$ and let f be an excessive function satisfying the equation*

$$f(x) = \max\{g(x), Tf(x)\}. \tag{2.52}$$

Set

$$\tau_\varepsilon = \inf\{n \geq 0 : f(x_n) \leq g(x_n) + \varepsilon\}, \qquad \varepsilon \geq 0.$$

[7] We assume here that all integrals $T^n g(x) = M_x g(x_n)$, $n = 1, 2, \ldots$ are defined.

Then, if a point $x \in E$ is such that $f(x) < \infty$, for this point and any $n = 0, 1, \ldots$

$$M_x f(x_{\tau_\varepsilon \wedge n}) = f(x). \tag{2.53}$$

PROOF. For the point $x \in E$ considered, $|f(x)| < \infty$ and $M_x f(x_0) = f(x)$ since $P_x\{x_0 = x\} = 1$. Therefore, each of the integrals $M_x f(x_0) I_{\{\tau_\varepsilon = 0\}}$, $M_x f(x_0) I_{\{\tau_\varepsilon > 0\}}$ is finite and

$$f(x) = M_x f(x_0) I_{\{\tau_\varepsilon = 0\}} + M_x f(x_0) I_{\{\tau_\varepsilon > 0\}}. \tag{2.54}$$

On the set $\{\omega : \tau_\varepsilon(\omega) > 0\}$ we have $f(x_0(\omega)) > g(x_0(\omega))$ and, therefore, by virtue of (2.52), $f(x_0) = Tf(x_0) = M_{x_0}[f(x_1) | \mathscr{F}_0]$. We find from the above and (2.54) that

$$\begin{aligned} f(x) &= M_x f(x_0) I_{\{\tau_\varepsilon = 0\}} + M_x f(x_1) I_{\{\tau_\varepsilon > 0\}} \\ &= M_x f(x_{\tau_\varepsilon}) I_{\{\tau_\varepsilon \le 1\}} + M_x f(x_1) I_{\{\tau_\varepsilon > 1\}}. \end{aligned}$$

Similar considerations show that

$$\begin{aligned} f(x) &= M_x f(x_{\tau_\varepsilon}) I_{\{\tau_\varepsilon \le 2\}} + M_x f(x_2) I_{\{\tau_\varepsilon > 2\}} \\ &\;\;\vdots \\ &= M_x f(x_{\tau_\varepsilon}) I_{\{\tau_\varepsilon \le n\}} + M_x f(x_n) I_{\{\tau_\varepsilon > n\}} \\ &= M_x f(x_{\tau_\varepsilon \wedge n}), \end{aligned}$$

thus proving the lemma. □

Lemma 8. *Let a function $g \in B(A^+)$ and let $v(x)$ be its smallest excessive majorant. Then*

$$\overline{\lim_n}\, v(x_n) = \overline{\lim_n}\, g(x_n) \qquad (P_x\text{-a.s.}),\ x \in E \tag{2.55}$$

and if $\tau_\varepsilon = \inf\{n \ge 0 : v(x_n) \le g(x_n) + \varepsilon\}$, then for any $\varepsilon > 0$

$$P_x(\tau_\varepsilon < \infty\} = 1, \qquad x \in E. \tag{2.56}$$

PROOF. Since $v(x) \ge g(x)$, then it is clear that

$$\overline{\lim_n}\, v(x_n) \ge \overline{\lim_n}\, g(x_n).$$

To prove the inverse inequality we shall fix a point $x \in E$ and denote

$$\psi_n = \sup_{j \ge n} g(x_j), \qquad \varphi_n = M_x[\psi_n | \mathscr{F}_n].$$

By virtue of Markovianness, $\varphi_n = \varphi(x_n)$ (P_x-a.s.) where $\varphi(x) = M_x \psi_0$. It is clear that $M_x \varphi(x_1) < \infty$, $\varphi(x) \ge g(x)$, and

$$T\varphi(x) = M_x \varphi(x_1) = M_x M_{x_1} \psi_0 = M_x \psi_1 \le M_x \psi_0 = \varphi(x).$$

Therefore, $\varphi(x)$ is the excessive majorant of the function $g(x)$ and, since $v(x)$ is the smallest excessive majorant of $g(x)$, then $v(x) \le \varphi(x)$; therefore, $\varphi_n = \varphi(x_n) \ge v(x_n)$ (P_x-a.s.)

Let m be fixed and let $n \geq m$. Then

$$M_x[\psi_m | \mathscr{F}_n] \geq M_x[\psi_n | \mathscr{F}_n] = \varphi_n \geq v(x_n)$$

and, therefore,

$$\varlimsup_n M_x[\psi_m | \mathscr{F}_n] \geq \varlimsup_n v(x_n).$$

For each $m \in N$ a sequence $(M_x[\psi_m | \mathscr{F}_n], \mathscr{F}_n, P_x)$, $n \geq m$, forms (by virtue of the condition $g \in B(\mathrm{A}^+)$), a generalized martingale (see Section 1.3). Then by virtue of Theorem 1.10.2° the limit $\lim_n M_x[\psi_m | \mathscr{F}_n]$ exists (P_x-a.s.) and

$$\lim M_x[\psi_m | \mathscr{F}_n] \leq M_x[\psi_m | \mathscr{F}_\infty] \qquad (P_x\text{-a.s.}),$$

where $\mathscr{F}_\infty = \sigma(\bigcup_n \mathscr{F}_n)$. We note that the random variable ψ_m is $\mathscr{F}_\infty$-measurable and, therefore,

$$\lim_n M_x[\psi_m | \mathscr{F}_n] \leq \psi_m \qquad (P_x\text{-a.s.}).$$

This together with $\varlimsup_n M_x[\psi_m | \mathscr{F}_n] \geq \varlimsup_n v(x_n)$ yields

$$\varlimsup_n v(x_n) \leq \psi_m \qquad (P_x\text{-a.s.})$$

and, therefore,

$$\varlimsup_n v(x_n) \leq \inf_m \psi_m = \inf_m \sup_{j \geq m} g(x_j) = \varlimsup_n g(x_n),$$

thus proving (2.55)

We shall show next that $P_x\{\tau_\varepsilon < \infty\} = 1, x \in E, \varepsilon > 0$. Let $A = \{\omega : \tau_\varepsilon(\omega) = \infty\}$; then for $\omega \in A$

$$v(x_n) > g(x_n) + \varepsilon$$

and, by virtue of the fact that $\varepsilon > 0$,

$$\varlimsup_n v(x_n) > \varlimsup_n g(x_n).$$

But $P_x\{\varlimsup_n g(x_n) < \infty\} = 1, x \in E$. Hence it follows from (2.55) that $P_x\{A\} = 0, x \in E$, thus proving the lemma.

2.4.7

The sequence $Q^n g(x)$ constructed for a function $g \in L$ in Lemma 4 is monotone, nondecreasing in n, and converges to $v(x)$—the smallest excessive majorant of the function $g(x)$.

In a number of cases the sequence of functions to be constructed below can also prove to be useful (as will be shown in Lemma 11) for functions $g \in L(\mathrm{A}^-, \mathrm{A}^+)$ decreasing monotonically (or—to be more precise—monotone, nonincreasing), is convergent to $v(x)$—the smallest excessive majorant of the function $g(x)$.

We shall associate each function $g \in B$ with an operator G (acting on functions $f \in B$ for which the expectations are $Tf(x)$, $x \in E$) defined by the formula

$$Gf(x) = \max\{g(x), Tf(x)\}.$$

Let us denote by G^n the nth power of the operator G, $G^0 f(x) \equiv f(x)$. (If $f = g$, then $Gg(x) = Qg(x)$; if $f = v$ is the smallest excessive majorant of the function $g \in L$, then $Gv(x) = v(x)$.)

Lemma 9. *Let $g \in B(A^+)$ and let $\varphi(x) = M_x[\sup_n g(x_n)]$. Then*

$$G^{n+1}\varphi(x) \leq G^n\varphi(x), \qquad n \in N,$$

and $\tilde{v}(x) = \lim_n G^n\varphi(x)$ satisfies

$$\tilde{v}(x) = \max\{g(x), T\tilde{v}(x)\}. \tag{2.57}$$

PROOF. The inequality $G^{n+1}\varphi(x) \leq G^n\varphi(x)$ can be verified by induction. We need only to show that $G\varphi(x) \leq \varphi(x)$. In fact,

$$\begin{aligned} G\varphi(x) &= \max\{g(x), T\varphi(x)\} \\ &= \max\left\{g(x), M_x\left[\sup_{j\geq 1} g(x_j)\right]\right\} \\ &\leq M_x\left\{\max\left[g(x), \sup_{j\geq 1} g(x_j)\right]\right\} = M_x\left[\sup_{j\geq 0} g(x_j)\right] = \varphi(x). \end{aligned}$$

Further, since

$$\varphi(x) \geq G^n\varphi(x) \downarrow \tilde{v}(x), \qquad n \to \infty,$$
$$T\varphi(x) < \infty, \quad x \in E,$$

by passing to the limit ($n \to \infty$) in the equality

$$G^n\varphi(x) = \max\{g(x), TG^{n-1}\varphi(x)\}$$

we have (by the Lebesgue theorem on monotone convergence) that $\tilde{v}(x)$ satisfies Equation (2.57).

The function $\tilde{v}(x)$, being the excessive majorant of the function $g \in B(A^+)$, need not be the smallest excessive majorant (see Example in Section 6). However, if $g \in B(A^-, A^+)$, the function $\tilde{v}(x)$ coincides with $v(x)$, which is the smallest excessive majorant of $g(x)$. To show this we shall prove first:

Lemma 10. *Let the function $g \in B(A^+)$ and let $\tilde{v}(x) = \lim_n G^n\varphi(x)$. We have*

$$\overline{\lim_n}\, \tilde{v}(x_n) = \overline{\lim_n}\, g(x_n) \qquad (P_x\text{-a.s.}),\ x \in E. \tag{2.58}$$

Proof. The inequality $\overline{\lim}_n \tilde{v}(x_n) \geq \overline{\lim}_n g(x_n)$ is obvious. On the other hand, for each $x \in E$, $n \in N$, and $m \leq n$, (P_x-a.s.)

$$\begin{aligned}\tilde{v}(x_n) &\leq G^n\varphi(x_n) \leq \varphi(x_n)\\ &= M_{x_n}\left[\sup_{j\geq 0} g(x_j)\right]\\ &= M_x\left[\sup_{j\geq n} g(x_j)\,|\,\mathscr{F}_n\right] \leq M_x\left[\sup_{j\geq m} g(x_j)\,|\,\mathscr{F}_n\right].\end{aligned} \tag{2.59}$$

It follows from (2.59) (as in Lemma 8) that

$$\overline{\lim_n}\, \tilde{v}(x_n) \leq \sup_{j\geq m} g(x_j)$$

and, therefore,

$$\overline{\lim_n}\, \tilde{v}(x_n) \leq \inf_m \sup_{j\geq m} g(x_j) = \overline{\lim_n}\, g(x_n),$$

thus proving the lemma.

Corollary (Compare with (2.56)). *Let $g \in B(A^+)$ and let*

$$\tilde{\tau}_\varepsilon = \inf\{n \geq 0 : \tilde{v}(x_n) \leq g(x_n) + \varepsilon\}, \qquad \varepsilon > 0. \tag{2.60}$$

Then

$$P_x\{\tilde{\tau}_\varepsilon < \infty\} = 1, \qquad x \in E. \tag{2.61}$$

Lemma 11

(1) *Let a function $g \in B(A^+)$ and let $\tilde{v}(x) = \lim_n G^n\varphi(x)$. Then for any $\varepsilon > 0$*

$$\tilde{v}(x) \leq M_x\tilde{v}(x_{\tilde{\tau}_\varepsilon}), \tag{2.62}$$

where the time $\tilde{\tau}_\varepsilon$ was defined in (2.60).

(2) *Let a function $g \in B(A^-, A^+)$. Then*

$$\tilde{v}(x) = M_x\tilde{v}(x_{\tilde{\tau}_\varepsilon}) \tag{2.63}$$

and $\tilde{v}(x) = v(x)$ where $v(x)$ is the smallest excessive majorant of the function $g(x)$.

Proof

(1) By virtue of Lemma 9 the function $\tilde{v}(x)$ satisfies the equation

$$\tilde{v}(x) = \max\{g(x), T\tilde{v}(x)\}.$$

Let us apply Lemma 7 to the function $f(x) = \tilde{v}(x)$. Then

$$\begin{aligned}\tilde{v}(x) &= M_x\tilde{v}(x_{\tilde{\tau}_\varepsilon \wedge n})\\ &= M_x[\tilde{v}(x_{\tilde{\tau}_\varepsilon})I_{\{\tilde{\tau}_\varepsilon \leq n\}}] + M_x[\tilde{v}(x_n I_{\{\tilde{\tau}_\varepsilon > n\}}].\end{aligned} \tag{2.64}$$

We get from (2.59) that

$$M_x[\tilde{v}(x_n)I_{\{\tilde{\tau}_\varepsilon > n\}}] \le M_x\left[M_x\left(\sup_{j\ge n} g^+(x_j)|\mathscr{F}_n\right)I_{\{\tilde{\tau}_\varepsilon > n\}}\right]$$

$$= M_x\left[\sup_{j\ge n} g^+(x_j)I_{\{\tilde{\tau}_\varepsilon > n\}}\right] \le M_x\left[\sup_{j\ge 0} g^+(x_j)I_{\{\tilde{\tau}_\varepsilon > n\}}\right].$$

But $g \in B(A^+)$, and by virtue of (2.61) $P_x\{\tilde{\tau}_\varepsilon > n\} \to 0$, $n \to \infty$. Hence

$$\overline{\lim_n}\, M_x[\tilde{v}(x_n)I_{\{\tilde{\tau}_\varepsilon > n\}}] \le \overline{\lim_n}\, M_x\left[\sup_{j\ge 0} g^+(x_j)I_{\{\tilde{\tau}_\varepsilon > n\}}\right] = 0,$$

and by passing in (2.64) to the limit ($n \to \infty$) we have from Fatou's lemma that

$$\tilde{v}(x) \le \overline{\lim_n}\, M_x[\tilde{v}(x_{\tilde{\tau}_\varepsilon})I_{\{\tilde{\tau}_\varepsilon \le n\}}] + \overline{\lim_n}\, M_x[\tilde{v}(x_n)I_{\{\tilde{\tau}_\varepsilon > n\}}]$$

$$\le M_x\tilde{v}(x_{\tilde{\tau}_\varepsilon})I_{\{\tilde{\tau}_\varepsilon < \infty\}} = M_x\tilde{v}(x_{\tilde{\tau}_\varepsilon}),$$

which fact proves the required inequality in (2.62).

(2) If the function $g \in B(A^-)$, then we can prove in a similar way that

$$\underline{\lim_n}\, M_x[\tilde{v}(x_n)I_{\{\tilde{\tau}_\varepsilon > n\}}] \ge -\ \underline{\lim_n}\, M_x\left[\sup_{j\ge 0} g^-(x_j)I_{\{\tilde{\tau}_\varepsilon > n\}}\right] = 0.$$

Hence, if $g \in B(A^-, A^+)$, then

$$\lim_n M_x[\tilde{v}(x_n)\, I_{\{\tilde{\tau}_\varepsilon > n\}}] = 0,$$

which fact together with (2.64) proves (2.63).

Finally, to prove the equality $v(x) = \tilde{v}(x)$ we note that for $\varepsilon > 0$

$$\tilde{v}(x_{\tilde{\tau}_\varepsilon}) \le g(x_{\tilde{\tau}_\varepsilon}) + \varepsilon \le v(x_{\tilde{\tau}_\varepsilon}) + \varepsilon.$$

Therefore, if $g \in B(A^-, A^+)$, then by virtue of (2.63) and (2.41) we have

$$\tilde{v}(x) = M_x\tilde{v}(x_{\tilde{\tau}_\varepsilon}) \le M_x g(x_{\tilde{\tau}_\varepsilon}) + \varepsilon \le M_x v(x_{\tilde{\tau}_\varepsilon}) + \varepsilon \le v(x) + \varepsilon,$$

which, because of arbitrariness of $\varepsilon > 0$, yields the inequality $\tilde{v}(x) \le v(x)$, this together with the obvious inequality $\tilde{v}(x) \ge v(x)$ proves that the functions $\tilde{v}(x)$ and $v(x)$ coincide. □

Corollary. *Let a function $g \in B(A^-, A^+)$ and let $v(x)$ be its smallest excessive majorant. Then for any $\varepsilon > 0$*

$$v(x) = M_x v(x_{\tau_\varepsilon}), \tag{2.65}$$

where $\tau_\varepsilon = \inf\{n \ge 0: v(x_n) \le g(x_n) + \varepsilon\}$.

Remark 4. We can prove (2.65) (without using the equality $\tilde{v}(x) = v(x)$ and (2.63)) if we apply Lemma 7 immediately to the function $f(x) = v(x)$.

2.5 The excessive characterization of the payoff and ε-optimal stopping rules (under the condition A^-)

2.5.1

Let a function $g \in B$ and let

$$s(x) = \sup_{\tau \in \mathfrak{M}_g} M_x g(x_\tau),$$

$$\bar{s}(x) = \sup_{\tau \in \overline{\mathfrak{M}}_g} M_x g(x_\tau)$$

be the payoffs introduced in Section 1. The theorem which follows describes (on the assumption that $g \in L(A^-)$) the structure of the payoffs $s(x)$ and $\bar{s}(x)$ and reveals the essence of the concept of excessive functions in the problems of optimal stopping of Markov random sequences $X = (x_n, \mathscr{F}_n, P_x)$, $n \in N$, $x \in E$.

Theorem 3. *Let a function $g \in L(A^-)$. Then:*

(1) *the payoff $s(x)$ is the smallest excessive majorant of the function $g(x)$;*
(2) *the payoff $s(x)$ and the payoff $\bar{s}(x)$ coincide, i.e., $s(x) = \bar{s}(x)$;*

(3)
$$s(x) = \max\{g(x), Ts(x)\}; \tag{2.66}$$

(4)
$$s(x) = \lim_n Q^n g(x) = \lim_b \lim_n Q^n g^b(x), \tag{2.67}$$

where $g^b(x) = \min(b, g(x))$, $b \geq 0$.

Proof. Let $v(x)$ be the smallest majorant of a function $g(x)$. Then, since $\overline{\lim}_n v(x_n) \geq \overline{\lim}_n g(x_n)$, by virtue of (2.41) for any[8] $\tau \in \overline{\mathfrak{M}}$

$$v(x) \geq M_x v(x_\tau) \geq M_x g(x_\tau). \tag{2.68}$$

(We recall that $f(x_\infty)$ is to be understood as $\overline{\lim}_n f(x_n)$.) Therefore,

$$v(x) \geq \sup_{\tau \in \overline{\mathfrak{M}}} M_x g(x_\tau) = \bar{s}(x) \geq s(x). \tag{2.69}$$

To prove that $v(x) \leq s(x)$ we shall assume first that $g \in L(A^-, A^+)$. Then by virtue of Lemma 8 the time

$$\tau_\varepsilon = \inf\{n \geq 0 : v(x_n) \leq g(x_n) + \varepsilon\}, \qquad \varepsilon > 0,$$

is a stopping time and by virtue of (2.65)

$$v(x) = M_x v(x_{\tau_\varepsilon}). \tag{2.70}$$

[8] If $g \in L(A^-)$, then $\mathfrak{M}_g = \mathfrak{M}$ and $\overline{\mathfrak{M}}_g = \overline{\mathfrak{M}}$.

Hence

$$v(x) = M_x v(x_{\tau_\varepsilon}) \leq M_x g(x_{\tau_\varepsilon}) + \varepsilon \leq s(x) + \varepsilon, \tag{2.71}$$

which fact, because of arbitrariness of $\varepsilon > 0$ and (2.68), proves the required equality $v(x) = s(x)$ (in the case $g \in L(A^-, A^+)$).

To prove the inequality $v(x) \leq s(x)$ in the general case we shall set $g^b(x) = \min(b, g(x))$, $b \geq 0$, let $v^b(x)$ be the smallest excessive majorant of the function $g^b(x)$, and write $s^b(x) = \sup_{\tau \in \mathfrak{M}} M_x g^b(x_\tau)$. Then, according to what has already been proved,

$$s(x) \geq s^b(x) = \sup_{\tau \in \mathfrak{M}} M_x g^b(x_\tau) = v^b(x).$$

The sequence $\{v^b(x), b \geq 0\}$ does not decrease. Let $v^*(x) = \lim_{b \to \infty} v^b(x)$. We shall prove that in fact $v^*(x) = v(x)$. We have

$$Tv^*(x) = T\left(\lim_b v^b\right)(x) = \lim_b Tv^b(x) \leq \lim_b v^b(x) = v^*(x),$$

i.e., the function $v^*(x)$ is excessive. Since $g^b(x) \uparrow g(x)$ and $v^b(x) \geq g^b(x)$, then $v^*(x) \geq g(x)$. Therefore, $v^*(x)$ is the excessive majorant of the function $g(x)$. We need only to prove that this excessive majorant is the smallest one. Let $f(x)$ be an excessive majorant of a function $g(x)$. Then $f(x) \geq g^b(x)$ and $f(x) \geq v^b(x)$, from which we have $f(x) \geq v^*(x)$. Therefore, $s(x) \geq v^*(x) = v(x)$ which was to be proved.

The recursion relation $s(x) = \max\{g(x), Ts(x)\}$ follows obviously from Lemma 5 and the equality $s(x) = v(x)$. Finally, (2.67) follows from the equalities $s(x) = \lim_b s^b(x)$ and $s(x) = v(x)$ and from Lemma 6.

2.5.2

Remark 1. Let $g \in L(A^-)$ and let a Markov time $\bar{\tau} \in \overline{\mathfrak{M}}$ be such that the corresponding gain $f(x) = M_x g(x_{\bar{\tau}})$ is an excessive function, with $f(x) \geq g(x)$. Then, since the payoff $\bar{s}(x)$ is the smallest excessive majorant of the function $g(x)$, and, obviously, $\bar{s}(x) \geq f(x)$, then $\bar{s}(x) = f(x)$. Therefore, the time $\bar{\tau}$ is a $(0, \bar{s})$-optimal time. We note that, in general, the $(0, \bar{s})$-optimal time needs not, in fact, be a $(0, s)$-optimal time. Section 6 contains an example illustrating the method of finding $(0, \bar{s})$-optimal times based on the given remark.

Remark 2. Let $\mathfrak{C}(\overline{\mathfrak{C}})$ be a set of stopping times τ (Markov times $\bar{\tau}$) from class $\mathfrak{M}$ $(\overline{\mathfrak{M}})$ which are times of first entry into Borel sets (i.e., $\tau = \inf\{n \geq 0 : x_n \in C\}$, $C \in \mathscr{B}$). Then

$$s(x) = \sup_{\tau \in \mathfrak{C}} M_x g(x_\tau) = \sup_{\tau \in \overline{\mathfrak{C}}} M_x g(x_\tau). \tag{2.72}$$

This result, which follows immediately from the proof of Theorem 3, reveals the significance of the class of times of first entry into Borel sets in the problems of optimal stopping of Markov chains. This fact does not imply that the optimal time, if its exists, is necessarily a time of first entry into some Borel set.

Remark 3. Let $F^X = \{\mathscr{F}_n{}^X\}$, $n \in N$, be a system of σ-algebras $\mathscr{F}_n{}^X = \sigma\{\omega : x_0, \ldots, x_n\}$. Denote by $\mathfrak{M}[F^X]$ the class of Markov times τ such that for each n the event $\{\tau = n\} \in \mathscr{F}_n^X$. Then it follows from Remark 2 that

$$s(x) = \sup_{\tau \in \mathfrak{M}[F^X]} M_x g(x_\tau). \tag{2.73}$$

In this sense the system of σ-algebras $F^X = \{\mathscr{F}_n^X\}$, $n \in N$, is naturally said to be a sufficient system in the problem "$s(x) = \sup M_x g(x_\tau)$." (For more detail about sufficient systems see Section 13).

Remark 4. In the further study of the structure of payoffs $s(x)$ (especially in the case where the condition A^- is violated and the excessive characterization is not, in general, true any longer) the remark which follows immediately from the proof of Theorem 3 proves to be useful: The payoff $s(x)$ is the smallest function from the class of functions $f \in L$ for which $f(x) \geq g(x)$ and $f(x) \geq M_x f(x_\tau)$ for any $\tau \in \mathfrak{M}_g$. (Compare this assertion with Remark 4 in Section 7 and with Theorem 7 in Section 8.)

Remark 5. It follows from the proof of the theorem that $s(x) \geq M_x s(x_\infty) \geq M_x g(x_\infty)$, and, therefore, along with the equation $s(x) = \max\{g(x), Ts(x)\}$ the payoff $s(x)$ also satisfies the equation

$$s(x) = \max\{g(x), M_x g(x_\infty), Ts(x)\}.$$

Furthermore, $s(x)$ is the smallest excessive majorant of the function $g(x)$ as well as the smallest excessive majorant of a function $G(x) = \max\{g(x), M_x g(x_\infty)\}$. In fact, the function $G \in L(A^-)$. Therefore, the payoff $S(x) = \sup M_x G(x_\tau)$ satisfies the equation $S(x) = \max\{G(x), TS(x)\}$ and is the smallest excessive majorant of the function $G(x)$. We shall show that $s(x) = S(x)$. The inequality $S(x) \geq s(x)$ is obvious. On the other hand, $s(x)$ satisfies the equation $s(x) = \max\{G(x), Ts(x)\}$ and, therefore, is the excessive majorant of the function $G(x)$. But $S(x)$ is the smallest excessive majorant of $G(x)$, hence $s(x) \geq S(x)$. From this we have $s(x) = S(x)$.

2.5.3

Theorem 3 contains the excessive characterization of the payoffs $s(x)$ and $\bar{s}(x)$. We know from this theorem that the payoff $s(x)$ coinciding with $\bar{s}(x)$ is the smallest excessive majorant of the function $g(x)$ under the condition $g \in L(A^-)$. In turn, Lemma 6 describes a constructive method of finding the smallest excessive majorant and, therefore, of finding the payoff $s(x)$ in the problem "$s(x) = \sup M_x g(x_\tau)$" with the function $g \in L(A^-)$.

We shall investigate next the question of the existence and structure of ε-optimal and optimal stopping rules.

Theorem 4. *Let a function $g \in L(A^-, A^+)$ and let $v(x)$ be the smallest excessive majorant of this function (coinciding with the payoff $s(x)$). Then:*

(1) *For any* $\varepsilon > 0$ *the time*

$$\tau_\varepsilon = \inf\{n \geq 0 : v(x_n) \leq g(x_n) + \varepsilon\} \tag{2.74}$$

is an (ε, s)*-optimal stopping time*;

(2) *The time*

$$\tau_0 = \inf\{n \geq 0 : v(x_n) = g(x_n)\} \tag{2.75}$$

is a $(0, \bar{s})$*-optimal Markov time*;

(3) *If the time* τ_0 *is a stopping time* ($\tau_0 \in \mathfrak{M}$), *it is a* $(0, s)$*-optimal time*;

(4) *If the set* E *is finite, the time* τ_0 *is a* $(0, s)$*-optimal stopping time.*

PROOF

(1) By virtue of (2.56), $P_x\{\tau_\varepsilon < \infty\} = 1$ for any $\varepsilon > 0$ and $x \in E$. Hence the fact that the stopping time τ_ε is an (ε, s)-optimal time follows from (2.71).

(2) By applying Lemma 7 to the function $f(x) = v(x)$, we find that for any $n \in N$

$$\begin{aligned}
v(x) &= M_x v(x_{\tau_0 \wedge n}) \\
&= M_x\{I_{\{\tau_0 < n\}} v(x_{\tau_0}) + I_{\{n \leq \tau_0 < \infty\}} v(x_n) + I_{\{\tau_0 = \infty\}} v(x_n)\} \\
&\leq M_x\{I_{\{\tau_0 < n\}} v(x_{\tau_0})\} + M_x\left\{I_{\{n \leq \tau_0 < \infty\}} M_{x_n}\left[\sup_j g(x_j)\right]\right\} \\
&\quad + M_x\left\{I_{\{\tau_0 = \infty\}} M_{x_n}\left[\sup_j g(x_j)\right]\right\} \\
&\leq M_x\{I_{\{\tau_0 < n\}} v(x_{\tau_0})\} + M_x\left\{I_{\{n \leq \tau_0 < \infty\}} \sup_j g^+(x_j)\right\} \\
&\quad + M_x\left\{I_{\{\tau_0 = \infty\}} \sup_{j \geq n} g(x_j)\right\}.
\end{aligned}$$

From this by virtue of the condition $g \in L(A^+)$ and Fatou's lemma we get

$$v(x) \leq M_x\{I_{\{\tau_0 < \infty\}} v(x_{\tau_0})\} + M_x\left\{I_{\{\tau_0 = \infty\}} \overline{\lim_n}\, g(x_n)\right\}. \tag{2.76}$$

But by the definition of the time τ_0

$$M_x\{I_{\{\tau_0 < \infty\}} v(x_{\tau_0})\} = M_x\{I_{\{\tau_0 < \infty\}} g(x_{\tau_0})\}.$$

Hence it follows from (2.76) that

$$v(x) \leq M_x\{I_{\{\tau_0 < \infty\}} g(x_{\tau_0})\} + M_x\left\{I_{\{\tau_0 = \infty\}} \overline{\lim_n}\, g(x_n)\right\} = M_x g(x_{\tau_0}), \tag{2.77}$$

i.e., the time τ_0 is a $(0, \bar{s})$-optimal time.

(3) Let $\tau_0 \in \mathfrak{M}$. Then (compare (2.76) and (2.77))

$$\begin{aligned} v(x) &\le M_x\{I_{\{\tau_0<\infty\}} g(x_{\tau_0})\} + M_x\{I_{\{\tau_0=\infty\}} \overline{\lim}\, g(x_n)\} \\ &\le M_x\{I_{\{\tau_0<\infty\}} g(x_{\tau_0})\} + M_x\left\{I_{\{\tau_0=\infty\}} \sup_n g^+(x_n)\right\} \\ &= M_x\{I_{\{\tau_0<\infty\}} g(x_{\tau_0})\} = M_x g(x_{\tau_0}), \end{aligned}$$

which fact proves that the time τ_0 is a $(0, s)$-optimal time.

(4) Set $\Gamma_\varepsilon = \{x : v(x) \le g(x) + \varepsilon\}$. It is clear that $\Gamma_\varepsilon \supseteq \Gamma_0$ and $\Gamma_\varepsilon \downarrow \Gamma_0$, $\varepsilon \downarrow 0$. If the set E is finite, there will be ε' such that $\Gamma_\varepsilon = \Gamma_0$ for all $\varepsilon \le \varepsilon'$. Therefore, the time $\tau_0 = \tau_\varepsilon$, $\varepsilon \le \varepsilon'$, with P_x-probability 1 is finite and, by virtue of (3), $s(x) = M_x g(x_{\tau_0})$, thus proving the theorem. □

2.5.4

Remark 1. Assertion (2) of Theorem 4 (see also (2.77)) explains why the value $\overline{\lim}_n g(x_n)$ is to be taken as $g(x_\infty)$ in defining the payoff $\bar{s}(x)$. The point is that for $\varepsilon > 0$ the times τ_ε defined in (2.74) are (ε, s)-optimal times. One might think that this result holds for $\varepsilon = 0$ as well. However, this is not true any longer due, first of all, to the fact that the times τ_0 are not, generally speaking, finite Markov times, i.e., stopping times. If we still try to find out whether these times are optimal in some reasonable sense we shall get a positive answer to this question by defining the payoff $\bar{s}(x)$ as $\sup_{\tau\in\mathfrak{M}} M_x g(x_\tau)$ where $g(x_\infty)$ is understood as $\overline{\lim}_n g(x_n)$.

Remark 2. The condition A^+ appearing in Theorem 4 cannot, in general, be weakened (see, in this connection, an example in the next section). We have, however, the following. Let $g \in L(A^-)$ and for a given $x_0 \in E$ let $M_{x_0}[\sup_n g^+(x_n)] < \infty$. Then the time τ_ε, $\varepsilon > 0$, is an (ε, s)-optimal time at a point x_0 (i.e., $M_{x_0} g(x_{\tau_\varepsilon}) \ge s(x_0) - \varepsilon$), and the time τ_0 is a $(0, \bar{s})$-optimal time at this point. If, in addition, $P_{x_0}\{\tau_0 < \infty\} = 1$, τ_0 will be an optimal stopping time (at the point x_0).

2.5.5

We shall consider some properties of the domains

$$C = \{x : v(x) > g(x)\}$$

and

$$\Gamma = \{x : v(x) = g(x)\}.$$

If the function $g \in L(A^-, A^+)$, then by Theorem 4 the $(0, \bar{s})$-optimal time

$$\tau_0 = \inf\{n \ge 0 : x_n \in \Gamma\},$$

with $\tau_0 = \infty$ if $x_n \notin \Gamma$ for all $n \ge 0$.

Hence the set C is naturally said to be a set of **continued observation**, and the set Γ is a set of **stopping** or a set of **terminal observation**.

Since $v(x) \geq Qg(x)$ it is clear that $C \supseteq \{x : Qg(x) > g(x)\} = \{x : Tg(x) > g(x)\}$.

The meaning of this relation is obvious: If a point x is such that the "prediction" of a gain is greater than the gain of "instant" stopping (equal to $g(x)$) by one step (i.e., the value $Tg(x) = M_x g(x_1)$), one should make in advance at least one observation at this point.

It is also clear that

$$\Gamma \subseteq \{x : Tg(x) \leq g(x)\};$$

this inclusion, in general, is strict.

We give a condition under which, in fact, the set

$$\Gamma = \{x : Tg(x) \leq g(x)\}.$$

Let a Markov chain $X = (x_n, \mathscr{F}_n, P_x)$, $n \in N$, and let a function g belonging to the class $L(A^-, A^+)$ be such that for each point $x \in \{x : Tg(x) \leq g(x)\}$

$$Tg(x_1) \leq g(x_1) \qquad (P_x\text{-a.s.}).$$

From this it follows (by virtue of the fact that the chain X is homogeneous) that for all $n \geq 0$ and $x \in \{x : Tg(x) \leq g(x)\}$

$$Tg(x_n) \leq g(x_n) \qquad (P_x\text{-a.s.})$$

or, equivalently,

$$Qg(x_n) = g(x_n) \qquad (P_x\text{-a.s.}).$$

Hence

$$\begin{aligned} Qg(x) &= g(x), \\ Q^2 g(x) &= \max\{g(x), TQg(x)\} \\ &= \max\{g(x), M_x Qg(x_1)\} \\ &= \max\{g(x), M_x g(x_1)\} \\ &= \max\{g(x), Tg(x)\} = g(x), \end{aligned}$$

and, similarly, for any $n \geq 1$

$$Q^n g(x) = g(x), \ x \in \{x : Tg(x) \leq g(x)\}.$$

But $v(x) = \lim_n Q^n g(x)$. Hence, if the point $x \in \{x : Tg(x) \leq g(x)\}$ and the condition indicated is satisfied, then $v(x) = g(x)$, i.e., the point $x \in \Gamma$ and, therefore, $\Gamma = \{x : Tg(x) \leq g(x)\}$.

Thus, in the case considered (in [22] referred to as monotone) the $(0, \bar{s})$-optimal time τ_0 can be expressed simply as follows:

$$\begin{aligned} \tau_0 &= \inf\{n \geq 0 : x_n \in \Gamma\} \\ &= \inf\{n \geq 0 : Tg(x_n) \leq g(x_n)\}. \end{aligned}$$

2.5.6

Let $\mathfrak{M}(m; \infty)$ ($\overline{\mathfrak{M}}(m; \infty)$) be a class of stopping times (Markov times) $\tau = \tau(\omega)$ for which $\tau(\omega) \geq m$ for all $\omega \in \Omega$. Set

$$\mathfrak{M}_g(m; \infty) = \{\tau \in \mathfrak{M}(m; \infty) : M_x g^-(x_\tau) < \infty, x \in E\},$$
$$\overline{\mathfrak{M}}_g(m; \infty) = \{\tau \in \overline{\mathfrak{M}}(m; \infty) : M_x g^-(x_\tau) < \infty, x \in E\},$$

and introduce the payoffs

$$s_{m,\infty}(x) = \sup_{\tau \in \mathfrak{M}_g(m;\infty)} M_x g(x_\tau), \tag{2.78}$$

$$\gamma_{m,\infty}(x; \omega) = \operatorname*{ess\,sup}_{\tau \in \mathfrak{M}_g(m;\infty)} M_x[g(x_\tau) | \mathcal{F}_m]. \tag{2.79}$$

In similar fashion we can define the functions $\bar{s}_{m,\infty}(x)$ and $\bar{\gamma}_{m,\infty}(x; \omega)$, taking sup and ess sup from the class $\overline{\mathfrak{M}}_g(m, \infty)$. The theorem which follows generalizes the results of Theorems 3 and 4 to the case considered (compare also with Theorem 2).

Theorem 5. *Let a function $g \in L(A^-)$. Then for any $m \geq 0$:*

(1)
$$s_{m,\infty}(x) = M_x s(x_m), \ x \in E; \tag{2.80}$$

(2)
$$\gamma_{m,\infty}(x; \omega) = s(x_m) \ (P_x\text{-a.s.}), \ x \in E; \tag{2.81}$$

(3) $\bar{s}_{m,\infty}(x) = s_{m,\infty}(x)$, $\bar{\gamma}_{m,\infty}(x; \omega) = \gamma_{m,\infty}(x; \omega)$ (P_x-a.s.), $x \in E$. *Let a function $g \in L(A^-, A^+)$. Then:*

(4) *The time*

$$\tau^\varepsilon_{m,\infty} = \inf\{n \geq m : s(x_n) \leq g(x_n) + \varepsilon\} \tag{2.82}$$

is for any $\varepsilon > 0$ an (ε, s)-optimal time in the sense that

$$M_x\{g(x_{\tau^\varepsilon_{m,\infty}}) | \mathcal{F}_m\} \geq \gamma_{m,\infty}(x; \omega) - \varepsilon \qquad (P_x\text{-a.s.}), \ x \in E \tag{2.83}$$

and

$$M_x\{g(x_{\tau^\varepsilon_{m,\infty}})\} \geq s_{m,\infty}(x) - \varepsilon, \qquad x \in E; \tag{2.84}$$

(5) *The Markov time $\tau^0_{m,\infty}$ is a $(0, \bar{s})$-optimal time in the sense that*

$$M_x\{g(x_{\tau^0_{m,\infty}}) | \mathcal{F}_m\} = \bar{\gamma}_{m,\infty}(x; \omega) \qquad (P_x\text{-a.s.}), \ x \in E, \tag{2.85}$$

$$M_x\{g(x_{\tau^0_{m,\infty}})\} = \bar{s}_{m,\infty}(x), \qquad x \in E; \tag{2.86}$$

(6) *If the time $\tau^0_{m,\infty}$ is a stopping time, then it is a $(0, s)$-optimal time.*

PROOF. Let $v(x)$ be the smallest excessive majorant of the function $g \in L(A^-)$. Then, by virtue of (2.40), for $\tau \in \overline{\mathfrak{M}}(m, \infty)$

$$v(x_m) \geq M_x[v(x_\tau) | \mathcal{F}_m] \geq M_x[g(x_\tau) | \mathcal{F}_m] \qquad (P_x\text{-a.s.}), \ x \in E.$$

Therefore,

$$v(x_m) \geq \operatorname*{ess\,sup}_{\tau \in \overline{\mathfrak{M}}(m;\infty)} M_x[g(x_\tau) | \mathcal{F}_m] = \bar{\gamma}_{m,\infty}(x; \omega) \geq \gamma_{m,\infty}(x; \omega) \tag{2.87}$$

and

$$M_x v(x_m) \geq \bar{s}_{m,\infty}(x) \geq s_{m,\infty}(x). \tag{2.88}$$

Hence to prove the first three assertions of the theorem we need only to show that $v(x_m) \leq \gamma_{m,\infty}(x;\omega)$ and $M_x v(x_m) \leq s_{m,\infty}(x)$.

To this end we assume first that a function $g \in L(A^+)$. Then the time $\tau_{m,\infty}^{\varepsilon} = \inf\{n \geq m : v(x_n) \leq g(x_n) + \varepsilon\}$ is a stopping time for $\varepsilon > 0$ and

$$v(x_m) = M_{x_m} v(x_{\tau_{0,\infty}^{\varepsilon}}).$$

Hence (P_x-a.s.)

$$\begin{aligned} v(x_m) &= M_{x_m} v(x_{\tau_{0,\infty}^{\varepsilon}}) \\ &\leq M_{x_m} g(x_{\tau_{0,\infty}^{\varepsilon}}) + \varepsilon \\ &= M_x[g(x_{\tau_{m,\infty}^{\varepsilon}}) | \mathscr{F}_m] + \varepsilon \\ &\leq \operatorname*{ess\,sup}_{\tau \in \mathfrak{M}(m;\infty)} M_x[g(x_\tau) | \mathscr{F}_m] + \varepsilon \\ &= \gamma_{m,\infty}(x;\omega) + \varepsilon, \end{aligned} \tag{2.89}$$

$$M_x v(x_m) \leq s_{m,\infty}(x) + \varepsilon \tag{2.90}$$

for $x \in E$.

Because of arbitrariness of $\varepsilon > 0$ we have from the above the required inequalities

$$v(x_m) \leq \gamma_{m,\infty}(x;\omega) \qquad (P_x\text{-a.s.}), \qquad M_x v(x_m) \leq s_{m,\infty}(x).$$

The general case can be reduced to one already considered ($g \in L(A^+)$) by the same method as that used for proving Theorem 3.

Assertion (4) follows immediately from (2.89) and (2.90). To prove assertions (5) and (6) we need to use the inequality

$$\begin{aligned} v(x_m) &\leq M_x\{I_{\{\tau_{m,\infty}^0 < \infty\}} g(x_{\tau_{m,\infty}^0}) | \mathscr{F}_m\} \\ &\quad + M_x\left\{I_{\{\tau_{m,\infty}^0 = \infty\}} \overline{\lim_n}\, g(x_n) | \mathscr{F}_m\right\} \\ &= M_x\{g(x_{\tau_{m,\infty}^0}) | \mathscr{F}_m\}, \end{aligned}$$

which can be proved in similar fashion to (2.77).

2.6 Examples

2.6.1

The examples which follow pursue two objectives. Some examples, based on Theorems 3 and 4, illustrate methods of finding payoffs and optimal stopping rules; other examples illustrate the fact that the conditions of these theorems cannot, in general, be weakened.

EXAMPLE 1. Let us obtain the solution of the problem of optimal stopping on the basis of Remark 1 to Theorem 3. Let $\xi_1, \xi_2, \ldots$ be a sequence of independent random variables uniformly distributed and given on a probability space $(\Omega, \mathscr{F}, P)$ and taking on the two values $+1$ and -1 with probabilities $P(\xi_i = +1) = p$, $P(\xi_i = -1) = q = 1 - p$. Set $x_0 = x$, $x_n = x + \xi_1 + \cdots + \xi_n$ where $x \in E = \{0, \pm 1, \pm 2, \ldots\}$. Then the process $X = (x_n, \mathscr{F}_n, P_x)$, $n \in N$, forms a Markov chain with values in E where $\mathscr{F}_n = \sigma\{\omega : x_0, \ldots, x_n\}$ and P_x is the distribution of probabilities on sets from $\mathscr{F}_\infty = \sigma(\bigcup_n \mathscr{F}_n)$ corresponding to an initial state x and induced naturally by the random variables $\xi_1, \xi_2, \ldots$.

Let a function $g(x) = \max\{0, x\}$. It is easy to understand that in the case $p \geq q$ the time $\tau^* \equiv \infty$ is a $(0, \bar{s})$-optimal time, with $\bar{s}(x) = \infty$ for all $x \in E$. The case $p < q$ is of greater interest. We shall show that in this situation there exists a $(0, \bar{s})$-optimal time τ^* which is not a stopping time.

We shall define Markov times $\tau_\gamma = \inf\{n \geq 0 : x_n \geq \gamma\}$ where $\gamma \in E$. It can easily be shown that probability $p_\gamma(x)$ for the set $\Gamma_\gamma = [\gamma, \infty]$ to be reached for various $x \in E$ can be described by

$$p_\gamma(x) = \begin{cases} \left(\dfrac{p}{q}\right)^{-x}, & x \leq \gamma, \\ 1, & x > \gamma. \end{cases}$$

Hence

$$f_\gamma(x) = M_x g(x_{\tau_\gamma}) = \begin{cases} \gamma\left(\dfrac{p}{q}\right)^{\gamma - x}, & x \leq \gamma, \\ x & x > \gamma. \end{cases}$$

Set $f^*(x) = \sup_\gamma f_\gamma(x)$. Then $f^*(x) = f_{\gamma^*}(x)$ where γ^* is a maximum point of the function $\gamma(p/q)^\gamma$ on the set E and

$$f^*(x) = \begin{cases} \gamma^*\left(\dfrac{p}{q}\right)^{\gamma^* - x}, & x \leq \gamma^*, \\ x, & x > \gamma^*. \end{cases}$$

It is easy to convince oneself that $f^*(x) \geq g(x)$ and $f^*(x) \geq Tf^*(x)$ for all $x \in E$, which fact implies that the time $\tau^* = \tau_{\gamma^*}$ is a $(0, \bar{s})$-optimal time:

$$M_x g(x_{\tau^*}) = \bar{s}(x).$$

We shall note that in the example considered $P_x(\overline{\lim}_n\, g(x_n) = 0) = 1$ for all $x \in E$ and $P_x(\tau^* = \infty) > 0$ for all $x < \gamma^*$. Hence the time τ^* being a $(0, \bar{s})$-optimal time, is not, at the same time, a $(0, s)$-optimal time since $\tau^* \notin \mathfrak{M}$ (see Figure 1).

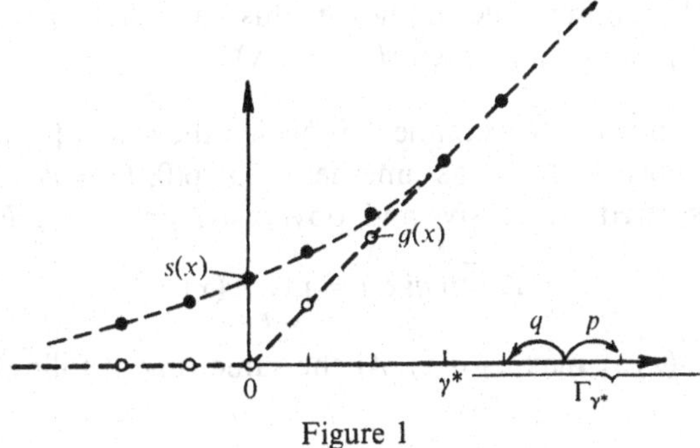

Figure 1

2.6.2

If the set E is finite, $|g(x)| < \infty$, $x \in E$, then, by (4) of Theorem 4 there always exists an optimal stopping time. Examples 2–4 which follow pertain to this very case.

EXAMPLE 2. Let $E = \{0, 1, \ldots, N\}$ and let a transition probability $p(x, y) = P_x(x_1 = y)$ be chosen so that

$$p(0, 0) = p(N, N) = 1, \qquad p(i, i + 1) = p(i, i - 1) = \tfrac{1}{2},$$
$$i = 1, \ldots, N - 1.$$

Excessiveness of the function $v = v(x)$ in the case considered implies that this function is convex upward:

$$v(x) \geq \frac{v(x + 1) + v(x - 1)}{2}, \qquad x = 1, \ldots, N - 1,$$

with $v(0) = g(0)$, $v(N) = g(N)$. Hence the excessive majorant $v(x)$ of the function $g(x)$ is the **smallest convex upward function,** "spanned from above" by $g(x)$ while satisfying the end-point conditions $v(0) = g(0)$, $v(N) = g(N)$ (see Figure 2).

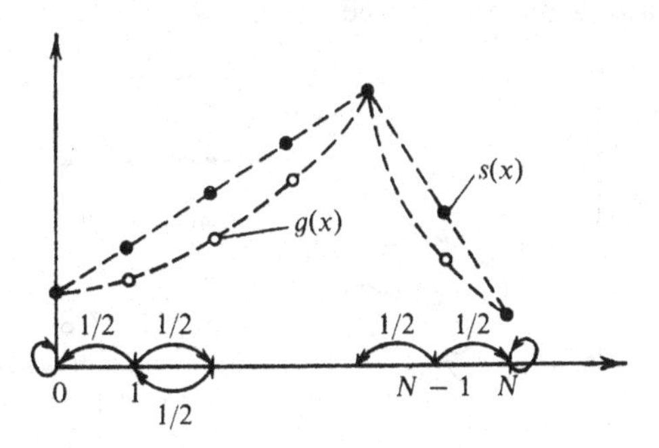

Figure 2

The optimal stopping rule implies in this case that one needs to stop observations at points x for which $v(x) = g(x)$.

EXAMPLE 3. In contrast to Example 2, in which the states $\{0\}$ and $\{N\}$ were assumed to be absorbent, we assume here that $p(0, 1) = p(N, N - 1) = 1$. The chain considered is recursive and, obviously, for any $x \in E$

$$M_x \overline{\lim}\, g(x_n) = \max_{x \in E} g(x).$$

It is clear that $s(x) \leq \max_{x \in E} g(x)$. At the same time it follows from (2.68) that

$$s(x) \geq M_x \overline{\lim}\, g(x_n) = \max_{x \in E} g(x).$$

Hence $s(x) = \max_{x \in E} g(x)$ and the optimal stopping rule implies in this case that one needs to stop observations at the first entry into one of the three points x where the function $g(x)$ attains a maximum (see Figure 3).

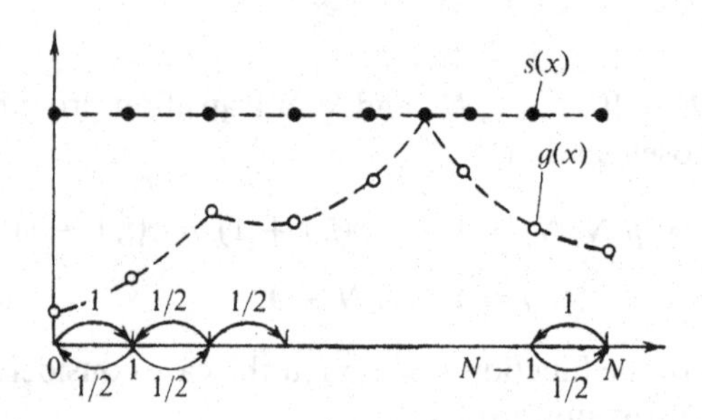

Figure 3

EXAMPLE 4. Again let $E = \{0, 1, \ldots, N\}$ and let $p(0, 0) = 1$, $p(N, N - 1) = 1$,

$$p(i, i + 1) = p(i, i - 1) = \tfrac{1}{2}, \qquad i = 1, \ldots, N - 1.$$

Then the smallest excessive majorant $v(x)$ of the function $g(x)$ is the smallest "convex" hull of the function $g(x)$ satisfying the restrictions: $v(0) = g(0)$, $v(x) \geq g(x_0)$, $x \geq x_0$ where x_0 is that (first) point at which the function $g(x)$ attains a maximum (see Figure 4).

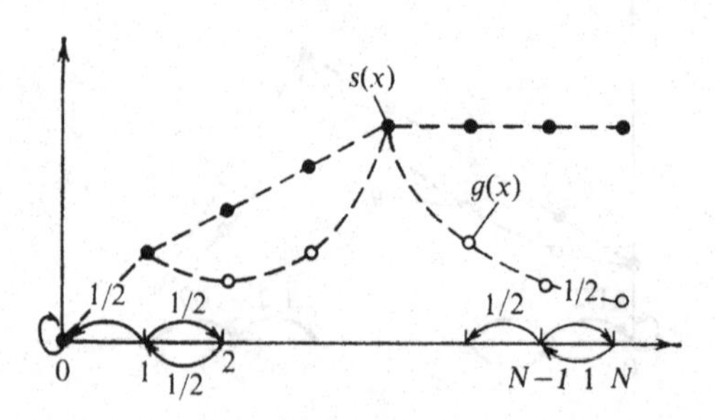

Figure 4

2.6.3

In the case of a finite number of states an optimal stopping time exists. If the set of states is countable this is, in general, not true any longer as is shown by the next example.

EXAMPLE 5. Let $E = \{0, 1, \ldots\}$, let $p(i, i + 1) = 1$ (a deterministic motion to the right), and let $g(x) \geq 0$ be a monotone increasing function with $\lim_{x\to\infty} g(x) = K < \infty$.

Since here $\varphi(x) = M_x[\sup g(x_n)] \equiv K$, then $G^n\varphi(x) \equiv K$ and, therefore, by virtue of Lemma 11 the smallest excessive majorant $v(x) \equiv K$ (which fact was obvious, in fact, *a priori*). It can be readily seen that there exists no optimal stopping time here whereas the time $\tau \equiv \infty$ is a $(0, \bar{s})$-optimal time. On the other hand, it is clear that the time $\tau_\varepsilon = \inf\{n \geq 0 : x_n \geq K - \varepsilon\}$ is an (ε, s)-optimal time for any $\varepsilon > 0$.

2.6.4

One could expect that in the case where the condition A^+ is violated the times

$$\tau_\varepsilon = \inf\{n \geq 0 : v(x_n) \leq g(x_n) + \varepsilon\}$$

would be still $(\varepsilon, \bar{s})$-optimal times. However, this is not, in general, true as follows from the next example.

EXAMPLE 6. Let $E = \{0, 1, 2, \ldots\}$, let $p(0, 0) = 1$, $p(i, i + 1) = p(i, i - 1) = \frac{1}{2}(i = 1, 2, \ldots)$, and let $g(0) = 1$, $g(i) = i(i = 1, 2, \ldots)$. We can show that in this case $M_x[\sup g(x_n)] = \infty, x = 1, 2, \ldots$. Taking advantage of Lemma 6, we can easily find the smallest excessive majorant $v(x)$ of the function $g(x) : v(0) = 1$, $v(x) = x + 1$, $x = 1, 2, \ldots$. For $0 \leq \varepsilon < 1$ the set $\Gamma_\varepsilon = \{x : v(x) \leq g(x) + \varepsilon\}$ consists of a single point $\{0\}$ and the time $\tau_\varepsilon = \inf\{n \geq 0 : x_n \in \Gamma_\varepsilon\}$ is finite with P_x-probability 1 for any $x \in E$. Hence $M_x g(x_{\tau_\varepsilon}) = 1, x \in E$. But, on the other hand, it is clear that the time $\tilde{\tau} \equiv 0$ prescribing instant stopping guarantees at any point $x = 2, 3, \ldots$ a gain $M_x g(x_{\tilde{\tau}}) = g(x)$ equal to 2, 3, ... respectively, larger than the gain of stopping guaranteed by the time τ_ε (see Figure 5).

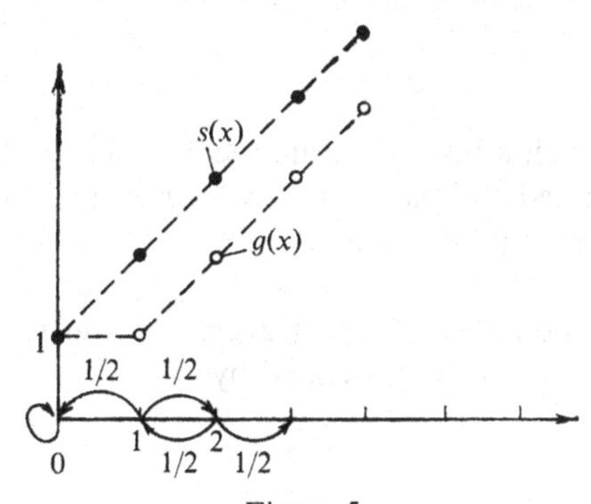

Figure 5

We note that this example shows that the payoff $s(x)$ can be finite as well for all $x \in E$ in the case where the condition A^+ is violated.

Generally speaking we cannot assert in the case where the condition A^+ is violated that there will be a time $\sigma_\varepsilon \in \mathfrak{M}$ such that it is also an (ε, s)-optimal time at least for points x at which $s(x) < \infty$, i.e., so that[9]

$$s(x) \leq M_x g(x_{\sigma_\varepsilon}) + \varepsilon, \qquad x \in \{x : s(x) < \infty\}.$$

Nevertheless, for each fixed x_0 where $s(x_0) < \infty$, and for any $\varepsilon > 0$ there exists (by the definition of sup) a stopping time $\sigma_\varepsilon(x_0)$ such that

$$s(x_0) \leq M_{x_0} g(x_{\sigma_\varepsilon(x_0)}) + \varepsilon.$$

This time can be constructed as follows. Set

$$g^b(x) = \min(b, g(x)), \qquad s^b(x) = \sup_{\tau \in \mathfrak{M}} M_x g^b(x_\tau).$$

We have $s^b(x) \uparrow s(x)$. Hence for the point x_0, where $s(x_0) < \infty$ and $\varepsilon > 0$, we can find $B = B(x_0, \varepsilon)$ such that for all $b \geq B$

$$s(x_0) - s^b(x_0) \leq \frac{\varepsilon}{2}. \tag{2.91}$$

It follows from the fact that the function $g^b(x)$ has an upper bound that the Markov time

$$\sigma_\varepsilon(x_0) = \inf\left\{n \geq 0 : s^{B(x_0, \varepsilon)}(x_n) \leq g^{B(x_0, \varepsilon)}(x_n) + \frac{\varepsilon}{2}\right\}$$

is such that $\sigma_\varepsilon(x_0) \in \mathfrak{M}$ and

$$M_x g^B(x_{\sigma_\varepsilon(x_0)}) \geq s^B(x) - \frac{\varepsilon}{2}, \qquad x \in E. \tag{2.92}$$

We get the required result from (2.92) and (2.91) since

$$s(x_0) \geq M_{x_0} g(x_{\sigma_\varepsilon(x_0)}) \geq M_{x_0} g^B(x_{\sigma_\varepsilon(x_0)}) \geq s^B(x_0) - \frac{\varepsilon}{2} \geq s(x_0) - \varepsilon.$$

2.6.5

The two examples which follow illustrate the fact that in the case where the condition A^- is violated the payoff $s(x)$ need not, in fact, be the smallest excessive majorant of the function $g(x)$.

EXAMPLE 7. Let the state space $E = \{0, 2, 2^2, \dots\}$ and let a Markov chain $X = (x_n, \mathscr{F}_n, P_x)$, $n \in N$, $x \in E$, be defined by

$$x_{n+1} = 2x_n \cdot \xi_{n+1},$$

[9] In this connection see also Remark 2 to Theorem 3.14.

where $\xi_1, \xi_2, \ldots$ is a sequence of independent random variables uniformly distributed and given on a probability space $(\Omega, \mathscr{F}, P)$, $P(\xi_n = 0) = P(\xi_n = 1) = \frac{1}{2}$, $\mathscr{F}_n = \sigma\{\omega : x_0, \ldots, x_n\}$, and measure P_x is defined naturally. In other words, a "particle" leaving a point $x \in E$, gets with equal probability either to a point $2x$ or to a point 0 at which it remains.

Set $g(x) = -x$ and consider the payoff

$$s(x) = \sup_{\tau \in \mathfrak{M}} M_x g(x_\tau), \qquad x \in E.$$

It can be easily verified that

$$M_x\left[\sup_n g^-(x_n)\right] = M_x\left[\sup_n x_n\right] = +\infty, \qquad x \neq 0,$$

thus violating the condition A$^-$.

Let us find the smallest excessive majorant $v(x)$ of the function $g(x)$ considered. We can easily verify that $Tg(x) = g(x)$, $x \in E$. Therefore, the function $v(x) = g(x)$ is the smallest excessive majorant of $g(x)$ and, if Theorem 3 remained true, then $s(x) = g(x) = -x$. However, in fact, $s(x) \equiv 0$. Indeed, let $\tilde{\tau} = \inf\{n : x_n = 0\}$. Since $P_x(\tilde{\tau} < \infty) = 1$, $x \in E$, then $\tilde{\tau}$ is a stopping time. But $M_x g(x_{\tilde{\tau}}) = 0$ and, obviously, $s(x) \leq 0$. Hence the time $\tilde{\tau}$ is an optimal time, $s(x) \equiv 0$, and $s(x) \neq v(x)$ where $v(x)$ is the smallest excessive majorant of $g(x)$.

It is easy to explain why here $s(x) \neq v(x)$. In proving Theorem 3 we relied essentially on Lemma 3 which, as this example illustrates, need not be true in the case where the condition A$^-$ is violated. (In the example considered, (2.41) is not satisfied for a function $f(x) = g(x)$ $(= -x)$ and times $\sigma \equiv 0$ and $\tau = \tilde{\tau}$.)

This example leads one conjecture that in the general case the characterization of the payoff $s(x)$ should be sought in the (more narrow) class of the functions for which Lemma 3 holds (at least, for times $\sigma \equiv 0$ and $\tau \in \overline{\mathfrak{M}}_g$) rather than in the class of excessive functions.

It is interesting to note that in the example considered the payoffs $s_n(x) = -x$, $n \in N$, whereas $s(x) \equiv 0$. Thus this example indicates that, generally speaking, the payoffs

$$s_n(x) \nrightarrow s(x), \qquad n \to \infty.$$

Example 8. Let $X = (x_n, \mathscr{F}_n, P_x)$, $n \in N$, where $x \in E = \{0, \pm 1, \pm 2, \ldots\}$, $x_n = x + \xi_1 + \cdots + \xi_n$, $n > 0$, $x_0 = x$, $\mathscr{F}_n = \sigma\{\omega : x_0, \ldots, x_n\}$, and let $\xi_1, \xi_2, \ldots$ be a sequence of independent random variables uniformly distributed taking on values ± 1 with probabilities $\frac{1}{2}$. Also let $g(x) = x$. Then $Qg(x) = g(x)$ and, therefore, for any $n \in N$, $s_n(x) = Q^n g(x) = x$. It is clear that the smallest excessive majorant of the function $g(x) = x$ is the function itself. It is clear also that $P_x(\overline{\lim}\, x_n = +\infty) = 1$ and hence $s(x) = +\infty$ for any $x \in E$.

2.7 The structure and methods of finding a payoff for a function $g \in B(\mathrm{a}^-)$

2.7.1

Examples 7 and 8 given in the previous section illustrate that the payoff $s(x)$ is not, in general, the smallest excessive majorant of the function $g(x)$ in the case where the condition A^- is violated. The structure of the payoff $s(x)$ in the general case will be discussed in Section 9. In the present section we shall consider the case where the function $g(x)$ satisfies the condition a^-. It turns out that in this case the payoff is the smallest excessive majorant of the function[10]

$$G(x) = \max\{g(x), M_x g(x_\infty)\} \tag{2.93}$$

rather than the function $g(x)$.

Thus, let the function $g(x)$ belong to the class $B(\mathrm{a}^-)$, i.e., the condition

$$\mathrm{a}^- : M_x[g^-(x_\infty)] < \infty, \qquad x \in E, \tag{2.94}$$

is satisfied where $g(x_\infty) = \overline{\lim}_n\, g(x_n)$. (Note that in Example 7 of Section 6 $M_x g(x_\infty) = 0$ and, therefore, the function $G(x) \equiv 0$.)

Theorem 6. *Let a function $g \in B(\mathrm{a}^-)$. Then:*

(1) *The payoff $\bar{s}(x)$ is*[11] *the smallest excessive majorant of the function* $G(x) = \max\{g(x), M_x g(x_\infty)\}$;

(2) *The payoff $\bar{s}(x) = \bar{S}(x)$ where*

$$\bar{S}(x) = \sup_{\tau \in \mathfrak{M}_G} M_x G(x_\tau); \tag{2.95}$$

(3) $\bar{s}(x) = \max\{g(x), M_x g(x_\infty), T\bar{s}(x)\}$.

To prove this theorem we shall need:

Lemma 12

(1) *The function*

$$V(x) = \lim_n Q^n G(x), \qquad x \in E, \tag{2.96}$$

is the smallest excessive majorant of the function $G(x)$;

(2) *The payoff $\bar{S}(x) = V(x)$.*

[10] Compare with Remark 5 to Theorem 3.

[11] It will follow from Theorem 9 that in the case considered the payoffs $\bar{s}(x)$ and $s(x)$ also coincide.

PROOF

(1) By virtue of the inequality $G(x) \geq M_x g(x_\infty)$ and Markovianness,

$$G(x_1) \geq M_{x_1} g(x_\infty) = M_x\{g(x_\infty)|\mathscr{F}_1\} \qquad (P_x\text{-a.s.}),\ x \in E.$$

Hence, since $M_x g(x_\infty) > -\infty$, $M_x G(x_1)$ is defined and is not equal to $-\infty$, i.e., $M_x G^-(x_1) < \infty$. Therefore, $G \in L$ and by virtue of Lemma 6 the function $V(x)$ is the smallest excessive majorant of $G(x)$.

(2) We shall show that $\bar{S}(x) = V(x)$. To this end we shall prove first the inequality $V(x) \leq \bar{S}(x)$.

Let $G(x) \leq b < \infty$ and let $\tilde{V}(x) = \lim_n G^n\varphi(x)$ where $\varphi(x) = M_x[\sup_n G(x_n)]$. It is clear that $V(x) \leq \tilde{V}(x)$, and by virtue of (1) of Lemma 11 we have

$$\tilde{V}(x) \leq M_x \tilde{V}(x_{\tilde{\tau}_\varepsilon}),$$

where the stopping time

$$\tilde{\tau}_\varepsilon = \inf\{n \geq 0 : \tilde{V}(x_n) \leq G(x_n) + \varepsilon\}.$$

Hence

$$V(x) \leq \tilde{V}(x) \leq M_x \tilde{V}(x_{\tilde{\tau}_\varepsilon}) \leq M_x G(x_{\tilde{\tau}_\varepsilon}) + \varepsilon \leq \bar{S}(x) + \varepsilon$$

and, therefore, $V(x) \leq \bar{S}(x)$.

In the general case this inequality can be proved, as in Theorem 3, by considering the functions $G^b(x) = \min(b, G(x))$ with the subsequent passage to the limit with respect to $b \uparrow \infty$.

We shall prove the inverse inequality $V(x) \geq \bar{S}(x)$. (Lemma 3 was used in Theorem 3 to prove the analogous inequality but it is unapplicable in this case since the function $G(x)$ does not satisfy, in general, the condition A^-.)

To this end we shall show that the family $\{V^-(x_n), n \in N\}$ is uniformly integrable (with respect to each measure P_x, $x \in E$).

We have $V(x) \geq G(x) \geq M_x g(x_\infty)$ from which, by virtue of Markovianness, we get

$$V(x_n) \geq M_{x_n} g(x_\infty) = M_x\{g(x_\infty)|\mathscr{F}_n\} \qquad (P_x\text{-a.s.}),\ x \in E,$$

and, therefore,

$$V^-(x_n) \leq M_x[g^-(x_\infty)|\mathscr{F}_n] = M_x\left[\varliminf_m g^-(x_m)|\mathscr{F}_n\right].$$

The sequence $\{M_x(\varliminf g^-(x_m)|\mathscr{F}_n), \mathscr{F}_n, P_x\}$, $n \in N$, is, obviously, a uniformly integrable martingale (for each $x \in E$). Hence the variables $\{V^-(x_n), n \in N\}$ are also uniformly integrable.

We shall consider the functions $V^b(x) = \min(b, V(x))$, $b \geq 0$, as well as $V(x)$. These functions are excessive and $M_x[V^b(x_{n+1})|\mathscr{F}_n] \leq V^b(x_n)$ P_x-a.s.), $x \in E$, by virtue of Properties VI and IV from Section 4. Since the variables $\{V^-(x_n), n \in N\}$ are uniformly integrable, the variables $\{V^b(x_n), n \in N\}$ are also uniformly integrable. Therefore, for each $x \in E$ the objects

$\{V^b(x_n), \mathscr{F}_n, P_x\}$, $n \in N$, form a uniformly integrable supermartingale, and by virtue of Theorem 1.9 there exists a limit $\lim_n V^b(x_n)$ $(= V^b(x_\infty))$, with

$$M_x[V^b(x_\infty)|\mathscr{F}_n] \leq V^b(x_n) \qquad (P_x\text{-a.s.}),\ x \in E. \tag{2.97}$$

This condition guarantees (Theorem 1.11) that for any Markov times σ and τ such that $P_x(\sigma \leq \tau) = 1$, $x \in E$,

$$M_x[V^b(x_\tau)|\mathscr{F}_\sigma] \leq V^b(x_\sigma)$$

and

$$M_x V^b(x_\tau) = M_x V^b(x_\sigma) \leq V^b(x).$$

We shall pass to the limit in these inequalities for $b \uparrow \infty$ and find that

$$M_x V(x_\tau) \leq M_x V(x_\sigma) \leq V(x) \tag{2.98}$$

and, therefore,

$$M_x G(x_\tau) \leq V(x), \tag{2.99}$$

since $V(x) \geq G(x) \geq M_x g(x_\infty)$ and $M_{x_\tau} g(x_\infty) = M_x[g(x_\infty)|\mathscr{F}_\tau] > -\infty$ $(P_x\text{-a.s.})$, $x \in E$. The required inequality follows from (2.99), thus proving the lemma. □

Proof of Theorem 6. We shall convince ourselves that $\bar{s}(x) = \bar{S}(x)$. It is clear that $\bar{s}(x) \leq \bar{S}(x)$, $x \in E$. The inverse inequality needs to be proved only for the $x \in E$ for which $\bar{s}(x) < \infty$ since if $\bar{s}(x) = \infty$, then $\bar{S}(x) = \infty$.

We shall show first that for all $x \in E$ where $\bar{s}(x) < \infty$,

$$g(x_\infty) = G(x_\infty) \qquad (P_x\text{-a.s.}). \tag{2.100}$$

In fact,

$$\begin{aligned} G(x_\infty) &= \overline{\lim_n}\, G(x_n) \\ &= \overline{\lim_n}\, \max\{g(x_n), M_{x_n} g(x_\infty)\} \\ &= \overline{\lim_n}\, \max\{g(x_n), M_x[g(x_\infty)|\mathscr{F}_n]\} \\ &= \max\left\{g(x_\infty), \overline{\lim_n}\, M_x[g(x_\infty)|\mathscr{F}_n]\right\}. \end{aligned} \tag{2.101}$$

By virtue of the assumption $\bar{s}(x) < \infty$ the expectation $M_x g(x_\infty) < \infty$. At the same time $g \in B(a^-)$, i.e., $M_x g^-(x_\infty) < \infty$. Hence $M_x|g(x_\infty)| < \infty$ and, therefore, by virtue of Theorem 1.10,

$$\lim_n M_x[g(x_\infty)|\mathscr{F}_n] = M_x[g(x_\infty)|\mathscr{F}_\infty] = g(x_\infty) \qquad (P_x\text{-a.s.}),$$

which fact together with (2.101) proves (2.100).

Furthermore, let $\bar{s}(x) < \infty$ at the point $x \in E$. For the arbitrary Markov time τ we set

$$\sigma_\tau(\omega) = \begin{cases} \tau(\omega) & \text{if } \tau(\omega) < \infty \text{ and } M_{x_\tau} g(x_\infty) \le g(x_\tau), \\ +\infty & \text{otherwise.} \end{cases}$$

It is clear that $\sigma_\tau(\omega)$ is a Markov time. We shall prove that if $M_x G(x_\tau)$ exists, then

$$M_x G(x_\tau) = M_x g(x_{\sigma_\tau}). \tag{2.102}$$

Since $g(x_\infty) = G(x_\infty)$, it suffices to show that

$$M_x I_{\{\tau < \infty\}} G(x_\tau) = M_x I_{\{\tau < \infty\}} g(x_{\sigma_\tau}). \tag{2.103}$$

Let

$$A = \{\omega : \tau(\omega) < \infty, M_{x_\tau} g(x_\infty) \le g(x_\tau)\},$$
$$B = \{\omega : \tau(\omega) < \infty, M_{x_\tau} g(x_\infty) > g(x_\tau)\}.$$

By virtue of Markovianness and the fact that the set B is $\mathscr{F}_\tau$-measurable,

$$\begin{aligned} M_x I_{\{\tau < \infty\}} G(x_\tau) &= M_x I_{\{\tau < \infty\}} \max[g(x_\tau), M_{x_\tau} g(x_\infty)] \\ &= M_x I_A g(x_\tau) + M_x I_B M_{x_\tau} g(x_\infty) \\ &= M_x I_A g(x_\tau) + M_x I_B M_x[g(x_\infty) | \mathscr{F}_\tau] \\ &= M_x I_A g(x_\tau) + M_x M_x[I_B g(x_\infty) | \mathscr{F}_\tau] \\ &= M_x I_A g(x_\tau) + M_x I_B g(x_\infty) \\ &= M_x I_{\{\tau < \infty\}} g(x_{\sigma_\tau}), \end{aligned}$$

which fact proves both (2.103) and (2.102).

From (2.102) we get the required inequality $\bar{S}(x) \le \bar{s}(x)$ which together with the obvious inequality $\bar{S}(x) \ge \bar{s}(x)$ proves that the payoffs $\bar{S}(x)$ and $\bar{s}(x)$ coincide here.

Finally, Equation (2.95) follows immediately from Lemmas 5 and 12, thus completing the proof of the theorem.

2.7.2

Remark 1. We shall again consider Example 7 from Section 6. It can be readily seen that the function $g \in B(a^-)$ and $M_x g(x_\infty) = 0$. Hence $G(x) \equiv 0$, and it is clear from Equation (2.95) that $\bar{s}(x) \equiv 0$.

Remark 2. The results of Section 9 imply that the payoffs $s(x)$ and $\bar{s}(x)$ coincide under the conditions of Theorem 6; therefore the payoff $s(x)$ satisfies the equation

$$s(x) = \max\{g(x), M_x g(x_\infty), Ts(x)\}. \tag{2.104}$$

Remark 3. It follows from (2.98) and irom (1) of Theorem 6 that if $P_x(\sigma \le \tau) = 1$, $x \in E$, then

$$-\infty < M_x \bar{s}(x_\infty) \le M_x \bar{s}(x_\tau) \le M_x \bar{s}(x_\sigma) \le \bar{s}(x). \tag{2.105}$$

Remark 4. It follows from (2.105) that the payoff $\bar{s}(x)$ satisfies for any $\tau \in \overline{\mathfrak{M}}$ the inequality

$$M_x \bar{s}(x_\tau) \le \bar{s}(x), \qquad x \in E.$$

Let $f(x)$ be some $\mathscr{B}$-measurable function majorizing $g(x)$ and satisfying for any $\tau \in \mathfrak{M}_g$ the inequality

$$M_x f(x_\tau) \le f(x), \qquad x \in E. \tag{2.106}$$

Since $M_x g(x_\tau) \le M_x f(x_\tau)$, $\bar{s}(x) \le f(x)$.

In other words, the payoff $\bar{s}(x)$ is the smallest function from among those satisfying (2.106) and majorizing $g(x)$ (compare with Remark 4 to Theorem 3 and with Theorem 7).

2.8 Regular functions: the structure of the payoff and ε-optimal stopping rules (under the condition A^+)

2.8.1

In investigating the structure of the payoffs $s(x)$ and $\bar{s}(x)$ in the general case we find it useful to make the following:

Definition 1. Let $\mathfrak{R}$ be some class of Markov times $\tau \in \overline{\mathfrak{M}}$. The function $f \in B$ is said to be an *$\mathfrak{R}$-regular function* if for any $\tau \in \mathfrak{R}$ and $x \in E$ the expectations $M_x f(x_\tau)$ are defined and for all $\tau, \sigma \in \mathfrak{R}$ with $P_x(\sigma \le \tau) = 1$, $x \in E$, the inequalities

$$M_x f(x_\tau) \le M_x f(x_\sigma), \qquad x \in E, \tag{2.107}$$

are satisfied where $f(x_\infty) = \overline{\lim}_n f(x_n)$.

If, in addition, $f(x) \ge g(x)$, the function $f(x)$ is said to be an *$\mathfrak{R}$-regular majorant of* $g(x)$. The smallest $\mathfrak{R}$-regular majorants of the function $g(x)$ can be defined in an obvious way.

Two classes of functions play a fundamental role: $\mathfrak{R} = \overline{\mathfrak{M}}$ and $\mathfrak{R} = \mathfrak{M}_g$. For simplicity the $\overline{\mathfrak{M}}$-regular functions will be said to be *regular functions.*

If the function $g \in B(A^-)$, the class of the smallest excessive majorants and that of the smallest regular majorants of the function g coincide (Lemma 3). Hence, we can formulate Theorem 3 as to say that the payoff $s(x)$ is the smallest regular majorant of $g(x)$. We note next that Examples 7 and 8 given in Section 6 illustrate the facts that where condition A^- is violated the excessive function need not be regular and that the payoff is in general not the smallest excessive majorant of the function $g(x)$.

The main result of this section consists in proving that the payoff $s(x)$ is the smallest regular majorant of the functions $g(x)$ of the classes $B(A^+)$. We shall consider in Section 9 the general case and show that the payoff $s(x)$ is the smallest $\mathfrak{M}_g$-regular majorant of $g(x)$.

2.8.2

Theorem 7. *Let the function $g \in B(A^+)$. Then:*

(1) *The payoff $s(x)$ is the smallest regular majorant of the function $g(x)$;*
(2) *The payoff $\bar{s}(x)$ and the payoff $s(x)$ coincide: $\bar{s}(x) = s(x)$;*
(3) $s(x) = \max\{g(x), Ts(x)\}$;
(4) $s(x) = \lim_{b\to\infty} \lim_{a\to-\infty} \lim_{n\to\infty} Q^n g_a^b(x)$;
where

$$g_a^b(x) = \begin{cases} b, & g(x) > b, \\ g(x), & a \le g(x) \le b, \\ a, & g(x) < a, \end{cases}$$

and $a \le 0, b \ge 0$.

PROOF. (a) Let

$$g_a(x) = \max\{a, g(x)\}, \qquad a \le 0,$$
$$s_a(x) = \sup_{\tau\in\overline{\mathfrak{M}}} M_x g_a(x_\tau)$$

and let $s_*(x) = \lim_{a\to-\infty} s_a(x)$. It is obvious that

$$s_*(x) \ge \bar{s}(x) \ge s(x) \ge g(x) > -\infty. \tag{2.108}$$

By virtue of Theorem 3

$$s_a(x) = \max\{g_a(x), Ts_a(x)\}, \tag{2.109}$$

and since $s_a(x) \downarrow s_*(x)$, $M_x[\sup_n g^+(x_n)] < \infty$, then by virtue of the Theorem on monotone convergence $Ts_a(x) \downarrow Ts_*(x)$ and, therefore,

$$s_*(x) = \max\{g(x), Ts_*(x)\}. \tag{2.110}$$

By virtue of Lemma 3, for any Markov times σ and τ with $P_x(\sigma \le \tau) = 1$, $x \in E$,

$$M_x s_a(x_\tau) \le M_x s_a(x_\sigma) \le s_a(x) < \infty.$$

This fact together with the inequality $s_*(x) \le s_a(x)$ shows that the expectations $M_x s_*(x_\tau)$ are defined ($\tau \in \overline{\mathfrak{M}}$, $x \in E$) and that

$$M_x s_*(x_\tau) \le M_x s_*(x_\sigma), \tag{2.111}$$

i.e., $s_*(x)$ is a regular function.

(b) We show next that $s_*(x) \le s(x)$. To this end we introduce the Markov times

$$\sigma_\varepsilon^a = \inf\{m \ge 0 : s_*(x_m) \le g_a(x_m) + \varepsilon\},$$
$$\tau_\varepsilon^a = \inf\{m \ge 0 : s_a(x_m) \le g_a(x_m) + \varepsilon\},$$
$$\tau_\varepsilon^* = \inf\{m \ge 0 : s_*(x_m) \le g(x_m) + \varepsilon\},$$

and show that for $\varepsilon > 0$, $P_x(\tau_\varepsilon^* < \infty) = 1$ and

$$s_*(x) \le M_x s_*(x_{\tau_\varepsilon^*}). \tag{2.112}$$

The above implies that

$$s_*(x) \le M_x g(x_{\tau_\varepsilon^*}) + \varepsilon$$

and, therefore, $s_*(x) \le s(x)$; this together with (2.108) proves $s_*(x) = \bar{s}(x) = s(x)$.

The function $g_a \in B(A^-, A^+)$, hence by virtue of Theorem 3 and the corollary to Lemma 11

$$P_x(\tau_\varepsilon^a < \infty) = 1, \qquad x \in E,$$

and, therefore,

$$s_a(x) = M_x s_a(x_{\tau_\varepsilon^a}). \tag{2.113}$$

It is clear that $\sigma_\varepsilon^\alpha \le \tau_\varepsilon^a$, $a \le \alpha \le 0$. Hence (Lemma 3)

$$M_x s_a(x_{\tau_\varepsilon^a}) \le M_x s_a(x_{\sigma_\varepsilon^\alpha}) \le s_a(x).$$

This together with (2.113) leads us to

$$s_a(x) = M_x s_a(x_{\sigma_\varepsilon^\alpha}), \qquad a \le \alpha, \tag{2.114}$$

where, obviously, $P_x(\sigma_\varepsilon^\alpha < \infty) = 1$, $x \in E$.

We take the limit in (2.114) as $a \to -\infty$. Then by virtue of the condition A^+ and the Theorem on monotone convergence we have

$$s_*(x) = M_x s_*(\tau_{\sigma_\varepsilon^\alpha}), \qquad \varepsilon > 0, \alpha \le 0.$$

By recalling the definition of the time $\sigma_\varepsilon^\alpha$ we can show from the above that

$$\begin{aligned} -\infty < s_*(x) \le M_x g_\alpha(x_{\sigma_\varepsilon^\alpha}) + \varepsilon &= M_x[g_\alpha(x_{\sigma_\varepsilon^\alpha}) \cdot I_{\{g(x_{\sigma_\varepsilon^\alpha}) \le \alpha\}}] \\ &\quad + M_x[g_\alpha(x_{\sigma_\varepsilon^\alpha}) \cdot I_{\{g(x_{\sigma_\varepsilon^\alpha}) > \alpha\}}] + \varepsilon \\ &\le \alpha P_x\{g(x_{\sigma_\varepsilon^\alpha}) \le \alpha\} + C_x, \end{aligned}$$

where $0 \le C_x = M_x[\sup_j g^+(x_j)] + \varepsilon < \infty$.

Hence for $\alpha > 0$

$$P_x\{g(x_{\sigma_\varepsilon^\alpha}) \le \alpha\} \le \frac{C_x - s_*(x)}{-\alpha}$$

and, therefore,

$$\lim_{\alpha \to -\infty} P_x\{g(x_{\sigma_\varepsilon^\alpha}) \le \alpha\} = 0. \tag{2.115}$$

We shall take advantage of this relation to prove that $P_x(\tau_\varepsilon^* < \infty) = 1$, $x \in E$. To this end we shall note that on the set $\{\omega : g(x_{\sigma_\varepsilon^\alpha}) > \alpha\}$

$$\tau_\varepsilon^*(\omega) = \sigma_\varepsilon^\alpha(\omega). \tag{2.116}$$

Hence

$$\begin{aligned} P_x(\tau_\varepsilon^* < \infty) &\ge P_x(\tau_\varepsilon^* < \infty, g(x_{\sigma_\varepsilon^\alpha}) > \alpha) \\ &= P_x(\sigma_\varepsilon^\alpha < \infty, g(x_{\sigma_\varepsilon^\alpha}) > \alpha) \\ &= P_x(g(x_{\sigma_\varepsilon^\alpha}) > \alpha) \to 1, \qquad \alpha \to -\infty. \end{aligned}$$

(c) Thus, $P_x(\tau_\varepsilon^* < \infty) = 1$, $x \in E$. We shall next prove (2.112) which, as was noted above, implies that $s_*(x)$ coincides with $s(x)$ and $\bar{s}(x)$.

It follows from (2.115) that for each fixed $x \in E$ there will be a subsequence $\{\alpha_i\}$, $\alpha_i \to -\infty$, $i \to \infty$, such that (P_x-a.s.)

$$\lim_{i\to\infty} I_{\{g(x_{\sigma_\varepsilon^{\alpha_i}}) \le \alpha_i\}}(\omega) = 0. \tag{2.117}$$

By making use of the equality established above

$$s_*(x) = M_x s_*(x_{\sigma_\varepsilon^{\alpha_i}}),$$

we find from (2.116) and (2.117) that:

$$\begin{aligned} s_*(x) &= M_x I_{\{g(x_{\sigma_\varepsilon^{\alpha_i}}) > \alpha_i\}} s_*(x_{\sigma_\varepsilon^{\alpha_i}}) + M_x I_{\{g(x_{\sigma_\varepsilon^{\alpha_i}}) \le \alpha_i\}} s_*(x_{\sigma_\varepsilon^{\alpha_i}}) \\ &= M_x I_{\{g(x_{\sigma_\varepsilon^{\alpha_i}}) > \alpha_i\}} s_*(x_{\tau_\varepsilon^*}) + M_x I_{\{g(x_{\sigma_\varepsilon^{\alpha_i}}) \le \alpha_i\}} s_*(x_{\sigma_\varepsilon^{\alpha_i}}). \end{aligned}$$

By virtue of the condition A^+ and Fatou's lemma we obtain from this

$$\begin{aligned} s_*(x) &\le M_x \overline{\lim_{i\to\infty}} I_{\{g(x_{\sigma_\varepsilon^{\alpha_i}}) > \alpha_i\}} s_*(x_{\tau_\varepsilon^*}) + M_x \overline{\lim_{i\to\infty}} I_{\{g(x_{\sigma_\varepsilon^{\alpha_i}}) \le \alpha_i\}} s_*(x_{\sigma_\varepsilon^{\alpha_i}}) \\ &= M_x s_*(x_{\tau_\varepsilon^*}). \end{aligned}$$

Therefore, $s_*(x) = \bar{s}(x) = s(x)$, which fact proves (2) of the theorem and, by virtue of (2.110), (3) as well.

(d) We shall show that the payoff $\bar{s}(x)$ is the smallest function in the class of all regular majorants of the function $g(x)$. If $f(x) \ge g(x)$, $M_x f(x_\tau)$ is defined for all $\tau \in \overline{\mathfrak{M}}$, $x \in E$, and $f(x) \ge M_x f(x_\tau)$, then $f(x) \ge M_x f(x_\tau) \ge M_x g(x_\tau)$ where the second inequality holds true since both $M_x f(x_\tau)$ and $M_x g(x_\tau)$ are defined. (We recall that $g \in B(A^+)$) and $f(x_\tau) \ge g(x_\tau)$. Hence $f(x) \ge \bar{s}(x)$, which proves (1) of the theorem (the existence of $M_x \bar{s}(x_\tau)$), while the inequality $M_x \bar{s}(x_\tau) \le M_x \bar{s}(x_\sigma)$ for $\sigma, \tau \in \overline{\mathfrak{M}}$ such that $P_x(\sigma \le \tau) = 1$ follows from the equality $\bar{s}(x) = s_*(x)$ and (2.111).)

The final assertion of the theorem is a particular case of Lemma 14 which will follow. □

Remark. If the function $g \in B(A^+)$, the payoff

$$s(x) = \lim_{a\to-\infty} s_a(x), \tag{2.118}$$

where $s_a(x) = \sup_{\tau\in\mathfrak{M}} M_x g_a(x_\tau)$ and $g_a(x) = \max\{a, g(x)\}$, $a \le 0$.

2.8.3

We shall investigate here some methods for constructing the payoffs $s(x)$ which are (under the condition A^+) the smallest regular majorants of the function $g(x)$. Lemma 13 which follows shows that in the case where the condition A^+ is satisfied the function $\tilde{v}(x) = \lim_{n\to\infty} G^n\varphi(x)$ (see Lemma 9) coincides with the payoff $s(x)$.

Lemma 13. *Let a function $g \in B(\mathrm{A}^+)$,*

$$\varphi(x) = M_x\left[\sup_n g(x_n)\right].$$

Then the payoff

$$s(x) = \lim_{n\to\infty} G^n\varphi(x), \tag{2.119}$$

where G^n is the nth power of the operator G defined in Subsection 4.7.

PROOF. Set

$$g_a(x) = \max\{a, g(x)\}, \qquad a \le 0,$$
$$\tilde{v}(x) = \lim_{n\to\infty} G^n\varphi(x),$$
$$\varphi_a(x) = M_x\left[\sup_j g_a(x_j)\right],$$
$$G_a\varphi_a(x) = \max[g_a(x), T\varphi_a(x)].$$

Then

$$G_a^N\varphi_a(x) \ge G^N\varphi(x),$$

and by virtue of Lemma 11 the payoff

$$s_a(x) = \sup_{\tau\in\mathfrak{M}} M_x g_a(x_\tau) = \lim_{N\to\infty} G_a^N\varphi_a(x).$$

According to (2.118), $s(x) = \lim_{a\to-\infty} s_a(x)$. Hence

$$s(x) = \lim_{a\to-\infty} s_a(x) \ge \lim_{N\to\infty} G^N\varphi(x) = \tilde{v}(x).$$

To prove the inverse inequality, $s(x) \le \tilde{v}(x)$, we shall note that by (2.110)

$$Gs(x) = \max\{g(x), Ts(x)\} = s(x).$$

Hence $G^N s(x) = s(x)$ and since $s(x) \le \varphi(x)$ then

$$s(x) = G^N s(x) \le G^N\varphi(x).$$

Therefore, $s(x) \le \tilde{v}(x)$. □

Corollary. *If $g \in B(\mathrm{A}^+)$, then*

$$s(x) = \lim_{n\to\infty}\lim_{N\to\infty} G_n^N\varphi_n(x).$$

For $a \le 0$ and $b \ge 0$ again let

$$g^b(x) = \min(b, g(x)), \qquad g_a(x) = \max(a, g(x))$$

and

$$g_a^b(x) = \begin{cases} b, & g(x) > b, \\ g(x), & a \le g(x) \le b, \\ a, & g(x) < a. \end{cases}$$

Lemma 14

(1) *If the function* $g \in B$, *then*

$$s(x) = \lim_{b \to \infty} s^b(x)$$

and

$$s(x) = \lim_{b \to \infty} \lim_{a \to -\infty} \lim_{N \to \infty} Q^N g_a^b(x). \tag{2.120}$$

(2) *If* $g \in B(\mathrm{A}^+)$, *then*

$$s(x) = \lim_{a \to -\infty} s_a(x)$$

and

$$s(x) = \begin{cases} \lim\limits_{a \to -\infty} \lim\limits_{b \to \infty} \lim\limits_{N \to \infty} Q^N g_a^b(x), \\ \lim\limits_{a \to -\infty} \lim\limits_{N \to \infty} \lim\limits_{b \to \infty} Q^N g_a^b(x), \\ \lim\limits_{b \to \infty} \lim\limits_{a \to -\infty} \lim\limits_{N \to \infty} Q^N g_a^b(x). \end{cases} \tag{2.121}$$

PROOF. Due to (2.118) and (2.51)

$$s_a(x) = \lim_{b \to \infty} \lim_{N \to \infty} Q^N g_a^b(x) = \lim_{N \to \infty} \lim_{b \to \infty} Q^N g_a^b(x). \tag{2.122}$$

This together with (2.118) proves the first two equalities in (2.121). To prove (1) of Lemma 14 and the last equality in (2.121) we shall first show that

$$s(x) = \lim_{b \to \infty} s^b(x). \tag{2.123}$$

Set $s^*(x) = \lim_{b \to \infty} s^b(x)$. It is seen that $s(x) \ge s^*(x)$. On the other hand, let $\tau \in \mathfrak{M}_g$. Then $M_x g^-(x_\tau) < \infty$, $M_x(g^b)^-(x_\tau) < \infty$ for all $x \in E$, and since $g^b(x) \uparrow g(x)$, $b \to \infty$, then by the Lebesgue theorem on monotone convergence

$$M_x g^b(x_\tau) \uparrow M_x g(x_\tau), \qquad b \to \infty.$$

But $M_x g^b(x_\tau) \le s^b(x) \le s^*(x)$. Hence $M_x g(x_\tau) \le s^*(x)$ and $s(x) \le s^*(x)$.

By virtue of the second equality in (2.121)

$$s^b(x) = \lim_{a \to -\infty} \lim_{N \to \infty} Q^N g_a^b(x),$$

from this we obtain

$$s(x) = \lim_{b\to\infty} s^b(x) = \lim_{b\to\infty} \lim_{a\to-\infty} \lim_{N\to\infty} Q^N g_a^b(x),$$

thus proving the lemma. □

Remark. Let us note that it is not possible, in general, to interchange the limits over a and N in the second equality in (2.121), i.e.,

$$s(x) = \lim_{a\to-\infty} \lim_{N\to\infty} \lim_{b\to\infty} Q^N g_a^b(x) \neq \lim_{N\to\infty} \lim_{a\to-\infty} \lim_{b\to\infty} Q^N g_a^b(x).$$

In fact, in Example 7 in Section 6

$$Q^N g_a^b(x) = Q^N g_a(x), \qquad \lim_{N\to\infty} Q^N g_a(x) = 0,$$

$$\lim_{a\to-\infty} Q^N g_a(x) = -x,$$

and, therefore,

$$0 = \lim_{a\to-\infty} \lim_{N\to\infty} Q^N g_a(x) \neq \lim_{N\to\infty} \lim_{a\to-\infty} Q^N g_a(x) = -x$$

for all $x \neq 0$.

2.8.4

We shall consider next the question of existence and structure of ε-optimal Markov times. The theorem which follows generalizes Theorem 4.

Theorem 8. *Let the function $g \in B(A^+)$ and let $s(x)$ be a payoff (the smallest regular majorant of $g(x)$). Then:*

(1) *For any $\varepsilon > 0$ the time*

$$\tau_\varepsilon = \inf\{n \geq 0 : s(x_n) \leq g(x_n) + \varepsilon\}$$

is an ε-optimal stopping time;

(2) *The time*

$$\tau_0 = \inf\{n \geq 0 : s(x_n) = g(x_n)\}$$

is an optimal Markov time;

(3) *If the time τ_0 is a stopping time ($\tau_0 \in \mathfrak{M}$), then it is an optimal time;*

(4) *For the time τ_0 to be $(0, s)$-optimal it suffices that*

$$\lim_n g(x_n) = -\infty \qquad (P_x\text{-a.s.}),\ x \in E. \tag{2.124}$$

PROOF. Proof of (1)–(3) is exactly the same as that of the corresponding assertions in Theorem 4. (Note that in Theorem 4 the condition A^- is necessary in fact only for the equality $s(x) = v(x)$ to hold.)

To prove the last assertion we shall assume that $P_{x_0}(\tau_0 = \infty) > 0$ for some $x_0 \in E$. But then, obviously, $\bar{s}(x_0) = -\infty$, which contradicts the inequalities $\bar{s}(x_0) \geq g(x_0) > -\infty$.

2.9 Regular characterization of the payoff (the general case)

2.9.1

Theorem 9. *Let the function* $g \in B$. *Then*:

(1) *The payoff* $s(x)$ *is the smallest* $\mathfrak{M}_g$*-regular majorant of the function* $g(x)$;
(2) $s(x) = \bar{s}(x)$;
(3)

$$s(x) = \lim_{b\to\infty} \lim_{a\to-\infty} \lim_{n\to\infty} Q^n g_a^b(x); \tag{2.125}$$

(4) *if* $g \in L$, *then*

$$s(x) = \max\{g(x), Ts(x)\}. \tag{2.126}$$

PROOF. Set $s^b(x) = \sup_{\tau\in\mathfrak{M}_g} M_x g^b(x_\tau)$, $\bar{s}^b(x) = \sup_{\tau\in\overline{\mathfrak{M}}_g} M_x g^b(x_\tau)$. Then by virtue of Lemma 14, $s(x) = \lim_{b\to\infty} s^b(x)$.

Let $\sigma, \tau \in \mathfrak{M}_g$ and let $P_g(\sigma \le \tau) = 1$, $x \in E$. By Theorem 7

$$M_x s^b(x_\tau) \le M_x s^b(x_\sigma). \tag{2.127}$$

Since $s^b(x) \ge -g^-(x)$ and $M_x g^-(x_\tau) < \infty$, then by the Theorem on monotone convergence we can take the limit in (2.127) as $b \uparrow \infty$ and obtain the inequality

$$M_x s(x_\tau) \le M_x s(x_\sigma). \tag{2.128}$$

Further, if $\tau \in \overline{\mathfrak{M}}_g$, then

$$M_x g^b(x_\tau) \le \bar{s}^b(x) = s^b(x) \le s(x)$$

($\bar{s}^b(x) = s^b(x)$ by Theorem 7) and, therefore (again by the Theorem on monotone convergence),

$$M_x g(x_\tau) \le s(x), \qquad \tau \in \overline{\mathfrak{M}}_g.$$

Hence $\bar{s}(x) \le s(x)$, which together with the obvious inequality $\bar{s}(x) \ge s(x)$ proves the equality $\bar{s}(x) = s(x)$.

Thus, the payoffs $s(x)$ and $\bar{s}(x)$ coincide and $s(x)$ is an $\mathfrak{M}_g$-regular majorant of the function $g(x)$. If $v(x)$ is also an $\mathfrak{M}_g$-regular majorant, then

$$M_x g(x_\tau) \le M_x v(x_\tau) \le v(x)$$

and, therefore, $s(x) \le v(x)$, i.e., $s(x)$ is the smallest $\mathfrak{M}_g$-regular majorant of the function $g(x)$.

Next let $g \in L$. Then the time $\tau(\omega) \equiv 1$ belongs to the class $\mathfrak{M}_g$ and, therefore, by virtue of (2.128) $s(x) \ge M_x s(x_1) = Ts(x)$. Hence

$$s(x) \ge \max\{g(x), Ts(x)\}. \tag{2.129}$$

On the other hand, by Theorem 7

$$s^b(x) = \max\{g^b(x), Ts^b(x)\}, \tag{2.130}$$

where $g^b(x) \le s^b(x) \le s(x)$. Since $M_x g^-(x_1) < \infty$, $M_x s^b(x_1) > -\infty$, which fact guarantees the existence of the integral $M_x s(x_1)$ and the validity of the inequality $M_x s^b(x_1) \le M_x s(x_1)$. Hence $s^b(x) \le \max\{g(x), Ts(x)\}$ and, therefore, $s(x) \le \max\{g(x), Ts(x)\}$; this together with (2.129) proves the validity of Equation (2.126).

Finally, (2.125) follows from Lemma 14. □

Remark 1. Let a time $\tau \in \mathfrak{M}_g$ be such that its associated gain $f(x) = M_x g(x_\tau)$ is an $\mathfrak{M}_g$-regular function. Then the time τ is optimal.

Remark 2. If $g \in B$, then Remarks 2 and 3 to Theorem 3 hold true (with obvious changes in the notation).

2.10 Convergence of the payoffs $s_n(x)$ and the optimal times τ_n^* as $n \to \infty$

2.10.1

Let the function $g \in L$. It follows from Theorem 1 that the payoffs $s_1(x)$, $s_2(x), \ldots$ can be found successively by the formulas

$$s_n(x) = Q^n g(x). \tag{2.131}$$

The knowledge of these payoffs enables us (at least, in principle) to find the optimal times as well:

$$\tau_n^* = \min\{0 \le m \le n : s_{n-m}(x_m) = g(x_m)\} \tag{2.132}$$

(in the classes $\mathfrak{M}_g(n)$).

Next let $n \to \infty$. Since $s_{n+1}(x) \ge s_n(x)$ and $\tau_{n+1}^* \ge \tau_n^*$ (P_x-a.s.), $x \in E$, the limits

$$s^*(x) = \lim_{n\to\infty} s_n(x), \qquad \tau^* = \lim_{n\to\infty} \tau_n^*$$

exist. In this case, if $g \in L$, then (by the Theorem on monotone convergence) from the equations

$$s_{n+1}(x) = \max\{g(x), Ts_n(x)\}$$

for $s^*(x)$ we get the equation

$$s^*(x) = \max\{g(x), Ts^*(x)\}. \tag{2.133}$$

It is natural to ask whether the limit $s^*(x)$ coincides with the payoff $s(x)$ (it is clear that $s^*(x) \le s(x)$ always)) and whether the time τ^* is a $(0, s)$-optimal time or, at least, a $(0, \bar{s})$-optimal time.

Examples 7 and 8 in Section 6 prove that this is not, generally speaking, true, which leads to another question: How should we characterize the difference between the functions $s^*(x)$ and $s(x)$? It follows from Theorem 9 that (assuming $g \in L$) the payoff $s(x)$ is the smallest $\mathfrak{M}_g$-regular majorant of the function $g(x)$. It follows from Equation (2.133) that the function $s^*(x) = \lim_{n\to\infty} s_n(x)$ is the excessive majorant of the function $g(x)$. We shall show that the function $s^*(x)$ is, in fact, the smallest excessive majorant of $g(x)$.

Indeed, let $f(x)$ be some excessive majorant of $g(x)$. Then

$$f(x) \geq \max\{g(x), Tf(x)\}$$

and

$$f(x) = Qf(x) = \max\{f(x), Tf(x)\} \geq \max\{g(x), Tg(x)\} = Qg(x).$$

We have from this

$$f(x) \geq Q^n g(x), \qquad n = 0, 1, \dots,$$

and, therefore, $f(x) \geq s^*(x) = \lim_{n\to\infty} Q^n g(x)$.

Therefore, the difference between the functions $s(x)$ and $s^*(x)$ can be characterized as follows.

Theorem 10. *Let the function $g \in L$. Then:*

(1) *The payoff $s(x)$ is the smallest $\mathfrak{M}_g$-regular majorant of the function $g(x)$;*
(2) *The function $s^*(x) = \lim_{n\to\infty} s_n(x)$ is the smallest excessive majorant of the function $g(x)$.*

2.10.2

We shall list the sufficient conditions guaranteeing the coincidence of the functions $s(x)$ and $s^*(x)$ and solve the problem of the optimalization of time $\tau^* = \lim_{n\to\infty} \tau_n^*$.

Theorem 11

(1) *If the function $g \in L(A^-)$, then*

$$\bar{s}(x) = s(x) = s^*(x).$$

(2) *If the function $g \in L(A^-, A^+)$, then the time $\tau^* = \lim_n \tau_n^*$ is $(0, \bar{s})$-optimal.*
(3) *If the function $g \in L(A^-, A^+)$ and if the time τ^* is a stopping time it will be $(0, s)$-optimal.*
(4) $\tau^* = \inf\{n \geq 0 : s^*(x_n) = g(x_n)\}$.

PROOF

(1) The coincidence of $s(x)$ and $\bar{s}(x)$ with $s^*(x)$ follows from Theorem 3 and Lemma 6.

(2) On the set $\{\omega : \tau^*(\omega) < \infty\}$

$$\lim_n g(x_{\tau_n^*}) = g(x_{\tau^*}).$$

Hence by Fatou's lemma

$$\begin{aligned}
s(x) &= \lim_n s_n(x) \\
&= \lim_n M_x g(x_{\tau_n^*}) \\
&\leq M_x \overline{\lim_n} g(x_{\tau_n^*}) \\
&= M_x \overline{\lim_n} g(x_{\tau_n^*}) I_{\{\tau^* < \infty\}} + M_x \overline{\lim_n} g(x_{\tau_n^*}) I_{\{\tau^* = \infty\}} \\
&= M_x g(x_{\tau^*}) I_{\{\tau^* < \infty\}} + M_x \overline{\lim} g(x_n) I_{\{\tau^* = \infty\}} \\
&= M_x g(x_{\tau^*}),
\end{aligned} \tag{2.134}$$

which proves that τ^* is a $(0, \bar{s})$-optimal time.

(3) If $P_x(\tau^* = \infty) = 0$, $x \in E$, then it follows from (2.134) that τ^* is a $(0, s)$-optimal time.

(4) To prove the last assertion of the theorem we shall set

$$\tilde{\tau} = \inf\{n \geq 0 : s^*(x_n) = g(x_n)\}$$

and show that $\tilde{\tau} = \tau^*$ (P_x-a.s.), $x \in E$.

Let $\omega_0 \in \{\omega : \tilde{\tau} = n\}$, $n < \infty$. Then $g(x_i) < s^*(x_i)$, $i = 0, 1, \ldots, n-1$, where $x_i = x_i(\omega_0)$ and, therefore, $g(x_i) < s_{N-i}(x_i)$ for sufficiently large N, i.e., $\tau_N^*(\omega_0) \geq n$. Hence $\tau^*(\omega_0) \geq \tau_N^*(\omega_0) \geq \tilde{\tau}(\omega_0)$. If $\tilde{\tau}(\omega_0) = \infty$, then $g(x_i) < s^*(x_i)$ for $x_i = x_i(\omega_0)$ for all $i \geq 0$ and, therefore, $\tau_N^*(\omega_0) > n$ for sufficiently large N and any $n < N$, and $\tau^*(\omega_0) = \lim_{N \to \infty} \tau_N^*(\omega_0) > n$, i.e., $\tau^*(\omega_0) = \infty$.

Thus, $\tilde{\tau} \leq \tau^*$. The inverse inequality is obvious. □

Remark. It is seen from the proof of Theorem 11 that (2) and (3) can be formulated as follows.

Let $g \in L(A^-)$ and at the point $x_0 \in E$ let

$$M_{x_0}\left[\sup_n g^+(x_n)\right] < \infty.$$

Then the time $\tau^* = \lim_n \tau_n^*$ is $(0, \bar{s})$-optimal at the point x_0, i.e.,

$$\bar{s}(x_0) = M_{x_0} g(x_{\tau^*}).$$

If, in addition, $P_{x_0}(\tau^* < \infty) = 1$, the time τ^* will be $(0, s)$-optimal at the point x_0.

2.10.3

It is sometimes difficult to use (3) of Theorem 11 due to the fact that it is difficult to verify whether the time τ^* is a stopping time. However, by some general considerations we are able to establish that there is an optimal time in the class of stopping times. It turns out that this fact alone implies that the time $\tau^* = \lim_n \tau_n^*$ is $(0, s)$-optimal. In this case the time τ^* will be the smallest time among all $(0, s)$-optimal times.

Theorem 12. *Let the function $g \in L(A^-)$ and let the stopping time σ^* be optimal: $M_x g(x_{\sigma^*}) = s(x)$ for all $x \in E$. Then if $s(x) < \infty$, $x \in E$, the time τ^* is an optimal stopping time and $\tau^* \leq \sigma^*$ (P_x-a.s.), $x \in E$.*

PROOF. If σ^* is an optimal stopping time, then $M_x g(x_{\sigma^*}) > -\infty$ and

$$s(x) = M_x g(x_{\sigma^*}) \leq M_x s(x_{\sigma^*}).$$

By Theorem 9, $M_x s(x_{\sigma^*}) \leq s(x)$. Hence it follows from the assumption $s(x) < \infty$ that

$$-\infty < M_x g(x_{\sigma^*}) = M_x s(x_{\sigma^*}) < \infty.$$

From this, taking into account the inequality $s(x_{\sigma^*}) \geq g(x_{\sigma^*})$ (P_x-a.s.), $x \in E$, we obtain

$$s(x_{\sigma^*}) = g(x_{\sigma^*}) \qquad (P_x\text{-a.s.}),\ x \in E, \tag{2.135}$$

which yields for $\tau^* = \inf\{n \geq 0 : s^*(x_n) = g(x_n)\} = \inf\{n \geq 0 : s(x_n) = g(x_n)\}$ the inequality $\tau^* \leq \sigma^*$ (P_x-a.s.), $x \in E$. We shall show next that the time τ^* is optimal. By virtue of Lemma 3

$$M_x s(x_{\sigma^*}) \leq M_x s(x_{\tau^*}) \tag{2.136}$$

and, therefore,

$$s(x) = M_x g(x_{\sigma^*}) \leq M_x s(x_{\sigma^*}) \leq M_x s(x_{\tau^*}) = M_x g(x_{\tau^*}), \tag{2.137}$$

where the last equality follows from the fact that for $x \in E$, $s(x_{\tau^*}) = g(x_{\tau^*})$ (P_x-a.s.) (see (1) and (4) of Theorem 11). Therefore, $s(x) \leq M_x g(x_{\tau^*})$, which fact proves that the stopping time $\tau^* = \lim_n \tau_n^*$ is optimal. □

Remark. The theorem still holds if one assumes that σ^* is an optimal Markov time (then, clearly, τ^* is also a Markov time and not a stopping time; cf. Theorem 3.10).

2.11 Solutions of recursive equations $f(x) = \max\{g(x), Tf(x)\}$

2.11.1

It follows from Theorem 3 that for the functions $g \in L(A^-)$ the payoff $s(x)$ is the smallest solution of the equation

$$f(x) = \max\{g(x), Tf(x)\}. \tag{2.138}$$

However, if $g \in L$ the smallest solution of this equation coincides with the function $s^*(x) = \lim_n s_n(x)$ which need not, generally speaking, coincide with the payoff $s(x)$ (Theorem 10).

Hence it is natural to investigate (under different assumptions on the classes of admissible functions $f(x)$) the cases where the solution of Equation (2.138) is unique, since then this solution will coincide automatically with the payoff. If there is no unique solution, it would be desirable to know how to distinguish, among all solutions of Equation (2.138), that solution which actually yields a payoff in the problem "$s(x) = \sup M_x g(x_\tau)$."

In addition to the investigation of these problems we shall show that each solution of Equation (2.138) can be regarded as the payoff in the problem of optimal stopping constructed in a specific way. In this sense we may say that Equation (2.138), being the simplest equation occurring in dynamic programming, is a characteristic feature of optimal stopping problems.

2.11.2

Let

$$P(n, x, \Gamma) = P_x\{x_n \in \Gamma\}, \qquad \mu(n, x, \Gamma) = \frac{1}{n}\sum_{i=1}^{n} P(i, x, \Gamma),$$

$$\Gamma \in \mathscr{B}, \qquad n \in N.$$

Suppose that on $(E, \mathscr{B})$ there exists a nonnegative measure μ such that for each $\mathscr{B}$-measurable bounded function $f = f(x)$, $x \in E$,

$$\int_E f(y)\mu(n, x, dy) \to \int_E f(y)\mu(dy), \qquad n \to \infty,$$

for all $x \in E$.

Theorem 13. *Let $f_1(x)$ and $f_2(x)$ be two solutions of Equation* (2.138) *belonging to the class L, coinciding on some measurable set $\Lambda \subseteq E$ and such that*

$$\sup_{x \in E} |f_1(x) - f_2(x)| < \infty.$$

If $\mu(E - \Lambda) < 1$, then $f_1(x) \equiv f_2(x)$.

PROOF. Set $r(x) = |f_1(x) - f_2(x)|$. Then it can be easily seen from (2.138) that

$$r(x) \leq Tr(x), \tag{2.139}$$

from which $r(x) \leq T^n r(x)$ and, therefore,

$$r(x) \leq \int_E r(y)\mu(n, x, dy). \tag{2.140}$$

Passing in (2.140) to the limit as $n \to \infty$, we find

$$r(x) \leq \int_E r(y)\mu(dy) \leq \sup_{y \in E} r(y) \cdot \mu(E - \Lambda)$$

and

$$\sup_{x \in E} r(x) \leq \sup_{y \in E} r(y) \cdot \mu(E - \Lambda).$$

Since, by assumption, $\mu(E - \Lambda) < 1$, then $r(x) \equiv 0$, i.e., $f_1(x) \equiv f_2(x)$.

Corollary 1. *If $P(1, x, E) = p < 1$ for all $x \in E$, then the solution of Equation (2.138) is unique in the class of measurable bounded functions.*

Corollary 2. *If the function $g(x)$ is bounded and if $f(x)$ is a bounded solution of Equation (2.138) coinciding with $g(x)$ on the set Λ, with $\mu(E - \Lambda) < 1$, then $f(x)$ is the smallest excessive majorant of the function $g(x)$ and, therefore, $f(x) = s(x)$.*

To prove the above it suffices to note that the payoff $s(x)$ also satisfies Equation (2.138) and $f(x)$ coincides with $s(x)$ on the set Λ.

2.11.3

A different criterion for the coincidence of two solutions of Equation (2.138) is:

Theorem 14. *Let $f_1(x)$ and $f_2(x)$ be two $\mathscr{B}$-measurable solutions of Equation (2.138) such that*

$$M_x\left\{\sup_n |f_1(x_n) - f_2(x_n)|\right\} < \infty, \qquad x \in E. \tag{2.141}$$

For any $\varepsilon > 0$, let there be a set $\Lambda_\varepsilon \in \mathscr{B}$ such that:

(1) $|f_1(x) - f_2(x)| < \varepsilon, \quad x \in \Lambda_\varepsilon$;
(2) $P_x\{x_n \in \Lambda_\varepsilon$ *for infinitely many* $n \in N\} = 1, x \in E$.

Then $f_1(x) \equiv f_2(x)$.

Proof. We shall form the process

$$R = (r(x_n), \mathscr{F}_n, P_x), \qquad n \in N, \qquad x \in E,$$

where $r(x) = |f_1(x) - f_2(x)|$. By virtue of (2.139), $0 \leq r(x) \leq T^n r(x)$. Hence the process R is a nonnegative submartingale. It follows from (2.141) and Theorem 1.9 that with P_x-probability 1 $\lim_n r(x_n)$ exists. By hypothesis (2) of Theorem 14, $x_n \in \Lambda_\varepsilon$ for infinitely many $n \in N$, hence since $\varepsilon > 0$ is arbitrary, $\lim_n r(x_n) = 0$ (P_x-a.s.), $x \in E$. From the inequality

$$0 \leq r(x) \leq M_x r(x_n)$$

by Fatou's lemma we obtain

$$0 \leq r(x) \leq \overline{\lim_n}\, M_x r(x_n) \leq M_x \overline{\lim_n}\, r(x_n) = 0,$$

thus proving the theorem. □

Corollary 1. *Let* $\Lambda = \bigcap_{\varepsilon > 0} \Lambda_\varepsilon$, i.e.,

$$\Lambda = \{x : |f_1(x) - f_2(x)| = 0\}$$

and $P_x(x_n \in \Lambda$ *for infinitely many* $n \in N) = 1$. *Then (under the hypotheses of the theorem)* $f_1(x) \equiv f_2(x)$.

Corollary 2. *If the solution* $f(x)$ *of Equation* (2.138) *coincides with the function* $g(x)$ *on a set* Λ *such that* $P_x\{x_n \in \Lambda$ *for infinitely many* $n \in N\} = 1$, $M_x\{\sup_n |f(x_n)|\} < \infty$, *and* $M_x\{\sup_n |g(x_n)|\} < \infty$, *then* $f(x)$ *coincides with the payoff* $s(x)$.

2.11.4

In the case where Equation (2.138) has many solutions it is useful to characterize the solution coinciding with the payoff $s(x) = \sup M_x g(x_\tau)$.

Theorem 15. *Let the function* $g \in L(A^-, A^+)$ *and let* $f(x)$ *be some solution of Equation* (2.138) *such that* $f \in L(A^+)$. *A necessary and sufficient condition for this solution to coincide with the payoff* $s(x)$ *is that the function* $f(x)$ *satisfy* (P_x-a.s.), $x \in E$, *the following "boundary condition at infinity"*:

$$\overline{\lim_n}\, f(x_n) = \overline{\lim_n}\, g(x_n). \tag{2.142}$$

PROOF

Necessity: If $f(x) = s(x)$, then (2.142) follows from Lemma 8 and Theorem 3.

Sufficiency: Set

$$\tau_\varepsilon = \inf\{n \geq 0 : f(x_n) \leq g(x_n) + \varepsilon\}, \qquad \varepsilon > 0.$$

Then (see the proof of Lemma 8) the probability $P_x(\tau_\varepsilon < \infty) = 1$, $x \in E$. Hence it follows from Lemma 7 (see also the proof of (2.63)) that

$$s(x) \geq M_x g(x_{\tau_\varepsilon}) \geq M f(x_{\tau_\varepsilon}) - \varepsilon = f(x) - \varepsilon.$$

Hence $s(x) \geq f(x)$. On the other hand, if $g \in L(A^-)$, then by Theorem 3 the payoff $s(x)$ is the smallest excessive majorant of the function $g(x)$. Hence, $f(x)$ being excessive, $s(x) = f(x)$.

2.11.5

Let the function $g \in L(A^-)$ and let $\eta = \eta(\omega)$ be some Φ_∞^x-measurable random variable ($\Phi_\infty^x = \bigcap_n \Phi_{n,\infty}^x$, $\Phi_{n,\infty}^x = \sigma\{\omega : x_n, x_{n+1}, \ldots\}$), $M_x \eta^- < \infty$, $x \in E$.

Set

$$\bar{s}_\eta(x) = \sup_{\tau \in \mathfrak{M}} M_x[g(x_\tau) I_{\{\tau < \infty\}} + \eta I_{\{\tau = \infty\}}], \tag{2.143}$$

i.e., in contrast to the payoff $\bar{s}(x)$ considered above we consider now a situation where the gain obtained on the set $\{\omega : \tau = \infty\}$ is equal to $\eta(\omega)$ (and it need not be equal to $\overline{\lim}_n\, g(x_n(\omega))$).

The payoff $\bar{s}_\eta(x)$ differs formally from the payoff investigated above:

$$\bar{s}(x) = \sup_{\tau \in \overline{\mathfrak{M}}} M_x\left[g(x_\tau)I_{\{\tau<\infty\}} + \overline{\lim_n}\, g(x_n)I_{\{\tau=\infty\}}\right]; \tag{2.144}$$

nevertheless, the problem of searching for the payoff $\bar{s}_\eta(x)$ can be reduced to the situation considered above. In fact, we have:

Theorem 16. *If the function $g \in L(A^-)$ and $M_x\eta^- < \infty$, $x \in E$,* then

$$\bar{s}_\eta(x) = \sup_{\tau \in \overline{\mathfrak{M}}} M_x\bar{g}(x_\tau), \tag{2.145}$$

where

$$\bar{g}(x) = \max\{g(x), M_x\eta\}. \tag{2.146}$$

In this case $\bar{s}_\eta(x)$ satisfies the equation

$$f(x) = \max\{g(x), Tf(x)\}. \tag{2.147}$$

Proof. Since the function $\bar{g} \in L(A^-)$, then by Theorem 3

$$\sup_{\tau \in \overline{\mathfrak{M}}} M_x\bar{g}(x_\tau) = \sup_{\tau \in \mathfrak{M}} M_x\bar{g}(x_\tau),$$

and, therefore, setting $\tilde{s}(x) = \sup_{\tau \in \mathfrak{M}} M_x\bar{g}(x_\tau)$ and

$$\sigma_\tau(\omega) = \begin{cases} \tau(\omega) & \text{if } \omega \in \{\omega : g(x_\tau) \geq M_{x_\tau}\eta\}, \\ +\infty & \text{if } \omega \in \{\omega : g(x_\tau) < M_{x_\tau}\eta\}, \end{cases}$$

it follows that

$$\begin{aligned} \tilde{s}(x) &= \sup_{\tau \in \mathfrak{M}} M_x[g(x_\tau)I_{\{g(x_\tau) \geq M_{x_\tau}\eta\}} + M_{x_\tau}\eta I_{\{g(x_\tau) < M_{x_\tau}\eta\}}] \\ &= \sup_{\tau \in \mathfrak{M}} M_x[g(x_\tau)I_{\{g(x_\tau) \geq M_{x_\tau}\eta\}} + \eta I_{\{g(x_\tau) < M_{x_\tau}\eta\}}] \\ &= \sup_{\tau \in \mathfrak{M}} M_x[g(x_{\sigma_\tau})I_{\{\sigma_\tau < \infty\}} + \eta I_{\{\sigma_\tau = \infty\}}] \\ &\leq \sup_{\tau \in \overline{\mathfrak{M}}} M_x[g(x_\tau)I_{\{\tau<\infty\}} + \eta I_{\{\tau=\infty\}}] = \bar{s}_\eta(x). \end{aligned}$$

Therefore, $\tilde{s}(x) \leq \bar{s}_\eta(x)$.

Further, since $\bar{g}(x) \geq M_x\eta$, by virtue of Theorem 1.10 and the condition $M_x\eta^- < \infty$, $x \in E$,

$$\overline{\lim_n}\, \bar{g}(x_n) \geq \overline{\lim_n}\, M_{x_n}\eta = \overline{\lim_n}\, M_x[\eta \mid \mathscr{F}_n] \geq M_x[\eta \mid \mathscr{F}_\infty] = \eta.$$

Hence

$$\begin{aligned} \tilde{s}(x) &= \sup_{\tau \in \overline{\mathfrak{M}}} M_x[\bar{g}(x_\tau)I_{\{\tau<\infty\}} + \overline{\lim_n}\, \bar{g}(x_n)I_{\{\tau=\infty\}}] \\ &\geq \sup_{\tau \in \overline{\mathfrak{M}}} M_x[g(x_\tau)I_{\{\tau<\infty\}} + \eta I_{\{\tau=\infty\}}] = \bar{s}_\eta(x). \end{aligned}$$

Therefore, $\tilde{s}(x) = \bar{s}_\eta(x)$, which proves (2.145).

We shall prove next Equation (2.147) for $\bar{s}_\eta(x)$.

The function $\bar{g} \in L(A^-)$, hence by Theorem 3

$$\bar{s}_\eta(x) = \tilde{s}(x) = \max[\bar{g}(x), T\tilde{s}(x)] = \max[g(x), M_x\eta, T\tilde{s}(x)].$$

But $T\tilde{s}(x) \geq TM_x\eta = M_x\eta$, and therefore,

$$\bar{s}_\eta(x) = \tilde{s}(x) = \max[g(x), T\tilde{s}(x)] = \max[g(x), T\bar{s}_\eta(x)].$$

The theorem is proved. □

This theorem shows that the payoffs $\bar{s}_\eta(x)$ satisfy the same equation, (2.147), for different η. We shall prove next that the converse holds as well—the functions $\bar{s}_\eta(x)$ for different η account for all the solutions of Equation (2.147). More precisely, we have:

Theorem 17. *Let $g \in L(A^-, A^+)$, and let $f \in L(A^+)$ be a solution of Equation (2.147). Then this solution can be represented as*

$$f(x) = \sup_{\tau \in \overline{\mathfrak{M}}} M_x[g(x_\tau)I_{\{\tau < \infty\}} + \eta I_{\{\tau = \infty\}}], \tag{2.148}$$

where $\eta = \lim_n f(x_n)$.

PROOF. Note first that $\lim_n f(x_n)$ (P_x-a.s.), $x \in E$ really exists by virtue of Property VII of excessive functions (see Section 4). Setting

$$\bar{g}(x) = \max\left[g(x), M_x \lim_n f(x_n)\right]$$

we shall show that the function $f(x)$ also satisfies the equation

$$f(x) = \max[\bar{g}(x), Tf(x)].$$

By virtue of Lemma 3

$$f(x) \geq M_x \lim_n f(x_n).$$

Hence

$$Tf(x) \geq M_x \lim_n f(x_n)$$

and

$$\max[\bar{g}(x), Tf(x)] = \max[g(x), Tf(x)] = f(x).$$

Let us show now that $\lim_n f(x_n) = \overline{\lim}_n \bar{g}(x_n)$ (P_x-a.s.), $x \in E$. We have

$$\overline{\lim_n} \bar{g}(x_n) = \overline{\lim_n}\left\{\max\left[g(x_n), M_{x_n} \lim_k f(x_k)\right]\right\}$$

$$= \max\left[\overline{\lim_n} g(x_n), \overline{\lim_n} M_{x_n} \lim_k f(x_k)\right].$$

Here $\overline{\lim}_n M_{x_n} \lim_k f(x_k) = \lim_k f(x_k)$. Hence

$$\overline{\lim_n}\, \bar{g}(x_n) = \max\left\{\overline{\lim_n}\, g(x_n),\ \lim_n f(x_n)\right\}.$$

But $\lim_n f(x_n) \geq \overline{\lim}_n g(x_n)$. Therefore,

$$\overline{\lim}\, \bar{g}(x_n) = \lim_n f(x_n).$$

It follows from this and Theorems 15 and 16 that

$$f(x) = \sup_{\tau \in \overline{\mathfrak{M}}} M_x \bar{g}(x_\tau) = \sup_{\tau \in \overline{\mathfrak{M}}} M_x \left[g(x_\tau) I_{\{\tau < \infty\}} + \lim_n f(x_n) I_{\{\tau = \infty\}} \right].$$

The theorem is proved. □

2.11.6

In the general case where $g \in B$, the payoff $s(x)$ is the smallest $\mathfrak{M}_g$-regular function dominating $g(x)$ (Theorem 9). This fact can be used (as an additional condition) for finding the necessary solution of Equation (2.147). However, it is not easy at all to verify the $\mathfrak{M}_g$- regularity. Hence it is of interest to find conditions (for the payoff) which characterize the necessary solution which can be easily verified as well.

In addition to Theorem 15 the following assumption, which is actually contained in Theorem 6, proves to be useful.

Theorem 18. *Let the function $g \in B(a^-)$. Any solution $f(x)$ of the equation*

$$f(x) = \max\{g(x), Tf(x)\} \tag{2.149}$$

that is the $\overline{\mathfrak{M}}_g$-regular majorant of the function $g(x)$ also satisfies the equation

$$f(x) = \max\{g(x), M_x g(x_\infty), Tf(x)\}, \tag{2.150}$$

where $g(x_\infty) = \overline{\lim}_n g(x_n)$.

Conversely, any solution of Equation (2.150) *satisfies Equation* (2.149) *and is the $\overline{\mathfrak{M}}_g$-regular majorant of $g(x)$.*

PROOF. If $f(x)$ is the $\overline{\mathfrak{M}}_g$-regular majorant of $g(x)$, then for any $\tau \in \overline{\mathfrak{M}}_g$, $M_x f(x_\tau)$ is defined and is greater than $-\infty$, and $M_x f(x_\tau) \leq f(x)$. In particular, by virtue of the condition $g \in B(a^-)$ the time $\tau \equiv \infty$ belongs to $\overline{\mathfrak{M}}_g$ and, therefore, $M_x g(x_\infty) \leq M_x f(x_\infty) \leq f(x)$. Hence Equation (2.150) follows from Equation (2.149).

Conversely, let $f(x)$ satisfy Equation (2.150). Then

$$f(x) \geq M_x g(x_\infty) = M_x \overline{\lim_n}\, g(x_n) \geq -M_x \underline{\lim_n}\, g^-(x_n).$$

Hence, if we write $\eta = -\underline{\lim}_n g^-(x_n)$, then $f(x) \geq M_x\eta$, and by virtue of Markovianness $f(x_n) \geq M_{x_n}\eta = M_x[\eta | \mathscr{F}_n]$ (P_x-a.s.), $x \in E$. This fact together with the inequality $f(x) \geq Tf(x)$ demonstrates that the supermartingale $(f(x_n), \mathscr{F}_n, P_x)$, $n \in N$, majorizes the martingale $(M_x(\eta | \mathscr{F}_n), \mathscr{F}_n, P_x)$, $n \in N$. Therefore, by virtue of Theorem 1.11 for any Markov time τ (and, in particular, for $\tau \in \overline{\mathfrak{M}}_g$) $M_x f(x_\tau)$ is defined and $M_x f(x_\tau) \leq M_x f(x_\sigma)$ $(P_x(\sigma \leq \tau) = 1, \ x \in E)$. Thus, $f(x)$ is the $\overline{\mathfrak{M}}_g$-regular majorant of the function $g(x)$.

We shall show next that the function $f(x)$ satisfies Equation (2.149) as well as Equation (2.150). In this case it is seen that we need only consider the cases where $M_x g(x_\infty) > \max\{g(x), Tf(x)\}$.

Thus, at a point $x \in E$ let $M_x g(x_\infty) > \max\{g(x), Tf(x)\}$ with $M_x g(x_\infty) < \infty$. Then by virtue of Theorem 1.11 we have again

$$M_x g(x_\infty) > Tf(x) = M_x f(x_1) \geq M_x f(x_\infty) \geq M_x g(x_\infty). \quad (2.151)$$

The contradiction thus obtained shows (in the case $M_x g(x_\infty) < \infty$) $M_x g(x_\infty) = \max\{g(x), Tf(x)\}$ and, therefore, $f(x) = \max\{g(x), Tf(x)\}$.

If $M_x g(x_\infty) = +\infty$ then, by virtue of (2.150), $f(x) = +\infty$ and, by virtue of (2.151), $Tf(x) = +\infty$. Therefore, the function $f(x)$ satisfies again Equation (2.149), thus proving the theorem.

To illustrate how this theorem can be exploited we shall consider Example 7 from Section 6. Since in this example $M_x g(x_\infty) = 0$, then by virtue of Theorems 9 and 18 the payoff $s(x)$ satisfies (along with (2.149)) the equation

$$f(x) = \max\{g(x), 0, Tf(x)\},$$

from which it can be seen that $f(x) \geq 0$. Hence $s(x) = 0$.

2.12 Criteria for the truncation of optimal stopping rules

2.12.1

Let $\mathfrak{M} = \{\tau\}$ be the class of stopping time where $P_x(\tau < \infty) = 1$, $x \in E$. We shall assume that there exists an optimal stopping rule $\tau^* \in \mathfrak{M}$, $M_x g(x_{\tau^*}) = s(x)$, $x \in E$. It may occur in this case that for some state $x \in E$ there is a finite $N(x)$ such that $P_x\{\tau^* \leq N(x)\} = 1$. In this case the optimal stopping time τ^* is said to be *truncated* at the point x. If there is a finite N such that $P_x\{\tau^* \leq N\} = 1$ for all $x \in E$, in other words, if $\tau^* \in \mathfrak{M}(N)$, then the stopping rule τ^* is said to be *truncated*.

The present section contains criteria enabling us to determine for which initial states optimal stopping rules are truncated, and also to determine whether the bound obtained by truncation $N(x)$ is exact (i.e., $P_x\{\tau^* \leq N(x)\} = 1$ and $P_x\{\tau^* = N(x)\} > 0$).

2.12.2

From now on we shall assume that the function $g \in L(A^-)$. By virtue of Theorem 3 the payoff

$$s(x) = \lim_n Q^n g(x). \tag{2.152}$$

Set

$$s_k(x) = \sup_{\tau \in \mathfrak{M}(k)} M_x g(x_\tau) \quad (= Q^k g(x)), \tag{2.153}$$

$$\alpha_k(x) = Q^k g(x) - TQ^k g(x), \tag{2.154}$$

$$\beta_k(x) = s_{k+1}(x) - s_k(x). \tag{2.155}$$

According to (2.155), (2.7), and (2.8), for all $n \geq k$

$$\beta_k(x_{n-k}) = \max\{g(x_{n-k}), TQ^k g(x_{n-k})\} - Q^k g(x_{n-k}) \qquad (P_x\text{-a.s.}),\ x \in E. \tag{2.156}$$

It can be seen from (2.154) and (2.156) that the condition $\beta_k(x_{n-k}) = 0$ is equivalent to the condition $\alpha_k(x_{n-k}) \geq 0$.

Theorem 19. *If, for a given state $x \in E$, for a certain $k \geq 0$ there is a finite $n_k = n_k(x)$ such that with the P_x-probability* 1

$$\beta_k(x_{n-k}) = 0, \qquad n \geq n_k, \tag{2.157}$$

then $s_n(x) = s(x)$ for all $n \geq n_k$.

If (2.157) is satisfied for k and l, $l < k$, and if $N_k = N_k(x)$ and $N_l = N_l(x)$ are the smallest numbers among $n_k = n_k(x)$ and $n_l = n_l(x)$ satisfying (2.157), then $N_k \leq N_l$.

To prove the theorem we shall need:

Lemma 15. *For all $n \geq 0$ and $x \in E$*

$$s_{n+2}(x) - s_{n+1}(x) \leq T(s_{n+1} - s_n)(x). \tag{2.158}$$

PROOF. The lemma follows from an analysis of the recursive relations

$$s_{k+1}(x) = \max\{g(x), Ts_k(x)\}, \quad k \geq 0. \tag{2.159}$$

(a) If $g(x) \geq Ts_{n+1}(x)$, then $g(x) \geq Ts_{n+1}(x) \geq Ts_n(x)$ and we obtain from (2.159) for $k = n + 1$ and $k = n$ that $s_{n+2}(x) = g(x)$, $s_{n+1}(x) = g(x)$. Therefore, (2.158) is satisfied.

(b) If $g(x) \leq Ts_n(x)$, then $g(x) \leq Ts_n(x) \leq Ts_{n+1}(x)$, and from (2.159) we have $s_{n+1}(x) = Ts_n(x)$, $s_{n+2}(x) = Ts_{n+1}(x)$. Therefore, $s_{n+2}(x) - s_{n+1}(x) = T(s_{n+1} - s_n)(x)$.

(c) If $Ts_n(x) \leq g(x) \leq Ts_{n+1}(x)$, then

$$s_{n+2}(x) - s_{n+1}(x) = Ts_{n+1}(x) - g(x) \leq Ts_{n+1}(x) - Ts_n(x) = T(s_{n+1} - s_n)(x),$$

which proves (2.158). □

PROOF OF THEOREM 19. By virtue of (2.158)

$$\begin{aligned} 0 &\leq s_{n_k+1}(x) - s_{n_k}(x) \\ &\leq T(s_{n_k} - s_{n_k-1})(x) \\ &\vdots \\ &\leq T^{n_k-k}(s_{k+1} - s_k)(x) \\ &= M_x[s_{k+1}(x_{n_k-k}) - s_k(x_{n_k-k})]. \end{aligned}$$

But $0 \leq s_{k+1}(x_{n_k-k}) - s_k(x_{n_k-k}) = \beta_k(x_{n_k-k})$; hence if, with P_x-probability 1, $\beta_k(x_{n_k-k}) = 0$ (or $\alpha_k(x_{n_k-k}) \geq 0$, which is the same), then $s_{n_k+1}(x) = s_{n_k}(x)$. Similarly, $s_n(x) = s_{n_k}(x)$ for all $n \geq n_k$, from which $s_n(x) = s(x)$, $n \geq n_k$.

The second assertion of the theorem follows from the inequality $N_k \leq N_{k-1}$, which can be easily deduced from (2.158).

Corollary 1. *If, for a given $x \in E$, for some $k \geq 0$ there is a finite $n_k = n_k(x)$ such that, with P_x-probability 1, $\beta_k(x_{n-k}) = 0$, $n \geq n_k$, then the stopping time*

$$\tau_{n_k}^* = \min\{m : s_{n_k-m}(x_m) = g(x_m)\}$$

is optimal (at the point x considered):

$$s(x) = M_x g(x_{\tau_{n_k}^*}).$$

Corollary 2. *If $M_k = \sup_x n_k(x) < \infty$ for some $k \geq 0$, then the stopping time*

$$\tau_{M_k}^* = \min\{m : s_{M_k-m}(x_m) = g(x_m)\}$$

is optimal:

$$s(x) = M_x g(x_{\tau_{M_k}^*}), \qquad x \in E.$$

Remark. It is very easy to construct the criteria for truncation of optimal stopping rules for small k. Thus, for $k = 0$

$$\alpha_0(x_n) = g(x_n) - Tg(x_n);$$

for $k = 1$

$$\alpha_1(x_{n-1}) = Qg(x_{n-1}) - TQg(x_{n-1}).$$

Hence, if for $k = 0$ there is $n_0 < \infty$ such that, with P_x-probability 1, for all $x \in E$ $g(x_n) \geq T_g(x_n)$, $n \geq n_0$, then the optimal stopping rule τ^* exists automatically and $P_x\{\tau^* \leq n_0\} = 1$ for all $x \in E$.

According to the second part of Theorem 19, $N_0(x) \geq N_1(x)$. Hence the criterion based on the analysis of the value $\alpha_1(x_{n-1})$ guarantees a more precise upper estimate for the truncation bound: $N(x) \le N_1(x) \le N_0(x)$.

Since $N(x) \le N_k(x)$, it is interesting to find out when $N(x) = N_k(x)$ for a certain k and all or certain $x \in E$.

Theorem 20. *If, for a given x, for some $k \ge 0$ there is a finite $N_k = N_k(x)$ such that, with P_x-probability 1,*

$$\beta_k(x_{n-k}) = 0, \qquad n \ge N_k, \tag{2.160}$$

and with positive P_x-probability the inequalities

$$Tg(x_i) \ge g(x_i), \qquad \beta_k(x_{N_k-k-1}) > 0, \qquad i = 0, 1, \ldots, N_k - k - 2, \tag{2.161}$$

are satisfied, then

$$s_{N_k-1}(x) < s_{N_k}(x) = s_{N_k+1}(x) = \cdots = s(x) \tag{2.162}$$

and $N(x) = N_k(x)$.

As a preliminary we shall prove:

Lemma 16. *If, for a given $x \in E$ in a space E^{n-m}, $n - m \ge 1$, there exists a set $A_1 \times \cdots \times A_{n-m}$ such that*

$$P_x\left[\bigcap_{i=0}^{n-m-1} [\{Tg(x_i) \ge g(x_i)\} \cap \{x_i \in A_i\}] \cap \{\beta_m(x_{n-m}) > 0\} \cap \{x_{n-m} \in A_{n-m}\}\right] > 0,$$

then

$$P_x\left[\bigcap_{i=0}^{n-m-1} \{\beta_{n-i}(x_i) = 0\} \cap \{x_i \in A_i\}\right] = 0. \tag{2.163}$$

Proof. The proof will be by induction. Let Equation (2.163) be satisfied for $i = j + 1, \ldots, n - m - 1$. We shall prove it for $i = j \ge 0$.

If $x_k \in A_k$, $0 \le k \le n - m - 1$, then

$$g(x_k) \le Tg(x_k) \le Ts_{n-k-1}(x_k) \le Ts_{n-k}(x_k).$$

For $x \in E$ follows

$$\begin{aligned}
\beta_{n-j}(x_j) &= s_{n-j+1}(x_j) - s_{n-j}(x_j) \\
&= T(s_{n-j} - s_{n-j-1})(x_j) \\
&= M_x[s_{n-j}(x_{j+1}) - s_{n-j-1}(x_{j+1}) | \mathscr{F}_j] \\
&\ge M_x[I_{A_{j+1}}(x_{j+1})[s_{n-j}(x_{j+1}) - s_{n-j-1}(x_{j+1})] | \mathscr{F}_j] \\
&= M_x[I_{A_{j+1}}(x_{j+1})\beta_{n-j-1}(x_{j+1}) | \mathscr{F}_j] > 0.
\end{aligned}$$

The lemma is proved. □

PROOF OF THEOREM 20. Note that $\beta_{N_k-1}(x_0) = s_{N_k}(x_0) - s_{N_k-1}(x_0)$. Hence, taking in (2.163) $i = 0$ and $n = N_k$, we shall get $\beta_{N_k-1}(x_0) > 0$, i.e., $s_{N_k-1}(x_0) < s_{N_k}(x_0)$.

By virtue of the preceding theorem, $s_{N_k}(x_0) = s_{N_k+1}(x_0) = \cdots = s(x_0)$, thus proving (2.162). It also follows from (2.162) and Corollary 1 to Theorem 19 that $N_k(x_0) = N(x_0)$. □

2.13 Randomized and sufficient classes of stopping times

2.13.1

Let $(\Omega, \mathscr{F})$ be a measurable space, and let $X = (x_n, \mathscr{F}_n, P_x)$, $n \in N$, be a Markov chain with values in a state space $(E, \mathscr{B})$. Denote by $\mathfrak{M}[F]$ the class of stopping times $\tau = \tau(\omega)$ (with respect to the system $F = \{\mathscr{F}_n\}$, $n \in N$) and consider the payoff

$$s(x) = \sup M_x g(x_\tau), \tag{2.164}$$

where the sup is taken over the times $\tau \in \mathfrak{M}[F]$ for which the mathematical expectations $M_x g(x_\tau)$ are defined for all $x \in E$.

Will the payoff $s(x)$ increase if we take the supremum in (2.164) over a wide class of stopping times rather than over times τ from the class $\mathfrak{M}[F]$?

To answer this question we shall give some necessary definitions.

Assume that in $\mathscr{F}$ there is distinguished a system $F^* = \{\mathscr{F}_n^*\}$, $n \in N$, of σ-algebras $\mathscr{F}_n^*$ having the property that

$$\mathscr{F}_n^* \subseteq \mathscr{F}_{n+1}^*, \qquad \mathscr{F}_n \subseteq \mathscr{F}_n^* \subseteq \mathscr{F}.$$

Assume that on the smallest σ-algebra containing all $\mathscr{F}_n^*$, $n \in N$, we are given probability measures P_x^*, $x \in E$, that are extensions of the measures P_x, $x \in E$ (i.e., $P_x^*(A) = P_x(A)$ if $A \in \sigma(\bigcup_n \mathscr{F}_n)$), and that the process $X^* = (x_n, \mathscr{F}_n^*, P_x^*)$, $n \in N$, is Markov.

Definition 1. The class of stopping times $\mathfrak{M}[F^*]$ (with respect to the system $F^* = \{\mathscr{F}_n^*\}$) is said to be the *class of randomized stopping times with respect to the system* $F = \{\mathscr{F}_n\}$.

Thus $\mathfrak{M}[F^*] \supseteq \mathfrak{M}[F]$; if we set

$$s^*(x) = \sup M_x g(x_\tau),$$

where the supremum is taken over the times from the class $\mathfrak{M}[F^*]$ for which $M_x g(x_\tau)$, $x \in E$, are defined, then $s^*(x) \geq s(x)$.

Nevertheless, $s^*(x) = s(x)$, i.e., the randomization does not imply increased payoff. Indeed, if $g \in B$ then by virtue of Theorem 9

$$s(x) = \lim_{b \to \infty} \lim_{a \to -\infty} \lim_{n \to \infty} Q^n g_a^b(x) \tag{2.165}$$

and

$$s^*(x) = \lim_{b\to\infty} \lim_{a\to-\infty} \lim_{n\to\infty} Q^{*n} g_a^b(x), \tag{2.166}$$

where

$$Q^* g(x) = \max\{g(x), M_x^* g(x_1)\}$$

(M_x^* is the mathematical expectation over the measure P_x^*).

But, obviously, $M_x^* g(x_1) = M_x g(x_1)$. Hence we conclude immediately from (2.165) and (2.166) that the payoffs $s^*(x)$ and $s(x)$ coincide.

Thus, we have proved:

Theorem 21. *Let the function $g \in B$. Then $s^*(x) = s(x)$, i.e., the additional introduction of randomized stopping times does not imply increased payoff.*

2.13.2

Even though randomization leads to no increase of the payoff, we can point out some useful applications of randomized stopping times.

For example, if for some $x \in E$ the payoff $s(x) = +\infty$, then the class $\mathfrak{M}[F] = \mathfrak{M}$ (as well as the class $\overline{\mathfrak{M}}(F) = \overline{\mathfrak{M}}$) need not contain an optimal time while the class $\mathfrak{M}[F^*]$ contains such a time.

In fact, let $s(x) = +\infty$ for some $x \in E$. Then, there will be a sequence of stopping times $\{\tau_i\}$, $i \in N$, $\tau_i \in \mathfrak{M}[F]$, such that $s(x) = \sup_i M_x g(x_{\tau_i})$. We may assume without loss of generality that $M_x g(x_{\tau_i}) \geq 2^i$.

Let $\rho = \rho(\omega)$ be an $\mathscr{F}$-measurable random variable taking on values $i = 1, 2, \ldots$ with the probability 2^{-i}. In this case[12]

$$P_x^*\{[\rho(\omega) = i] \cap A\} = P_x(A) \cdot 2^{-i}$$

for all $x \in E$, $A \in \sigma(\bigcup_n \mathscr{F}_n)$. We shall define the randomized time $\tau^* = \tau^*(\omega)$ by

$$\tau^*(\omega) = \tau_i(\omega) \qquad \text{if } \omega \in \{\omega : \rho(\omega) = i\}.$$

Then obviously,

$$M_x^* g(x_{\tau^*}) = \sum_{i=1}^{\infty} M_x g(x_{\tau_i}) \cdot 2^{-i} = \infty.$$

It is especially useful to consider the randomized stopping times in solving variational problems of optimal stopping. For example, let it be required to find $\sup M_{x^0} g(x_\tau)$, $x^0 \in E$, on the assumption that only those stopping times τ are to be considered for which $M_{x^0} f(x_\tau) = c$ where c is some constant and $f, g \in B$.

[12] Thereby we assume that the initial space $(\Omega, \mathscr{F})$ is sufficiently "rich." Otherwise, instead of $(\Omega, \mathscr{F})$ we should consider a new space $(\tilde{\Omega}, \tilde{\mathscr{F}})$ where $\tilde{\Omega} = \Omega \times \Omega^*$, $\tilde{\mathscr{F}} = \mathscr{F} \times \mathscr{F}^*$, and $(\Omega^*, \mathscr{F}^*)$ is some measurable space of the "randomized" outcomes $\omega^* \in \Omega^*$.

Even in the cases where there exist τ_1 and τ_2 belonging to the class $\mathfrak{M}[F]$ and such that $M_{x^0} f(x_{\tau^1}) = a < c$, $M_{x^0} f(x_{\tau^2}) = b > c$, in the class $\mathfrak{M}[F]$ there may be no time τ for which $M_{x^0} f(x_\tau) = c$. However, in the class $\mathfrak{M}[F^*]$ the time

$$\tau^*(\omega) = \tau_i(\omega) \qquad \text{if } \omega \in \{\omega : \rho(\omega) = i\},$$

where $i = 1, 2$ and $\rho(\omega)$ is an $\mathscr{F}$-measurable random variable such that

$$P_x^*\{\rho(\omega) = 1\} = \frac{b - c}{b - a}, \qquad P_x^*\{\rho(\omega) = 2\} = \frac{c - a}{b - a},$$

$$P_x^*\{[\rho(\omega) = i] \cup A\} = P_x(A) \cdot P_x^*\{\rho(\omega) = i\},$$

yields $M_{x^0}^* f(x_{\tau^*}) = c$.

2.13.3

We shall discuss next the question of sufficiency in optimal stopping problems.

It was noted above (see Remark 3 to Theorem 3) that in considering the payoff $s(x) = \sup M_x g(x_\tau)$ it suffices to take the supremum not over the class $\mathfrak{M} = \mathfrak{M}[F]$ but over a more narrow class $\mathfrak{M}[F^*]$. In this sense the system of σ-algebras $F^X = \{\mathscr{F}_n^X\}$, $n \in N$, where $F_n^X = \sigma(\omega : x_0, x_1, \ldots, x_n\}$ is sufficient. Hence it is natural to introduce:

Definition 2. The nondecreasing system $G = \{\mathscr{G}_n\}$, $n \in N$, of σ-algebras $\mathscr{G}_n \subseteq \mathscr{F}_n$, is said to be *sufficient* if

$$\sup_{\tau \in \mathfrak{M}[G]} M_x g(x_\tau) = \sup_{\tau \in \mathfrak{M}[F]} M_x g(x_\tau).$$

Thus, while randomization leads to expanding of the class $\mathfrak{M}[F]$, sufficiency narrows down admissible class of stopping times without decreasing the payoff.

In connection with the remark, made above regarding the sufficiency of the class $\mathfrak{M}[F^X]$, it is useful to note that the process $Y = (x_n, \mathscr{F}_n^X, P_x)$ as well as the process $X = (x_n, \mathscr{F}_n, P_x)$ will be Markov. The above fact implies that in solving the optimal stopping problem we can consider immediately the process Y instead of the process X.

Is it feasible to narrow down the class $\mathfrak{M}[F^X]$ even more without decreasing the payoff? From this point of view the simplest class is that of a stopping time τ identical to some time n, $\tau(\omega) \equiv n$, $n \in N$. It is seen that this class coincides with the class $\mathfrak{M}[G^0]$ where $G^0 = \{\mathscr{G}_n^0\}$, $n \in N$, and each σ-algebra $\mathscr{G}_n^0$ is trivial, i.e., $\mathscr{G}_n^0 = \{\varnothing, \Omega\}$, where $\varnothing$ is the empty set.

There exist nontrivial cases where the class $\mathfrak{M}[G^0]$ is sufficient, but these cases are the exception rather than the rule.

We give one general result which is useful for the determination of sufficient σ-algebras.

Theorem 22. *Let $X = (x_n, \mathscr{F}_n, P_x)$, $n \in N$, be a Markov process and let the function $g \in B(\mathrm{A}^+)$. The nondecreasing system $G = \{\mathscr{G}_n\}$, $n \in N$, of the σ-algebras $\mathscr{G}_n \subseteq \mathscr{F}_n$ is sufficient if:*

(1) *$g(x_n)$ is $\mathscr{G}_n$-measurable, $n \in N$;*
(2) *For the arbitrary $\mathscr{G}_{n+1}$-measurable variable $\varkappa = \varkappa(\omega)$ with*

$$M_x|\varkappa| < \infty, \qquad x \in E,$$

$$M_x(\varkappa|\mathscr{F}_n) = M_x(\varkappa|\mathscr{G}_n) \ (P_x\text{-a.s.}),\ x \in E,\ n \in N.$$

PROOF. By virtue of Theorem 7

$$s(x) = \lim_{b\to\infty} \lim_{a\to-\infty} \lim_{n\to\infty} Q^n g_a^b(x).$$

From (1) and (2) it follows that $s(x_n)$ is $\mathscr{G}_n$-measurable. Next, by Theorem 8, for any $\varepsilon > 0$ the time

$$\tau_\varepsilon = \inf\{n \geq 0 : s(x_n) \leq g(x_n) + \varepsilon\}$$

is (ε, s)-optimal. Since $g(x_n)$ and $s(x_n)$ are $\mathscr{G}_n$-measurable, $\tau_\varepsilon \in \mathfrak{M}[G]$ and, therefore, by virtue of arbitrariness of $\varepsilon > 0$,

$$\sup_{\tau\in\mathfrak{M}[F]} M_x g(x_\tau) = \sup_{\tau\in\mathfrak{M}[G]} M_x g(x_\tau),$$

i.e., the system G is sufficient.

Corollary. *Let $X = (X', X'') = ((x_n', x_n''), \mathscr{F}_n, P_{x',x''})$, $n \in N$, be a Markov process with state space $(E' \times E'', \mathscr{B}' \times \mathscr{B}'')$. Assume that the function $g(x', x'')$ is independent of x'' (more precisely, is $\mathscr{B}' \times \{\varnothing, E''\}$-measurable), and belongs to the class $B(\mathrm{A}^+)$. Assume also that the component X' is itself a Markov process. Then the system $F' = \{\mathscr{F}_n'\}$ with $\{\mathscr{F}_n'\} = \sigma\{\omega : x_0', \ldots, x_n'\}$ is sufficient:*

$$s(x', x'') = \sup_{\tau\in\mathfrak{M}[F]} M_{x',x''} g(x_\tau', x_\tau'') = \sup_{\tau\in\mathfrak{M}[F']} M_{x',x''} g(x_\tau', x_\tau'').$$

In this case the function $s(x', x'')$ is independent of x''.

2.14 Optimal stopping of a Markov sequence allowing for the cost of observation

2.14.1

Many statistical problems (compare with the problems considered in Chapter 4) are such that each observation of the Markov sequence $X = (x_n, \mathscr{F}_n, P_x)$ involves some cost to be taken into account in calculating the gain.

We now assume that at the time n of stopping the observation we obtain a gain equal to

$$G(n, x_0, \ldots, x_n) = \alpha^n g(x_n) - \sum_{s=0}^{n-1} \alpha^s c(x_s) \tag{2.167}$$

for $n \geq 1$ and equal to $G(0, x_0) = g(x_0)$ for $n = 0$. In (2.167) α is some constant, $0 < \alpha \leq 1$, and the functions $g(x)$ and $c(x)$ belonging to the class $\mathscr{B}$ are assumed (to simplify the discussion) to satisfy the conditions $c(x) \geq 0$ and

$$|g(x)| \leq G < \infty, \; M_x c(x_n) < \infty, \quad n = 0, 1, \ldots. \tag{2.168}$$

It is natural to interpret $c(x)$ as the cost for the opportunity to make a next observation (in the state x), and to interpret α as the parameter which accounts for the variation of the "values" with time.

By the payoff we shall mean the variable

$$s(x) = \sup M_x\left\{\alpha^\tau g(x_\tau) - \sum_{s=0}^{\tau-1}\alpha^s c(x_s)\right\}, \tag{2.169}$$

where the supremum is taken over the class of stopping times

$$\mathfrak{M}_{(\alpha, c)} = \left\{\tau \in \mathfrak{M} : M^x \sum_{s=0}^{\tau-1} \alpha^s c(x_s) < \infty, x \in E\right\}.$$

To describe the structure of the payoff $s(x)$ and the methods of finding (ε, s)-optimal stopping times, it is convenient to introduce the following:

Definition. The function $f \in B$ is said to be the (α, c)-*excessive majorant of the function* $g \in B$ if, for all $x \in E$, the mathematical expectations $Tf(x) = M_x f(x_1)$ (where $g(x) \leq f(x)$) and

$$\alpha Tf(x) - c(x) \leq f(x), \qquad x \in E, \tag{2.170}$$

are defined.

Theorem 23. *Let the functions* $g(x)$ *and* $c(x)$ *satisfy* (2.168), $0 < \alpha \leq 1$. *Then*:

(1) *The payoff* $s(x)$ *is the smallest* (α, c)-*excessive majorant of the function* $g(x)$;

(2) $$s(x) = \max\{g(x), \alpha Ts(x) - c(x)\};$$

(3) $$s(x) = \lim_{N \to \infty} Q^N_{(\alpha, c)} g(x), \tag{2.171}$$

where $Q^N_{(\alpha, c)}$ *is the* Nth *power of the operator*

$$Q(\alpha, c) f(x) = \max\{f(x), \alpha Tf(x) - c(x)\};$$

(4) *For any* $\varepsilon > 0$ *the time*

$$\tau_\varepsilon = \inf\{n \geq 0 : \alpha^n s(x_n) \leq \alpha^n g(x_n) + \varepsilon\}$$

is an ε-*optimal time from the class* $\mathfrak{M}_{(\alpha, c)}$;

(5) *If* $P_x(\tau_0 < \infty\} = 1$, $x \in E$, *the time* τ_0 *will be an optimal stopping time from the class* $\mathfrak{M}_{(\alpha, c)}$;

(6) *If* $P_x\{\sum_{s=0}^{\infty} \alpha^s c(x_s) = \infty\} = 1$, $x \in E$, *then* $P_x\{\tau_0 < \infty\} = 1$ *and the time* τ_0 *is an optimal stopping time in the class* $\mathfrak{M}_{(\alpha, c)}$.

PROOF. The theorem can be proved in several ways. For instance, the problem in question can be reduced to that already investigated (with $\alpha = 1$ and $c(x) \equiv 0$) with the aid of a new Markov chain which, unfortunately, entails a rather complex state space).[13] We prefer here a method based on the fact that the solutions developed above with $\alpha = 1$ and $c(x) \equiv 0$ can be readily carried over to the general case as well. We note first that the function

$$v(x) = \lim_{N \to \infty} Q^N_{(\alpha, c)} g(x)$$

is the smallest (α, c)-excessive majorant of the function $g(x)$, which fact can be proved in the same way as Lemma 6. Using the same proof as that of Lemma 5, we can establish that the function $v(x)$ satisfies the equation

$$v(x) = \max\{g(x), \alpha T v(x) - c(x)\}.$$

In Theorem 3 (in the case $\alpha = 1$, $c(x) \equiv 0$) the inequality $M_x v(x_\tau) \leq v(x)$, $\tau \in \overline{\mathfrak{M}}$, was crucial for proving the equality $s(x) = v(x)$. In the case being considered the inequality $M_x v(x_\tau) \leq v(x)$, $\tau \in \overline{\mathfrak{M}}$, must be replaced by the inequality

$$M_x\left[\alpha^\tau v(x_\tau) - \sum_{s=0}^{\tau-1} \alpha^s c(x_s)\right] \leq v(x), \tag{2.172}$$

whose validity for $\tau \in \mathfrak{M}_{(\alpha, c)}$ follows from the arguments given below.

The sequence $(v(x_n), \mathscr{F}_n, P_x)$, $n = 0, 1, \ldots,$ forms an (α, c)-supermartingale for each $x \in E$, i.e., $M_x|v(x_n)| < \infty$, $n = 0, 1, \ldots$ and

$$\alpha M_x[v(x_{n+1}) | \mathscr{F}_n] - c(x_n) \leq v(x_n), \qquad n = 0, 1, \ldots. \tag{2.173}$$

The inequalities $M_x|v(x_n)| < \infty$ follow from the fact that $|Q^N_{(\alpha, c)} g(x)| \leq G < \infty$ and, therefore, $|v(x)| \leq G < \infty$. Relation (2.172) follows immediately from Markovianness and the inequality

$$\alpha T v(x) - c(x) \leq v(x).$$

Let $\tau = \tau(\omega)$ be a stopping time (with respect to $F = \{\mathscr{F}_n\}$). Then for the (α, c)-supermartingales we have the inequality

$$M_x\left[\alpha^{\tau \wedge N} v(x_{\tau \wedge N}) - \sum_{s=0}^{(\tau \wedge N)-1} \alpha^s c(x_s)\right] \leq v(x), \qquad N < \infty,$$

which can be proved as in the case $\alpha = 1$, $c(x) = 0$ (see the proof in [69], theor. 2.1).

Since $c(x) \geq 0$, it follows from this inequality that for any $\tau \in \mathfrak{M}_{(\alpha, c)}$

$$M_x\left[\alpha^{\tau \wedge N} v(x_{\tau \wedge N}) - \sum_{s=0}^{\tau-1} \alpha^s c(x_s)\right] \leq v(x),$$

[13] See [94], Chap. 2, §8.

from which, making use of the Lebesgue theorem on bounded convergence we obtain the required inequality, (2.172). Hence

$$M_x\left[\alpha^\tau g(x_\tau) - \sum_{s=0}^{\tau-1}\alpha^s c(x_s)\right] \le M_x\left[\alpha^\tau v(x_\tau) - \sum_{s=0}^{\tau-1}\alpha^s c(x_s)\right] \le v(x),$$

and, therefore, $s(x) \le v(x)$.

To prove the inverse inequality we introduce the time

$$\tau_\varepsilon = \inf\{n \ge 0 : \alpha^n v(x_n) \le \alpha^n g(x_n) + \varepsilon\}.$$

Taking advantage of Lemma 8 it is easy to show[14] that

$$\overline{\lim_n}\,\alpha^n v(x_n) = \overline{\lim_n}\,\alpha^n g(x_n) \qquad (P_x\text{-a.s.}),\ x \in E,$$

i.e., $P_x\{\tau_\varepsilon < \infty\} = 1$, $x \in E$, if $\varepsilon > 0$. Let us show that, moreover, $\tau_\varepsilon \in \mathfrak{M}_{(\alpha, c)}$. In fact, by analogy with (2.53), we obtain

$$\begin{aligned} v(x) &= M_x\left[\alpha^{\tau_\varepsilon \wedge n} v(x_{\tau_\varepsilon \wedge n}) - \sum_{s=0}^{(\tau_\varepsilon \wedge n)-1}\alpha^s c(x_s)\right] \\ &\le M_x[\alpha^{\tau_\varepsilon} v(x_{\tau_\varepsilon}) I_{\{\tau_\varepsilon < n\}}] - M_x\left[\sum_{s=0}^{\tau_\varepsilon - 1}\alpha^s c(x_s) I_{\{\tau_\varepsilon < n\}}\right] + M_x[\alpha^n v(x_n) I_{\{\tau_\varepsilon \ge n\}}]. \end{aligned}$$

From this, taking into account that $|v(x)| \le G$, we find

$$M_x \sum_{s=0}^{\tau_\varepsilon - 1}\alpha^s c(x_s) I_{\{\tau_\varepsilon < n\}} \le 2G < \infty$$

and, therefore, due to the fact that $P_x\{\tau_\varepsilon < \infty\} = 1$, $x \in E$,

$$M_x \sum_{s=0}^{\tau_\varepsilon - 1}\alpha^s c(x_s) < \infty.$$

From this inequality and the relation

$$v(x) = M_x\left[\alpha^{\tau_\varepsilon \wedge n} v(x_{\tau_\varepsilon \wedge n}) - \sum_{s=0}^{(\tau_\varepsilon \wedge n)-1}\alpha^s c(x_s)\right]$$

we have (for $\varepsilon > 0$)

$$\begin{aligned} v(x) &= M_x\left[\alpha^{\tau_\varepsilon} v(x_{\tau_\varepsilon}) - \sum_{s=0}^{\tau_\varepsilon - 1}\alpha^s c(x_s)\right] \\ &\le M_x\left[\alpha^{\tau_\varepsilon} g(x_{\tau_\varepsilon}) - \sum_{s=0}^{\tau_\varepsilon - 1}\alpha^s c(x_s)\right] + \varepsilon, \end{aligned}$$

and, therefore, $v(x) \le s(x)$.

[14] In Lemma 8 the function $\psi_n = \sup_{j \ge n}[\alpha^j g(x_j) - \sum_{s=0}^{j-1}\alpha^s c(x_s)]$ has to be considered instead of the function $\psi_n = \sup_{j \ge n} g(x_j)$.

Thus, we have proved (1)–(4). (5) follows from the fact that setting $P_x\{\tau_0 < \infty\} = 1$, $x \in E$, yields

$$v(x) = M_x\left[\alpha^{\tau_0}v(x_{\tau_0}) - \sum_{s=0}^{\tau_0-1}\alpha^s c(x_s)\right]$$
$$= M_x\left[\alpha^{\tau_0}g(x_{\tau_0}) - \sum_{s=0}^{\tau_0-1}\alpha^s c(x_s)\right].$$

Finally, since $\overline{\lim}_n \alpha^n v(x_n) = \overline{\lim}_n \alpha^n g(x_n)$ (P_x-a.s.), $x \in E$, it follows from the relation

$$v(x) = M_x\left[\alpha^{\tau_0 \wedge n}v(x_{\tau_0 \wedge n}) - \sum_{s=0}^{(\tau_0 \wedge n)-1}\alpha^s c(x_s)\right]$$

by Fatou's lemma that

$$v(x) \le M_x\left[\alpha^{\tau_0}v(x_{\tau_0}) - \sum_{s=0}^{\tau_0-1}\alpha^s c(x_s)\right].$$

Since $\tau_0 = \inf\{n : \alpha^n v(x_n) = \alpha^n g(x_n)\}$, we have

$$v(x) \le M_x\left[\alpha^{\tau_0}g(x_{\tau_0}) - \sum_{s=0}^{\tau_0-1}\alpha^s c(x_s)\right].$$

If we assume that for a certain $x_0 \in E$ that $P_{x_0}\{\tau_0 = \infty\} > 0$ and $P_{x_0}\{\sum_{s=0}^{\infty}\alpha^s c(x_s) = \infty\} = 1$ we get $v(x_0) = -\infty$. This contradicts the obvious inequality $v(x_0) \ge g(x_0) > -\infty$. We can also see that

$$M_x \sum_{s=0}^{\tau_0-1}\alpha^s c(x_s) \le M_x \alpha^{\tau_0}g(x_{\tau_0}) - v(x) < \infty.$$

Therefore, $\tau_0 \in \mathfrak{M}_{(\alpha, c)}$ is an optimal time:

$$s(x) = M_x\left[\alpha^{\tau_0}g(x_{\tau_0}) - \sum_{s=0}^{\tau_0-1}\alpha^s c(x_s)\right]. \qquad \square$$

2.14.2

In several cases the optimal stopping problem considered in the preceding subsection can be easily reduced to a similar problem but with $c(x) \equiv 0$.

Theorem 24. *Let a nonnegative function $c(x)$ be such that*

$$M_x \sum_{s=0}^{\infty}\alpha^s c(x_s) < \infty, \qquad x \in E. \tag{2.174}$$

Then

$$s(x) = \sup M_x \alpha^\tau G(x_\tau) - f(x), \tag{2.175}$$

where

$$f(x) = M_x \sum_{s=0}^{\infty} \alpha^s c(x_s),$$

$$G(x) = g(x) + f(x).$$

PROOF. Set

$$\xi = \sum_{n=0}^{\infty} \alpha^n c(x_n).$$

Then for $\tau \in \mathfrak{M}$

$$\theta_\tau \xi = \sum_{n=0}^{\infty} \alpha^n c(x_{n+\tau}) = \alpha^{-\tau} \sum_{n=\tau}^{\infty} \alpha^n c(x_n),$$

and by virtue of the strong Markov property

$$\begin{aligned}
f(x) &= M_x \xi \\
&= M_x \sum_{n=0}^{\infty} \alpha^n c(x_n) \\
&= M_x \left[\sum_{n=0}^{\tau-1} \alpha^n c(x_n) + \sum_{n=\tau}^{\infty} \alpha^n c(x_n) \right] \\
&= M_x \sum_{n=0}^{\tau=1} \alpha^n c(x_n) + M_x \alpha^\tau \theta_\tau \xi \\
&= M_x \sum_{n=0}^{\tau-1} \alpha^n c(x_n) + M_x \alpha^\tau M_{x_\tau} \xi \\
&= M_x \sum_{n=0}^{\tau-1} \alpha^n c(x_n) + M_x \alpha^\tau f(x_\tau),
\end{aligned}$$

which proves (2.175). □

Remark 1. The theorem holds true for arbitrary functions $c(x) \in B$ satisfying the condition

$$M_x \sum_{s=0}^{\infty} \alpha^s |c(x_s)| > \infty. \tag{2.176}$$

Remark 2. In the cases when the conditions of (2.174) and (2.176) cannot be satisfied the method for reducing the problem involving $c(x) \neq 0$ to that involving $c(x) \equiv 0$ may be useful.

Let $|g(x)| \leq C$, $|c(x)| \leq C$ and let $f(x)$ be a bounded ($|f(x)| \leq K < \infty$) solution of the equation

$$\alpha T f(x) - f(x) = c(x). \tag{2.177}$$

Set $\mathfrak{M}^1 = \{\tau \in \mathfrak{M} : M_x \tau < \infty, x \in E\}$. Then for any time $\tau \in \mathfrak{M}^1$

$$f(x) = M_x\left[\alpha^\tau f(x_\tau) - \sum_{s=0}^{\tau-1} \alpha^s c(x_s)\right] \tag{2.178}$$

(which can be proved by analogy with the proof of Theorem 1.12), and therefore

$$\begin{aligned} s(x) &= \sup_{\tau \in \mathfrak{M}^1} M_x\left[\alpha^\tau g(x_\tau) - \sum_{s=0}^{\tau-1} \alpha^s c(x_s)\right] \\ &= f(x) + \sup_{\tau \in \mathfrak{M}^1} M_x[\alpha^\tau G(x_\tau)] \end{aligned} \tag{2.179}$$

with $G(x) = g(x) - f(x)$.

2.14.3

EXAMPLE. Let $\xi, \xi_1, \xi_2, \ldots$ be a sequence of independent uniformly distributed random variables given on a probability space $(\Omega, \mathscr{F}, P)$ with $M|\xi| < \infty$. For $x \in R$ set

$$x_n = \max\{x, \xi_1, \ldots, \xi_n\}, \qquad x_0 = x,$$

and let

$$s(x) = \sup M_x\left[\alpha^\tau x_\tau - c\sum_{s=0}^{\tau-1} \alpha^s\right],$$

where c is a nonnegative constant, $0 < \alpha \leq 1$, and sup is taken over all the Markov times τ for which

$$M_x\left[\alpha^\tau x_\tau - c\sum_{s=0}^{\tau-1} \alpha^s\right]^- < \infty, \qquad x \in E.$$

It is clear that $X = (x_n, G_n, P_x)$, $n \geq 0$, where $\mathscr{G}_n = \sigma\{\omega : \xi_1, \ldots, \xi_n\}$, $n \geq 1$, $\mathscr{G}_0 = \{\varnothing, \Omega\}$, and the P_x-measure on $\mathscr{G} = \sigma(\bigcap_n \mathscr{G}_n)$ induced by the distributions of the variables $\xi_1, \xi_2, \ldots$, is a Markov sequence.

Denote by γ the (unique) root of the equation

$$M(\xi - \gamma)^+ = \frac{(1-\alpha)\gamma + c}{\alpha} \tag{2.180}$$

and show that the example considered illustrates the monotone case noted in Subsection 5.5 (with obvious changes in notation).

In fact, let $g(x) = x$; then

$$\begin{aligned} [\alpha Tg(x) - c] - g(x) &= \alpha M \max\{x, \xi) - x - c \\ &= \alpha M[\max\{x, \xi\} - x] - x(1-\alpha) - c \\ &= \alpha M(\xi - x)^+ - x(1-\alpha) - c \end{aligned}$$

and, therefore,

$$\{x : [\alpha Tg(x) - c] - g(x) \leq 0\} = \{x : x \geq \gamma\}.$$

It follows from this that if the initial point $x \in \{x : x \geq \gamma\}$, then $x_n \geq \gamma$ for any n (since $x_n \geq x_{n-1} \geq \cdots \geq x$), and, therefore, if the time $\tau_0 = \inf\{n \geq 0 : x_n \geq \gamma\}$ is such that $P_x\{\tau_0 < \infty\} = 1, x \in R$, it will be an optimal stopping time (compare with Subsection 5.5).

It is interesting to note that if

$$s_N(x) = \sup_{\tau \in \mathfrak{M}(N)} M_x\left[\alpha^\tau x_\tau - c\sum_{s=0}^{\tau-1} \alpha^s\right],$$

then in the class $\mathfrak{M}(N)$ there exists an optimal stopping time τ_N^*, with

$$\tau_N^* = \min\{0 \leq n \leq N : x_n \geq \gamma\},$$

where the constant γ is independent of N and is the same as that for τ_0.

2.15 Reduction of the optimal stopping problem for arbitrary random sequences to the corresponding problem for Markov processes

2.15.1

Many optimal stopping problems (including the problems considered in Chapter 4) are usually formulated in terms that are not Markov, and to apply the theory described above we need to reduce the primary problem to a Markov problem.

The optimal stopping problem is frequently formulated as follows.

Let $(\Omega, \mathscr{F}, P)$ be a probability space, let $F = \{\mathscr{F}_n\}$, $n = 0, 1, \ldots$, be a nondecreasing system of σ-algebras $\mathscr{F}_n$ characterizing the information available to the observer before time n, and let $Z = (z_n, \mathscr{F}_n)$ be a random sequence where z_n is treated as the gain obtained at the final observation time n.

Set

$$V = \sup Mz_\tau, \tag{2.181}$$

where the supremum is taken over all stopping times τ (with respect to the system F), for which the mathematical expectations Mz_τ are defined.

The problem of searching for the payoff V and the ε-optimal times τ_ε (i.e., for which

$$V \leq Mz_{\tau_\varepsilon} + \varepsilon)$$

can be solved in different ways.

First, we can make use of the general theory for solving such problems discussed in detail, for example, in [22]. Second, this problem can be solved by the Markov methods suggested above, if we consider instead of the primary sequence $\{z_n, n \geq 0\}$ a Markov sequence (more precisely, a Markov random function) $\{\tilde{z}_n, n \geq 0\}$ with $\tilde{z}_n = (z_0, \ldots, z_n)$. Unfortunately, the above reduction (in which the whole past $(z_0, \ldots, \tilde{z}_n)$ is taken as the state $\tilde{z}_n$ at time n) may turn out to be of no real value due to the fact that the state space of the Markov chain $\{\tilde{z}_n, n \geq 0\}$ is very complex.

Hence, in order to apply efficiently the above theory of optimal stopping rules for Markov sequences the state space of the Markov chain resulting from reduction must be sufficiently simple.

We shall also emphasize the fact that in many cases the reduction of general problems of optimal stopping to Markov problems is justified because such a reduction makes it feasible to exploit the powerful analytical device of the Markov process theory. This follows also from the fact that the majority of familiar problems in which explicit solutions are obtained can be reformulated in Markov terms.

We consider in the present subsection the reduction of the optimal stopping problem for arbitrary random sequences to problems formulated in Markov terms. (In this case the concept of transitive statistics introduced below proves to be crucial.) However, before discussing this problem we shall consider "problem (2.181)" in more detail.

2.15.2

Let

$$\mathfrak{M}_z[F] = \{\tau \in \mathfrak{M}; Mz_\tau^- < \infty\}.$$

Then (compare with Section 1)

$$V = \sup_{\tau \in \mathfrak{M}_z[F]} Mz_\tau. \tag{2.182}$$

Assume now that the values z_n can be represented as

$$z_n = z_n' + z_n''$$

where z_n' and z_n'' are $\mathscr{F}_n$-measurable, $M|z_n''| < \infty$, $n = 0, 1, \ldots$, and for any $\tau \in \mathfrak{M}_z[F]$

$$Mz_\tau'' = Mz_0''. \tag{2.183}$$

(Equation (2.183) is satisfied, for instance, if the sequence $(z_n'', \mathscr{F}_n, P)$ forms a uniformly integrable martingale, i.e., $z_n'' = M(\eta | \mathscr{F}_n)$ where $M|\eta| < \infty$.) Then, obviously,

$$V = \sup_{\tau \in \mathfrak{M}_z[F]} Mz_\tau = \sup_{\tau \in \mathfrak{M}_z[F]} Mz_\tau' + Mz_0'', \tag{2.184}$$

and, therefore, instead of the problem "$V = \sup Mz_\tau$" is suffices to consider the problem "$V = \sup Mz_\tau'$." (We shall deal with this kind of situation in Section 4.3 in considering a so-called *disruption* problem.)

Therefore, the solution of the problem of finding the payoff $V = \sup Mz_\tau$ can be simplified if instead of the sequence $\{z_n, n \geq 0\}$ one considers another sequence $\{z'_n, n \geq 0\}$ such that the values $z''_n = z_n - z'_n$ possess the properties indicated above.

Frequently, another way to simplify the solution of the problem becomes feasible because a new sequence of (sufficient) σ-algebras $F' = \{\mathscr{F}'_n\}, n \geq 0$, can be found such that

$$\mathscr{F}'_n \subseteq \mathscr{F}'_{n+1}, \qquad \mathscr{F}'_n \subseteq \mathscr{F}_n$$

and possessing the property that the z'_n are $\mathscr{F}'_n$-measurable, where

$$\sup_{\tau \in \mathfrak{M}_{z'}[F]} Mz'_\tau = \sup_{\tau \in \mathfrak{M}_{z'}[F']} Mz'_\tau.$$

(Compare with Section 13 and with Remark 3 to Theorem 3, where $\mathscr{F}'_n = \mathscr{F}^X_n$.)

2.15.3

Assume next that in the problem "$V = \sup_{\tau \in \mathfrak{M}_z[F]} Mz_\tau$" the family of σ-algebras $F = \{\mathscr{F}_n\}, n \geq 0$, is of a special structure:

$$\mathscr{F}_n = \mathscr{F}^\xi_n = \sigma\{\omega : \xi_0, \xi_1, \ldots, \xi_n\}, \qquad n \geq 0,$$

where $\xi = (\xi_0, \xi_1, \ldots)$ is some sequence of random variables (or, in a more general form, a sequence of random elements[15] with values on some measure space $(X, \mathscr{X})$ given on $(\Omega, \mathscr{F}, P)$). We shall regard $\xi = (\xi_0, \xi_1, \ldots)$ as a sequence of observations, based on which we wish to stop the sequence $z_0, z_1, \ldots,$ so as to obtain the maximum possible (or close to it) gain $V = \sup Mz_\tau$.

The reduction given below of the problem $V = \sup Mz_\tau$ to the Markov problem requires the following:

Definition. Let $F^\xi = \{\mathscr{F}^\xi_n\}, n = 0, 1, \ldots,$ where $\mathscr{F}^\xi_n = \sigma\{\omega : \xi_0, \ldots, \xi_n\}$. The system of random elements $\eta = (\eta_0, \eta_1, \ldots)$ with values in $(Y, \mathscr{Y})$ is referred to as a *system of transitive statistics* (*with respect to* F^ξ), if:

(1) η_n is $\mathscr{F}^\xi_n/\mathscr{Y}$-measurable, $n = 0, 1, \ldots;$

(2) For each $n = 1, 2, \ldots$ there exists a $\mathscr{Y} \times \mathscr{X}/\mathscr{Y}$-measurable function $\varphi_n = \varphi_n(y, x)$ such that with probability 1

$$\eta_n(\omega) = \varphi_n(\eta_{n-1}(\omega), \xi_n(\omega)). \tag{2.185}$$

Lemma 17. *Let $\eta = (\eta_0, \eta_1, \ldots)$ be the system of transitive statistics (with respect to F^ξ) with values in $(Y, \mathscr{Y})$. If, for each $n = 0, 1, \ldots$ with probability 1*

$$P\{\xi_{n+1} \in B | \mathscr{F}^\xi_n\} = P\{\xi_{n+1} \in B | \eta_n\}, \qquad B \in \mathscr{X}, \tag{2.186}$$

[15] See Section 1.1.

then the elements $Y = (\eta_n, \mathscr{F}_n^\xi, P)$, $n = 0, 1, \ldots,$ *form a Markov random function*:

$$P\{\eta_{n+1} \in A \,|\, \mathscr{F}_n^\xi\} = P\{\eta_{n+1} \in A \,|\, \eta_n\}, \qquad A \in \mathscr{Y}. \tag{2.187}$$

PROOF. It suffices to show that for any bounded ($\mathscr{Y}$-measurable) function $f = f(y)$, $n = 0, 1, \ldots,$

$$M[f(\eta_{n+1}) \,|\, \mathscr{F}_n^\xi] = M[f(\eta_{n+1}) \,|\, \eta_n] \qquad (P\text{-a.s.}). \tag{2.188}$$

By virtue of (2.185) $f(y_{n+1}) = \varphi_{n+1}(y_n, x_{n+1})$ where $\varphi_{n+1}(y, x)$ is a bounded, $\mathscr{Y} \times \mathscr{X}/\mathscr{Y}$-measurable function. If $\varphi_{n+1}(y, x)$ is a function of the form

$$\varphi_{n+1}^N(y, x) = \sum_{k=1}^N g_{k,n+1}(y) I_{B_k}(x), \qquad B_k \in \mathscr{X}, \tag{2.189}$$

then we immediately obtain from (2.186)

$$M[\varphi_{n+1}^N(\eta_n, \xi_{n+1}) \,|\, \mathscr{F}_n^\xi] = M[\varphi_{n+1}^N(\eta_n, \xi_{n+1}) \,|\, \eta_n], \tag{2.190}$$

which proves (2.187) for the functions of the form (2.189).

To prove the lemma for the general case we need first to construct a sequence of functions of the form (2.189) monotonically convergent (as $N \to \infty$) to $\varphi_{n+1}(y, x)$, and second to make in (2.190) the passage to the limit as $N \to \infty$.

Let now $z_n = g(y_n)$ where $g(y)$ is a $\mathscr{Y}$-measurable function. Then

$$V = \sup Mg(\eta_\tau)$$

and, if the condition of (2.186) is satisfied we shall have a Markov problem of optimal stopping which differs from that considered above in the fact that the system $Y = (\eta_n, \mathscr{F}_n^\xi, P)$, $n \geq 0$, forms a Markov random function rather than a Markov process (in the sense of the definitions given in Chapter 1). This difference is not, however, essential in nature, and, as will be seen from what follows, we have all the machinery necessary to solve this problem.

Let $\pi(A) = P\{\eta_0 \in A\}$, $A \in \mathscr{Y}$, and let us assume that the Markov random function Y has regular stationary transition probabilities

$$P(A \,|\, y) = P\{\eta_{n+1} \in A \,|\, \eta_r = y\}.$$

These transition probabilities and initial distributions define completely the sequences $\eta = (\eta_0, \eta_1, \ldots)$, with

$$P\{\eta_0 \in A_0, \ldots, \eta_n \in A_n\}$$

$$= \int_{A_{n-1}} \cdots \int_{A_0} P(A_n \,|\, y_{n-1}) P(dy_{n-1} \,|\, y_{n-2}) \cdots P(dy_1 \,|\, y_0) \pi(dy_0). \tag{2.191}$$

To avoid nonessential questions of integrability we shall assume that $|g(y)| \le C < \infty$, and introduce the functions (compare with (2.20) and (2.79))

$$\Gamma_{m,n}(\omega) = \operatorname*{ess\,sup}_{\tau \in \mathfrak{M}(m;n)} M[g(\eta_\tau) | \mathscr{F}_m], \tag{2.192}$$

$$\Gamma_{m,\infty}(\omega) = \operatorname*{ess\,sup}_{\tau \in \mathfrak{M}(m;\infty)} M[g(\eta_\tau) | \mathscr{F}_m], \tag{2.193}$$

where $\mathscr{F}_m = \mathscr{F}_m^\xi$, $\mathfrak{M}(m;n) = \{\tau \in \mathfrak{M} : m \le \tau \le n\}$.

Also let

$$V_{m,n} = \sup_{\tau \in \mathfrak{M}(m;n)} Mg(\eta_\tau), \tag{2.194}$$

$$V_{m,\infty} = \sup_{\tau \in \mathfrak{M}(m;\infty)} Mg(\eta_\tau), \qquad V_{0,\infty} = V. \tag{2.195}$$

(Compare with (2.19) and (2.78).)

If we examine the proof of Lemma 1 (see also Remark 1 to Lemma 1), we shall easily see that instead of (2.17) we have for the case in question (P-a.s.) the inequality

$$M[g(\eta_\tau) | \mathscr{F}_m] \le Q^{n-m} g(\eta_m), \tag{2.196}$$

where $\tau \in \mathfrak{M}(m;n)$ and $Qg(y)$ is to be understood as

$$\max\{g(y), M[g(\eta_1) | \eta_0 = y]\}.$$

Similarly we can easily verify (compare with the proof of Theorem 2), that for the time

$$\tau_{m,n}^* = \min\{m \le j \le n : Q^{n-j} g(\eta_j) = g(\eta_j)\}$$

we have the equality

$$M[g(\eta_{\tau_{n,m}^*}) | \mathscr{F}_m] = Q^{n-m} g(\eta_m) \qquad (P\text{-a.s.}). \tag{2.197}$$

Thus,

$$\Gamma_{m,n}(\omega) = Q^{n-m} g(\eta_m(\omega)) \qquad (P\text{-a.s.})$$

and it follows from (2.196) and (2.197) that the time $\tau_{m,n}^*$ is optimal in the sense that (P-a.s.)

$$M[g(\eta_{\tau_{m,n}^*}) | \mathscr{F}_m] = \operatorname*{ess\,sup}_{\tau \in \mathfrak{M}(m;n)} M[g(\eta_\tau) | \mathscr{F}_m] \quad (= \Gamma_{m,n}(\omega))$$

and

$$M[g(\eta_{\tau_{m,n}^*})] = \sup_{\tau \in \mathfrak{M}(m;n)} Mg(\eta_\tau) \quad (= V_{m,n}).$$

In this case

$$\operatorname*{ess\,sup}_{\tau \in \mathfrak{M}(m;n)} M[g(\eta_\tau) | \mathscr{F}_m] = \operatorname*{ess\,sup}_{\tau \in \mathfrak{M}(m;n)} M[g(\eta_\tau) | \eta_m] \qquad (P\text{-a.s.}).$$

It can be seen from the above that, in fact, the function $\Gamma_{m,n}(\omega)$ depends on ω through $\eta_m(\omega)$ (to avoid new notation we shall write $\Gamma_{m,n}(\omega) = \Gamma_{m,n}(\eta_m(\omega))$ and it suffices to take the supremum in (2.192) and (2.194) not over all Markov times $\tau \in \mathfrak{M}(m; n) = \mathfrak{M}(m; n)[F^{\xi}]$, but over the class $\mathfrak{M}(m; n)[F^{\eta}] \subseteq \mathfrak{M}(m; n)[F^{\xi}]$.

Let us now show that the Markov process $W = (w_n, \mathscr{G}_n, P_w)$, $n \geq 0$, can be constructed with values in a state space $(Y, \mathscr{Y})$ and defined on a space $(\tilde{\Omega}, \tilde{\mathscr{F}})$ for which the functions $s_{n-m}(\omega)$ (see Section 2) are associated with the functions $\Gamma_{m,n}(\eta_m(\omega))$ in a simple way:

$$\Gamma_{m,n}(\eta_m(\omega)) = s_{n-m}(\eta_m(\omega)) \qquad (P\text{-a.s.}).$$

It follows from the above, in particular, that the solution of the optimal stopping problems in the classes $\mathfrak{M}(m; n)$ for the Markov random function $Y = (\eta_n, \mathscr{F}_n^{\xi}, P)$ can be obtained from the solution of this problem for the Markov process $W = (w_n, \mathscr{F}_n^{w}, P_w)$, $n \geq 0$, since the optimal stopping times can be constructed from the functions $\Gamma_{m,n}$ and $V_{m,n} = M\Gamma_{m,n}(\eta_m(\omega))$.

To construct the process W we shall proceed as follows.

We fix a point $y \in Y$ and construct a Markov random function leaving a point y at an initial time. We set

$$\tilde{\Omega} = Y \times Y \times \ldots, \qquad \tilde{\mathscr{F}} = \mathscr{F} \times \mathscr{F} \ldots,$$

and construct the sequence $\tilde{\eta}_0(\tilde{\omega}), \tilde{\eta}_1(\tilde{\omega}), \ldots$, assuming, $\tilde{\eta}_i(\tilde{\omega}) = y_i$ for $\tilde{\omega} = (y_0, y_1, \ldots) \in \tilde{\Omega}$.

Using the regular transition function $P(A \mid y)$ we construct the measure $\tilde{P}^y$ on the sets of the form

$$\{\tilde{\omega} : \tilde{\eta}_0(\tilde{\omega}) \in A_0, \tilde{\eta}_1(\tilde{\omega}) \in A_1, \ldots, \tilde{\eta}_n(\tilde{\omega}) \in A_n\}$$

with the aid of the formula (compare with (2.91))

$$\begin{aligned}&\tilde{P}^y\{\tilde{\eta}_0(\tilde{\omega}) \in A_0, \ldots, \tilde{\eta}_n(\tilde{\omega}) \in A_n\} \\ &\qquad = \int_{A_{n-1}} \cdots \int_{A_0} P(A_n \mid y_{n-1}) P(dy_{n-1} \mid y_{n-2}) \cdots P(dy_1 \mid y_0) \pi^y(dy_0),\end{aligned}$$

where π^y is the measure concentrated at the point y.

The system

$$\tilde{Y}^y = (\tilde{\eta}_n, \tilde{\mathscr{F}}_n, \tilde{P}^y), \qquad n \geq 0,$$

where $\tilde{\mathscr{F}}_n = \sigma\{\omega : \tilde{\eta}_0, \ldots, \tilde{\eta}_n\}$ is, as can be easily verified, a Markov random function with the same transition probability as in the primary Markov function $Y = (\eta_n, \mathscr{F}_n^{\xi}, P)$, $n \geq 0$. In this case $\tilde{P}^y\{\tilde{\eta}_0(\tilde{\omega}) = y\} = 1$.

Finally, let us consider the system of Markov random functions $\{\tilde{Y}^y, y \in Y\}$. It is easy to verify that this system forms a Markov family of random functions (in the sense of Definition 7, Section 1.4).

It was noted in Subsection 1.4.5 that we could always construct by the concommitant expansion of the space of elementary events) a Markov process (denoted by $W = (w_n, \mathcal{G}_n, P_w)$, $n \geq 0$, $w \in Y$) which has the same transition probabilities as those in the Markov family of the random functions $\{\tilde{Y}^y, y \in Y\}$.

Then, if

$$s_n(w) = \sup M_w g(\omega_\tau), \tag{2.198}$$

it follows from the above that (P-a.s.)

$$\Gamma_{m,n}(\eta_m(\omega)) = s_{n-m}(\eta_m(\omega)),$$
$$V_{m,n} = M\Gamma_{m,n}(\eta_m(\omega)) = Ms_{n-m}(\eta_m(\omega)).$$

In particular,

$$V_{0,n} = \sup_{\tau \in \mathfrak{M}(0,n)} Mg(\eta_\tau) = Ms_n(\eta_0) = \int_Y s_n(y)\pi(dy). \tag{2.199}$$

Having solved the optimal stopping problem (see (2.198)) for the Markov process $W = (w_n, \mathcal{G}_n, P_w)$, it can be seen that we obtain as well the solution of the corresponding problem for a Markov random function with transition function $P(A \mid y)$ and any initial distribution $\pi(A)$, $A \in \mathcal{Y}$.

The arguments given above are for the class $\mathfrak{M}(m; n)$. The appropriate passage to the limit as $n \to \infty$ justifies the previous conclusions (in the case $|g(x)| \leq C < \infty$ and for the classes $\mathfrak{M}(m; \infty)$). These conclusions can be extended to the case with the functions $g \in B$ satisfying the condition $Mg^-(\eta_0) = \int g^-(y)\pi(dy) < \infty$, by considering the "truncated" functions g_a^b introduced; first one has to go to the limit as $a \downarrow -\infty$, and then as $b \uparrow \infty$ (compare with (2.125)).

We shall note that the functions $\Gamma_{m,\infty}(\eta_m(\omega))$ introduced in (1.193) were found with the aid of the functions $s(x)$ by the formulas

$$\Gamma_{m,\infty}(\eta_m(\omega)) = s(\eta_m(\omega)). \tag{2.200}$$

In this case

$$\Gamma_{m,\infty}(\eta_m) = \max\{g(\eta_m), M[\Gamma_{m+1,\infty} \mid \eta_m]\} \tag{2.201}$$

and

$$V_{m,\infty} = M\Gamma_{m,\infty}. \tag{2.202}$$

In particular, it follows from the above for $m = 0$ that the payoff is given by:

$$V = \int_Y s(y)\pi(dy). \tag{2.203}$$

(2.203) implies that the following nontrivial relation is satisfied:

$$\sup_{\tau \in \mathfrak{M}} Mg(\eta_\tau) = \int_Y \operatorname*{ess\,sup}_{\tau \in \mathfrak{M}} M[g(\eta_\tau) \mid \eta_0 = y]\pi(dy).$$

The following fact should be emphasized: Letting

$$\tau = \inf\{n \geq 0 : \eta_n \in A_n\}, \qquad A_n \in \mathscr{Y},$$

and

$$\tilde{\tau} = \inf\{n \geq 0 : w_n \in A_n\}, \qquad A_n \in \mathscr{Y},$$

be the times of first entry of the sequences $\eta = (\eta_0, \eta_1, \ldots)$ and $w = (w_0, w_1, \ldots)$ into the same Borel sets $A_0, A_1, \ldots$, then, since the probability distributions

$$P\{\eta_0 \in B_0, \eta_1 \in B_1, \ldots | \eta_0 = w\}, \qquad B_n \in \mathscr{Y},$$

and

$$P_w\{w_0 \in B_0, w_1 \in B_1, \ldots\}, \quad B_n \in \mathscr{Y},$$

coincide (π-a.s.), it follows that

$$Mg(\eta_\tau) = \int_Y M_w g(w_{\tilde{\tau}}) \pi(dw). \tag{2.204}$$

In particular, it follows from the above that if the time $\tilde{\tau}$ is ε-optimal in the problem "$s(w) = \sup M_w g(w_{\tilde{\tau}})$," then the time τ will be ε-optimal in the primary problem "$V = \sup Mg(\eta_\tau)$." □

The problem "$V = \sup Mg(\eta_\tau)$" was reduced above to the standard problem "$s(w) = \sup M_w g(w_\tau)$" for the case of homogeneous processes and for the gain functions independent of time. The case of the nonhomogeneous process as well as the cases which allow for the cost of the observation, and where the function g depends on both the state coordinate and the time parameter can be considered in a similar way.

Without dwelling on these questions in detail, we shall discuss a few examples.

2.15.4

EXAMPLE 1. Let $\xi = (\xi_0, \xi_1, \ldots)$ be a sequence of independent, uniformly distributed random variables and let

$$z_n = \max(\xi_0, \xi_1, \ldots, \xi_n) - c(n).$$

If we set

$$\eta_n = \max(\xi_0, \xi_1, \ldots, \xi_n),$$

the system $Y = (\eta_n, \mathscr{F}_n^\xi, P)$, $n \geq 0$, will form a Markov random function. The problem of searching for the payoff

$$V = \sup Mz_\tau$$

is reduced to the problem of finding the payoff

$$s(x) = \sup M_x g(\tau, x_\tau)$$

with the function $g(n,x) = x - c(n)$ for the pertinent Markov process $X = (x_n, \mathscr{F}_n^x, P_x)$ with $x_n = \max(x, \xi_1, \ldots, \xi_n)$, $x_0 = x$.

In this case $V = \int s(x)\pi(dx)$ where $\pi(A) = P\{\xi_0 \in A\}$.

The example discussed at the end of Section 14 illustrated that (under certain assumptions) the optimal time (in the problem "$s(x) = \sup M_x g(\tau, x_\tau)$") exists and is the time of first entry of x_n into the set $[\gamma, \infty)$. It follows from the above (compare with (2.204)) that in the primary problem also the time of first entry of the variables η_n into the set $[\gamma, \infty)$ will be optimal.

EXAMPLE 2. Assume that on a certain probability space $(\Omega, \mathscr{F}, P)$ an unobservable random process $\theta = (\theta_0, \theta_1, \ldots)$ and an observable (at times $n = 1, 2, \ldots$) process $\xi = (\xi_1, \xi_2, \ldots)$ are given. At the stopping time n the observer obtains the gain $G_n(\theta_0, \ldots, \theta_n; \xi_1, \ldots, \xi_n)$ if $n \geq 1$, and $G_0(\theta_0)$ if $n = 0$. The magnitude of this gain in each particular case is not, in general, known to the observer since $\theta_0, \ldots, \theta_n$ are unknown.

The problem consists in finding the payoff

$$V = \sup MG_\tau(\theta_0, \ldots, \theta_\tau; \xi_1, \ldots, \xi_\tau) \tag{2.205}$$

(the supremum is taken over the times τ from $\mathfrak{M}(F^\xi)$ for which the mathematical expectations are defined in (2.205); $F^\xi = \{\mathscr{F}_n^\xi\}$, $\mathscr{F}_n^\xi = \sigma\{\omega: \xi_1, \ldots, \xi_n\}$, $\mathscr{F}_0^\xi = \{\varnothing, \Omega\}$) and in describing the structure of optimal or ε-optimal stopping times.

Let us assume for simplicity that the mathematical expectations $MG_n(\theta_0, \ldots, \theta_n; \xi_1, \ldots, \xi_n)$ are defined for $n = 0, 1, \ldots$. Then, setting $z_n = M(G_n | \mathscr{F}_n^\xi)$, we find that the payoff $V = \sup Mz_\tau$ where the function z_n is (unlike G_n) $\mathscr{F}_n^\xi$-measurable.

Let us assume further that the unobservable process $\theta = (\theta_0, \theta_1, \ldots)$ is a Markov chain with a finite number of states $\{0, 1, \ldots, N\}$, initial distribution $\pi = \{\pi(l) = P\{\theta_0 = l\}, l = 1, \ldots, N\}$,[16] and transition probabilities

$$P_{ij}(n) = P\{\theta_{n+1} = j | \theta_n = i\}. \tag{2.206}$$

We assume that the joint distributions of the sequence (θ, ξ) are arranged so that

$$\begin{aligned} P\{\theta_{n+1} &= j, \xi_{n+1} \leq x | \theta_0, \ldots, \theta_n; \xi_1, \ldots, \xi_n\} \\ &= P\{\xi_{n+1} \leq x | \theta_0, \ldots, \theta_n, \theta_{n+1} = j; \xi_1, \ldots, \xi_n\} \\ &= P\{\theta_{n+1} = j | \theta_0, \ldots, \theta_n; \xi_1, \ldots, \xi_n\} \\ &= P\{\xi_{n+1} \leq x | \theta_{n+1} = j\} P\{\theta_{n+1} = j | \theta_n\}. \end{aligned} \tag{2.207}$$

Let G_n be functions of the following structure:

$$G_0(\theta_0) = g_0(\theta_0),$$

$$G_n(\theta_0, \ldots, \theta_n) = \sum_{k=1}^{n-1} c_k(\theta_k) + g_n(\theta_n). \tag{2.208}$$

[16] We do not include $\pi(0)$ in the vector π since $\pi(0) = 1 - \sum_{l=1}^{n} \pi(l)$.

Now write

$$\pi_n^\pi(j) = P\{\theta_n = j | \mathscr{F}_n^\xi\},$$
$$\pi_{k,n}^\pi(j) = P\{\theta_k = j | \mathscr{F}_n^\xi\}.$$

Then $z_0 = M(G_0 | \mathscr{F}_0^\xi) = \sum_{j=0}^N g_0(j)\pi_0^\pi(j)$ and

$$z_n = M(G_n | \mathscr{F}_n^\xi) = \sum_{k=0}^{n-1} \sum_{j=0}^{N} c_k(j)\pi_{k,n}^\pi(j) + \sum_{j=0}^{N} g_n(j)\pi_n^\pi(j). \tag{2.209}$$

Setting $z_0' = z_0$,

$$z_n' = \sum_{k=0}^{n-1} \sum_{j=0}^{N} c_k(j)\pi_k^\pi(j) + \sum_{j=0}^{N} g_n(j)\pi_n^\pi(j), \qquad n \geq 1, \tag{2.210}$$

$$z_n'' = z_n - z_n' = \sum_{k=0}^{n-1} \sum_{j=0}^{N} c_k(j)[\pi_{k,n}^\pi(j) - \pi_k^\pi(j)]. \tag{2.111}$$

Since

$$\begin{aligned} M(z_{n+1}'' | \mathscr{F}_n^\xi) &= M\left\{\sum_{k=0}^{n} \sum_{j=0}^{N} c_k(j)[P(\theta_k = j | \mathscr{F}_{n+1}^\xi) - P(\theta_k = j | \mathscr{F}_k^\xi)] | \mathscr{F}_n^\xi\right\} \\ &= \sum_{k=0}^{n} \sum_{j=0}^{N} c_k(j)[P(\theta_k = j | \mathscr{F}_n^\xi) - P(\theta_k = j | \mathscr{F}_k^\xi)] \\ &= \sum_{k=0}^{n-1} \sum_{j=0}^{N} c_k(j)[P(\theta_k = j | \mathscr{F}_n^\xi) - P(\theta_k = j | \mathscr{F}_k^\xi)] = z_n'', \end{aligned}$$

the sequence $z'' = (z_n'', \mathscr{F}_n^\xi, P)$, $n = 0, 1, \ldots,$ forms a martingale. If this martingale were uniformly integrable, then for all $\tau \in \mathfrak{M}$ it would follow that (Theorem 1.11)

$$Mz_\tau'' = Mz_0'' = 0, \tag{2.212}$$

where the payoff $V^\pi = \sup Mz_\tau = \sup\, Mz_\tau'$ (we introduced the index π in the payoff V to emphasize the fact that this payoff is dependent on the *a priori* distribution π).

Let us consider in more detail the vector of *a posteriori* probabilities

$$\pi_n^\pi = (\pi_n(1), \ldots, \pi_n(N)), \qquad n = 0, 1, \ldots, \tag{2.213}$$

with $\pi_0^\pi = \pi$. By the Bayes formula

$$\left(p_j(x) = \frac{dP\{\xi_n \leq x | \theta_n = j\}}{dP\{\xi_n \leq x\}}\right),$$

we have

$$\pi_{n+1}^\pi(j) = \frac{p_j(\xi_{n+1}) \sum_{i=0}^N \pi_n^\pi(i)p_{ij}(n)}{\sum_{i,j=0}^N p_j(\xi_{n+1})\pi_n^\pi(i)p_{ij}(n)}. \tag{2.214}$$

This means that the system of (vector) statistics $(\pi_0^\pi, \pi_1^\pi, \ldots)$ is transitive.

Let us verify that (for a given π) the system[17] $\Pi^\pi = (\pi_n^\pi, \mathscr{F}_n^\xi, P^\pi), n \geq 0,$ forms a Markov random function. Using Lemma 17 we need only to verify Equation (2.187), whose validity in the case in question follows from the following chain of equations:

$$\begin{aligned} P^\pi\{\xi_{n+1} \in B | \mathscr{F}_n^\xi\} &= \sum_{k=0}^N P^\pi\{\xi_{n+1} \in B | \mathscr{F}_n^\xi, \theta_n = k\}\pi_n^\pi(k) \\ &= \sum_{i=0}^N \sum_{k=0}^N P^\pi\{\xi_{n+1} \in B | \mathscr{F}_n^\xi, \theta_n = k, \theta_{n+1} = l\} \\ &\quad \times P^\pi\{\theta_{n+1} = l | \theta_n = k, \mathscr{F}_n^\xi\}\pi_n^\pi(k) \\ &= \sum_{i=0}^N \sum_{k=0}^N P\{\xi_{n+1} \in B | \theta_{n+1} = l\} p_{ki}(n)\pi_n^\pi(k) \\ &= P^\pi\{\xi_{n+1} \in B | \pi_n^\pi\}. \end{aligned} \tag{2.215}$$

Therefore, the system $\Pi^\pi = (\pi_n^\pi, \mathscr{F}_n^\xi, P^\pi)$ forms a Markov random function and, if Equation (2.212) is satisfied for any $\tau \in \mathfrak{M}$, then

$$V^\pi = \sup_{\tau \in \mathfrak{M}} M^\pi g(\tau, \pi_\tau^\pi), \tag{2.216}$$

where $g(n, \pi_n^\pi) = z_n'$, the values of z_n' are defined in (2.210), and M^π is the expectation over the measure P^π.

Consider next a new space $(\hat{\Omega}, \hat{\mathscr{F}})$ where

$$\hat{\Omega} = S \times \Omega, \qquad \hat{\mathscr{F}} = \mathscr{B}_S \times \mathscr{F},$$

and

$$S = \{\pi : \pi = (\pi(1), \ldots, \pi(N)), \pi(i) \geq 0, \pi(1) + \cdots + \pi(N) \leq 1\}.$$

For $\hat{\omega} = (\pi, \omega)$ set

$$\begin{aligned} \hat{\pi}_0(\hat{\omega}) &= \pi, & \hat{\pi}_n(\hat{\omega}) &= \pi_n^\pi(\omega), \\ \hat{\xi}_0(\hat{\omega}) &= \pi, & \hat{\xi}_n(\hat{\omega}) &= \xi_n(\omega), \end{aligned}$$

$$\hat{\mathscr{F}}_n = \sigma\{\omega : \hat{\xi}_0, \ldots, \hat{\xi}_n\}, \qquad \hat{\mathscr{F}} = \sigma\left(\bigcup_n \hat{\mathscr{F}}_n\right).$$

If $A \in \hat{\mathscr{F}}$, we shall write

$$A_\pi = \{\omega : (\pi, \omega) \in A\}, \qquad A^\omega = \{\pi : (\pi, \omega) \in A\}.$$

We shall define on $(\hat{\Omega}, \hat{\mathscr{F}})$ the measure

$$\hat{P}_\pi(A) = P^\pi(A_\pi) I_{A^\omega}(\pi), \qquad A \in \hat{\mathscr{F}}.$$

[17] We denote by P^π the restriction of the measure P on the σ-algebra

$$\mathscr{F}^{\theta, \xi} = \sigma(\omega : \theta_0, \theta_1, \ldots; \xi_0, \xi_1, \xi_2, \ldots\}$$

where $\xi_0(\omega) \equiv \pi$.

If the transition functions $p_{ij}(n)$ are independent of n (i.e., the Markov chain θ is homogeneous), then it is easy to verify that the elements

$$\hat{\Pi} = (\hat{\pi}_n, \hat{\mathscr{F}}_n, \hat{P}_{\hat{\pi}}), \qquad n = 0, 1, \ldots, \tag{2.217}$$

form a homogeneous Markov process.

The results of Subsection 15.3 imply that the magnitude of the payoff

$$s(\hat{\pi}) = \sup_{\hat{\tau} \in \mathfrak{M}[\hat{\mathscr{F}}]} \hat{M}_{\hat{\pi}}\, g(\hat{\tau}, \hat{\pi}_{\hat{\tau}}) \tag{2.218}$$

at the point $\hat{\pi} \in S$ coincides with $V^{\hat{\pi}} = \sup_{\tau \in \mathfrak{M}[\mathscr{F}]} M^{\hat{\pi}} g(\tau, \pi_\tau^{\hat{\pi}})$.

In this case, if the time $\hat{\tau}_\varepsilon = \inf\{n \geq 0 : \hat{\pi}_n \in B_n\}$ is ε-optimal (in (2.218)) then the time

$$\tau_\varepsilon = \inf\{n \geq 0 : \pi_n^{\hat{\pi}} \in B_n\}$$

will be ε-optimal (in (2.216)) for any given $\hat{\pi}$.

Thus, (2.216) is a particular case of (2.218).

Notes to Chapter 2

2.1. The problem of existence and the construction of optimal stopping times in Bayes decision procedures were investigated first in Wald [105] and [106], Wald and Wolfowitz [104], and Arrow, Blackwell, and Girshick [3]. Soon after these publications and under their influence Snell [97] formulated the general problem of optimal stopping of random processes in discrete time. Chow and Robbins [17], Haggström [54], and Siegmund [95] developed further the results obtained by Snell. The studies in this direction which can be naturally called "martingale techniques" are summarized (for the case of discrete time) in Chow, Robbins, and Siegmund [22].

The investigations made in another direction in the theory of optimal stopping rules are based on the assumption that the observable process is Markov.

Dynkin, who started investigating problems of this kind, considered the payoff $s(x) = \sup \mathrm{M}_x g(x_\tau)$ for nonnegative functions $g(x)$ (assuming $g(x_\infty) = 0$).

The payoffs $s(x)$ and $\bar{s}(x)$ for arbitrary functions $g(x)$ make it convenient to define $g(x_\infty) = \overline{\lim}_{t\to\infty} g(x_t)$; these payoffs were introduced in the first edition of this book.

This Markov process formally fits Snell's scheme; nevertheless, the assumption about Markovianness enables us to obtain more meaningful and constructive results (particularly, for the case of continuous time). On the other hand, it is a well-known fact that any process can be considered a Markov process (by augmenting the states by the past time history). Hence the results of the Markov theory of optimal stopping rules can lead us to pertinent results for arbitrary random sequences as well.

2.2. The equations $s_n(x) = \max\{g(x), T_{s_{n-1}}(x)\}$ appeared in Wald [106], Arrow, Blackwell, and Girshick [3], Bellman [9], and other works in connection with various problems and under various general assumptions. These equations are often referred to as Wald–Bellman equations and can be deduced with the aid of dynamic programming principles ("backward" induction). The proofs of Theorem 1 and Theorem 2 are close to the proofs given in Chow, Robbins, and Siegmund [22], and in Haggström [54].

2.3. The problem of choosing the best object, also known as the "secretary problem," was investigated (in different formulations) by Gardner [42], Dynkin [32], Chow, Moriguti, Robbins, and Samuels [19], Gilbert and Mosteller [46], Gusein–Zade [53], Presman and Sonin [79], and De Groot [25].

2.4. The fact that excessive functions play a definite role in describing the structure of the payoff $s(x)$ was noted first by Dynkin [32]. Lemma 3, which is fundamental in determining the structure of the payoff $s(x)$, follows from the fact that the sequence $(f(x_n), \mathscr{F}_n, P_x)$ forms a supermartingale and $f \in L(A^-)$ (see Meyer [72]). Lemma 4 can be found in Dynkin [32]. Lemma 5, proved in Grigelionis and Shiryayev [49], demonstrates the fact that the smallest excessive majorant $v(x)$ of the function $g(x)$ satisfies the Wald–Bellman equation $v(x) = \max\{g(x), Tv(x)\}$. The technique for constructing the smallest excessive majorants $v(x)$ for (nonnegative) functions $g(x)$ (Lemma 6) was described by A. D. Ventsel. Lemma 7 is, in essence, a variant of the theorem from martingale theory on the transformation of optional sampling (Doob [28], theor. 2.2, chap. 7). The proof of Lemma 8 was taken from Snell [97]. The technique of constructing the smallest excessive majorants given in Lemmas 9 and 11 has been described here for the first time. Similar construction techniques can be found also in Siegmund [95].

2.5. Another proof of the assertion in Theorem 3 that the payoff $s(x)$ is the smallest excessive majorant of $g(x)$ for the case $g(x) \geq 0$ is due to Dynkin [32]. $(\varepsilon, \bar{s})$—optimal times have been considered here for the first time. $(\varepsilon, \bar{s})$—optimality of the time τ_ε (see Theorem 4) for the case $0 \leq g(x) \leq C < \infty$ was proved by Dynkin [32]. (2) and (3) of Theorem 4 are similar to the corresponding results given in Siegmund [95] and in Chow and Robbins [21].

2.6. Examples similar to those given in the present subsection can be found in Dynkin and Yushkevich [33]. Example 7 is given in Haggström [54].

2.7. The payoff structure for functions of the class $g \in B(a^-)$ is discussed in G. Yu. Engelbert [40] and Lazrijeva [65].

2.8–9. $\mathscr{R}$-regular functions in optimal stopping problems were investigated by Chow and Robbins [21], and by Shiryayev [94] (in the first edition of this book). The proofs of Theorems 7 and 9 are in Shiryayev [94] (chap. 2, following the proof of theor. 3), and in G. Yu. Engelbert [38].

2.10. The assertions of Theorem 11 for arbitrary random sequences are due to Chow and Robbins [17], [18], and [21], and Haggström [54].

2.11. The problem of uniqueness of the solution of recursive equations $f(x) = \max\{g(x), Tf(x)\}$ was investigated by Bellman [9], Grigelionis and Shiryayev [49], and Grigelionis [51]. Theorem 15 was proved by Siegmund [95]. Theorems 16 and 17 are due to Lazrijeva [65]. Theorem 18 can be found in G. Yu. Engelbert [40].

2.12. The results discussed in this section were obtained by Ray [80], and by Grigelionis and Shiryayev [48]. Examples illustrating the applications of the truncation criteria can be found in these papers.

2.13. The randomized and sufficient classes of stopping times were discussed in Siegmund [95], Shiryayev [88], Dynkin [34], and Grigelionis [52].

2.14. Functionals of the type (2.169) were investigated in Krylov [60]. The example for the case $\alpha = 1$ given at the end of this section was examined in Chow, Robbins, and Siegmund [22].

2.15. The reduction problem and the properties of transitive statistics were studied in Bahadur [4], Shiryayev [88] and [91], Grigelionis [52], and Chow, Robbins, and Siegmund [22].

Optimal stopping of Markov processes 3

3.1 The statement of the problem and main definitions

3.1.1

As in the case of discrete time, it would be natural to expect that for a wide class of Markov processes with continuous time the payoff permits the excessive or regular characterization (compare with Theorems 2.3, 2.7, and 2.9). This is true; however, to prove it, as well as to investigate the problems of existence and structure of optimal and ε-optimal times, we shall need rather deep results from the general theory of Markov processes and martingales.

Throughout this chapter the Markov process $X = (x_t, \mathscr{F}_t, P_x)$ will be understood as a (homogeneous, nonterminating) standard Markov process with continuous time $t \geq 0$ with values in a state space $(E, \mathscr{B})$ where E is semicompact.

3.1.2

We shall denote by $\mathbb{B}$ the aggregate of almost Borel and lower $\mathscr{C}_0$-continuous[1] functions $g = g(x)$ given on $(E, \mathscr{B})$ taking on values in $(-\infty, \infty]$ such that the process $\{g(x_t), t \geq 0\}$ is separable ([72]).

Let $\overline{\mathfrak{M}}$ be the class of all Markov times $\tau = \tau(\omega)$ (with respect to the system $F = \{\mathscr{F}_t\}, t \geq 0)$, and let $\mathfrak{M}$ be the class of finite Markov times $(P_x\{\tau < \infty\} = 1, x \in E)$, which are referred to as *stopping times.*

[1] The function $g(x)$ is referred to as lower (upper) $\mathscr{C}_0$-continuous if $P_x\{\underline{\lim}_{t\downarrow 0}\, g(x_t) \geq g(x)\} = 1$, $x \in E$ $(P_x\{\overline{\lim}_{t\downarrow 0}\, g(x_t) \leq g(x)\} = 1, x \in E)$.

Let us associate each function $g \in \mathbb{B}$ and Markov time $\tau \in \overline{\mathfrak{M}}$ with the random variable $g(x_\tau)$ by setting

$$g(x_\tau) = \begin{cases} g(x_{\tau(\omega)}(\omega)), & \text{if } \omega \in \{\omega : \tau(\omega) < \infty\}, \\ \overline{\lim}_{t \to \infty} g(x_t(\omega)), & \text{if } \omega \in \{\omega : \tau(\omega) = \infty\}. \end{cases}$$

By analogy with the case of discrete time we shall interpret the variable $g(x_\tau)$ as the gain obtained in the state x_τ with the observations being stopped at time τ. If the mathematical expectation $M_x g(x_\tau)$ is definable, it will naturally be referred to as the *average gain corresponding to the time* τ *and the initial state* $x \in E$.

Let the function $g \in \mathbb{B}$. We shall form the classes of Markov times

$$\overline{\mathfrak{M}}_g = \{\tau \in \overline{\mathfrak{M}} : M_x g^-(x_\tau) < \infty, x \in E\},$$
$$\mathfrak{M}_g = \{\tau \in \mathfrak{M} : M_x g^-(x_\tau) < \infty, x \in E\}$$

and set

$$\bar{s}(x) = \sup_{\tau \in \overline{\mathfrak{M}}_g} M_x g(x_\tau), \tag{3.1}$$

$$s(x) = \sup_{\tau \in \mathfrak{M}_g} M_x g(x_\tau). \tag{3.2}$$

As in the case of discrete time, each of the above functions is referred to as a *payoff*. The time $\tau_\varepsilon \in \mathfrak{M}_g$ is said to be (ε, s)-*optimal* or simply ε-*optimal* if for all $x \in E$

$$s(x) - \varepsilon \le M_x g(x_{\tau_\varepsilon}).$$

The 0-optimal stopping time is said to be *simply optimal*. Similarly, the time $\tau_\varepsilon \in \overline{\mathfrak{M}}_g$ is said to be (ε, $\bar{s}$)-*optimal* or simply ε-*optimal* if for all $x \in E$

$$\bar{s}(x) - \varepsilon \le M_x g(x_{\tau_\varepsilon}).$$

It is seen from the definitions of the payoffs $s(x)$ and $\bar{s}(x)$ that $s(x) \le \bar{s}(x)$. It will be shown that these payoffs coincide: $s(x) \equiv \bar{s}(x)$.

It is useful to note that the payoffs $\bar{s}(x)$ and $s(x)$ will not change if we take the suprema in (3.1) and (3.2) not over the classes $\overline{\mathfrak{M}}_g$ and $\mathfrak{M}_g$ but over the (wider) classes of times from $\overline{\mathfrak{M}}$ and $\mathfrak{M}$ (respectively) for which the mathematical expectations $M_x g(x_\tau)$ are defined for all $x \in E$ (compare with Section 2.1).

3.1.3

The subclasses of functions of the class $\mathbb{B}$ play an essential role in investigating the properties of payoffs and ε-optimal Markov times.

Let us denote by $\mathbb{L}$ the set of all functions $g \in \mathbb{B}$ having the property that for each of them there exists $\tau \in \overline{\mathfrak{M}}_g$ such that $P_x\{\tau > 0\} = 1$, $x \in E$.

The introduction of this class of functions can be easily justified: this is the "reserve" of functions $g(x)$ for which the optimal stopping problems considered are nontrivial in the sense that we can find at least one (nontrivial) time $\tau \in \overline{\mathfrak{M}}_g$ for which $P_x\{\tau > 0\} = 1, x \in E$, and $M_x g(x_\tau) > -\infty$ to compete with the (trivial) time $\tau \equiv 0$.

Further, let us denote by $\mathbb{B}(\mathrm{A}^-)$ and $\mathbb{B}(\mathrm{A}^+)$ the classes of the functions $g \in \mathbb{B}$ for which the conditions

$$\mathrm{A}^- : M_x\left[\sup_{t \geq 0} g^-(x_t)\right] < \infty, \qquad x \in E,$$

and

$$\mathrm{A}^+ : M_x\left[\sup_{t \geq 0} g^+(x_t)\right] < \infty, \qquad x \in E,$$

are (respectively) satisfied.

Also let

$$\mathbb{B}(\mathrm{A}^-, \mathrm{A}^+) = \mathbb{B}(\mathrm{A}^-) \cap \mathbb{B}(\mathrm{A}^+), \qquad \mathbb{L}(\mathrm{A}^-) = \mathbb{L} \cap \mathbb{B}(\mathrm{A}^-),$$
$$\mathbb{L}(\mathrm{A}^+) = \mathbb{L} \cap \mathbb{B}(\mathrm{A}^+), \qquad \mathbb{L}(\mathrm{A}^-, \mathrm{A}^+) = \mathbb{L}(\mathrm{A}^-) \cap \mathbb{L}(\mathrm{A}^+).$$

It is obvious that

$$\mathbb{B}(\mathrm{A}^-) = \mathbb{L}(\mathrm{A}^-) \subseteq \mathbb{L} \subseteq \mathbb{B}. \tag{3.3}$$

In addition to the "trivial" time $\tau \equiv 0$ let us consider the time $\tau \equiv \infty$. If the function $g \in \mathbb{B}$ is such that the condition

$$\mathrm{a}^- : M_x[g^-(x_\infty)] < \infty, \qquad x \in E, \tag{3.4}$$

is satisfied where $g(x_\infty) = \overline{\lim}_{t \to \infty} g(x_t)$, we shall say that the function $g(x)$ *belongs to the class* $\mathbb{B}(\mathrm{a}^-)$. It is clear that

$$\mathbb{B}(\mathrm{A}^-) \subseteq \mathbb{B}(\mathrm{a}^-) \subseteq \mathbb{B}.$$

Next, let $\mathbb{K}$ be some subset of the set $\mathbb{B}$. We denote by $\mathbb{K}_0$ the aggregate of $\mathscr{C}_0$-continuous functions (i.e., the functions which are upper continuous and lower continuous at the same time) from $\mathbb{K}$.

The main objective of this chapter is to study the structure of the payoffs and ε-optimal times for different classes of the functions $g(x)$. The systematic investigation of these questions starts with the investigation of the excessive and regular functions, in terms of which the characterization of the payoffs $s(x)$ and $\bar{s}(x)$ will be expressed.

3.2 Regular and excessive functions: excessive majorants

3.2.1

Let $X = (x_t, \mathscr{F}_t, P_x)$ be a Markov process. If the function $g \in \mathbb{B}$ is such that for any $\tau \in \overline{\mathfrak{M}}_g$

$$M_x g(x_\tau) \leq g(x), \qquad x \in E, \tag{3.5}$$

then it is clear that $\bar{s}(x) = g(x)$, and the optimal stopping rule depends on $\tau_0 \equiv 0$. Therefore, the optimal stopping problem can obviously be solved for the functions $g(x)$ satisfying (3.5). To a certain extent this fact explains the significance of the $\mathfrak{R}$-regular function to be introduced.

Definition 1. Let $\mathfrak{R} \subseteq \overline{\mathfrak{M}}$. The function $f \in \mathbb{B}$ is said to be $\mathfrak{R}$-*regular* if for any $\tau \in \mathfrak{R}$ and $x \in E$ the expectations $M_x f(x_\tau)$ are defined[2] and for all $\sigma, \tau \in \mathfrak{R}$ with $P_x(\sigma \leq \tau) = 1, x \in E$, the inequalities

$$M_x f(x_\tau) \leq M_x f(x_\sigma), \qquad x \in E, \tag{3.6}$$

are satisfied.

The two classes of functions $\mathfrak{R} = \overline{\mathfrak{M}}$ and $\mathfrak{R} = \overline{\mathfrak{M}}_g$ are considered to be crucial from now on. Let us agree to call the $\overline{\mathfrak{M}}$ regular functions *regular*.

Definition 2. The function $f \in \mathbb{B}$ is said to be *excessive* (*for the process X or with respect to the semigroup* $\{T_t\}$, $t \geq 0$) if for all $t \geq 0$ and $x \in E$ the mathematical expectations $T_t f(x)$ $(= M_x f(x_t))$ are defined and

$$T_t f(x) \leq f(x), \qquad x \in E, \qquad t \geq 0. \tag{3.7}$$

Before discussing how these concepts are related (we shall prove in Lemma 1 that any excessive function $f \in \mathbb{B}(A^-)$ is regular) and explaining their roles in optimal stopping times, we shall consider some properties of excessive functions.

I. The function $f(x) \equiv$ const is excessive.
II. If the excessive functions f and g are nonnegative, the function $af + bg$ where a and b are nonnegative constants is also excessive.
III. Let the sequence of excessive functions $\{f_n(x), n \in N\}$ be such that $M_x f_0^-(x_t) < \infty$, $t \geq 0$, and $f_n(x) \leq f_{n+1}(x)$. Then the function $f(x) = \lim_n f_n(x)$ is also excessive.
IV. Let $f(x)$ be an excessive function satisfying (for a given $x^0 \in E$) the condition

$$M_{x^0} f^-(x_t) < \infty, \qquad t \geq 0. \tag{3.8}$$

Then the system $(f(x_t), \mathscr{F}_t, P_{x^0})$ forms (for a given $x \in E$) a generalized supermartingale:

$$M_{x^0}[f(x_t) \mid \mathscr{F}_s] \leq f(x_s) \qquad (P_{x^0}\text{-a.s.}),\ s \leq t. \tag{3.9}$$

If, in addition, for some P_{x^0}-integrable random variable η

$$f(x_t) \geq M_{x^0}(\eta \mid \mathscr{F}_t) \qquad (P_{x^0}\text{-a.s.}),\ t \geq 0, \tag{3.10}$$

[2] This implies, in particular, that the function $f(x_\tau)$ is measurable with respect to that σ-algebra on which measures P_x, $x \in E$, are defined.

then for any $t \in (0, \infty)$ there exists a left limit $\lim_{u \uparrow t} f(x_u)$ (P_{x^0}-a.s.), the process $\{f(x_t), t \geq 0\}$ is right continuous.[3]

V. The excessive function $f(x)$ satisfying (3.10) for all $x^0 \in E$ is $\mathscr{C}_0$-continuous:

$$\lim_{t \downarrow 0} f(x_t) = f(x) \qquad (P_{x^0}\text{-a.s.}),\ x^0 \in E.$$

VI. A necessary and sufficient condition for the almost Borel function $f(x)$ with values in $(-\infty, \infty]$ satisfying (3.10) to be excessive (in the sense of Definition 2) is that (3.7) be satisfied and that for all $x \in E$

$$\lim_{t \downarrow 0} T_t f(x) = f(x). \tag{3.11}$$

VII. If the excessive function $f(x)$ satisfies (3.10), then for any $t \geq 0$ the function $f_t(x) = T_t f(x)$ is also excessive, with

$$T_t f(x) \leq T_s f(x), \qquad s \leq t.$$

VIII. If f and g are excessive functions satisfying (3.8), then the function $f \wedge g = \min(f, g)$ is also excessive.

IX. If f is an excessive function satisfying the condition

$$\sup_{t \geq 0} M_x f^-(x_t) < \infty,$$

then with P_x-probability 1 there exists a (finite or equal to $+\infty$) limit $\lim_{t \to \infty} f(x_t(\omega))$.

(Properties I–III are obvious. For the proof of Property IV, see, for example: [55], §5; [31], Theorems 12.4 and 12.6; [73], chap. 14; and [12], chap. 2. As regards Property VI see [31]. Properties VII and VIII can be verified in an elementary way. Finally, Property IX follows from Theorem 1.9.)

3.2.2

The following lemma, which plays a fundamental role in investigating the properties of the payoffs $s(x)$ and $\bar{s}(x)$ as well as in investigating the case of discrete time, shows, in particular, that any excessive function of the class $\mathbb{B}(A^-)$ is regular.

Lemma 1. *Let the excessive function $f \in \mathbb{B}(A^-)$. Then for any two Markov times σ and τ such that $P_x\{\sigma \leq \tau\} = 1$, $x \in E$, we have the inequalities*

$$M_x[f(x_\tau) | \mathscr{F}_\sigma] \leq f(x_\sigma) \qquad (P_x\text{-a.s.}),\ x \in E. \tag{3.12}$$

[3] Since the excessive function may take on the value $+\infty$, the continuity is to be defined in the topology of the expanded number line.

In particular,

$$M_x f(x_\tau) \le M_x f(x_\sigma) \le f(x), \tag{3.13}$$

i.e., any excessive function satisfying the condition A^- *is regular.*

The proof is analogous to that of Lemma 2.1.

Lemma 2. *If f is a regular function, then for any Markov times σ and τ satisfying the condition $P_x\{\sigma \le \tau\} = 1$, $x \in E$, we have the inequalities*

$$M_x[f(x_\tau)|\mathscr{F}_\sigma] \le f(x_\sigma) \qquad (P_x\text{-a.s.}),\ x \in E. \tag{3.14}$$

PROOF. Suppose (3.14) is not satisfied for any $x \in E$ and Markov times σ and τ. Set

$$A = \{\omega : M_x[f(x_\tau)|\mathscr{F}_\sigma] > f(x_\sigma)]\} \tag{3.15}$$

and

$$\rho = \begin{cases} \tau & \text{if } \omega \in A, \\ \sigma & \text{if } \omega \in \bar{A}. \end{cases}$$

The set $A \in \mathscr{F}_\sigma \subseteq \mathscr{F}_\tau$, and the time $\rho = \tau I_A + \sigma I_{\bar{A}}$ is a Markov time. Let $P_x(A) > 0$. Then

$$\begin{aligned} M_x f(x_\rho) &= M_x I_A f(x_\tau) + M_x I_{\bar{A}} f(x_\sigma) \\ &= M_x I_A M_x[f(x_\tau)|\mathscr{F}_\sigma] + M_x I_{\bar{A}} f(x_\sigma) > M_x f(x_\sigma), \end{aligned}$$

which fact contradicts (3.6) since $P_x\{\sigma \leqslant \rho\} = 1$.

Therefore, $P_x(A) = 0$ for all $x \in E$, thus proving (3.14). □

Remark 1. It follows from the lemma proved that (3.6), a requirement involved in the definition of the regular function, is equivalent to the property given by (3.14).

Remark 2. Lemma 2 remains valid if in defining the regular function we reject the assumption that it is lower $\mathscr{C}_0$-semicontinuous.

3.2.3

Definition 3. The excessive function $f = f(x)$ is referred to as the *excessive majorant of the function $g \in \mathbb{B}$ if $f(x) \ge g(x)$, $x \in E$.*

The function $f = f(x)$ is called the *smallest excessive majorant of the function* $g \in \mathbb{B}$ if $f(x)$ is the excessive majorant, $f(x) \le h(x)$, where $h(x)$ is an arbitrary excessive majorant of the function $g(x)$.

Let the function $g \in \mathbb{B}(\mathrm{A}^-)$. Set

$$Q_n g(x) = \max\{g(x), T_{2^{-n}} g(x)\}. \tag{3.16}$$

Lemma 3. *If $g \in \mathbb{B}(\mathrm{A}^-)$, then the function*

$$v(x) = \lim_n \lim_N Q_n^N g(x), \tag{3.17}$$

where Q_n^N is the Nth power of the operator Q_n, is the smallest excessive majorant of the function $g(x)$.

PROOF. Set $v_n(x) = \lim_N Q_n^N g(x)$. By virtue of Theorem 2.3 and Lemma 2.6

$$v_n(x) = \sup_{\tau \in \mathfrak{R}(n)} M_x g(x_\tau) \tag{3.18}$$

where $\mathfrak{R}(n) \subseteq \mathfrak{M}$ is the class of stopping times with the values $k \cdot 2^{-n}$, $k \in N$, and such that

$$\{\tau = k \cdot 2^{-n}\} \in \sigma\{\omega : x_0, x_{2^{-n}}, \ldots, x_{k \cdot 2^{-n}}\}.$$

Since $\mathfrak{R}(n+1) \supseteq \mathfrak{R}(n)$, $v_{n+1}(x) \geq v_n(x)$. Therefore, $\lim_n v_n(x)$ exists and is denoted by $v(x)$ in (3.17).

It is seen that $v(x) \geq g(x)$, $v_n(x) \geq T_{2^{-n}} v_n(x)$, and for any $m \in N$

$$v_n(x) \geq T_{m \cdot 2^{-n}} v_n(x).$$

Let us take $m = l \cdot 2^{n-k}$, $l \in N$. Then $v_n(x) \geq T_{l \cdot 2^{-k}} v_n(x)$ and

$$v(x) \geq T_{l \cdot 2^{-k}} v(x). \tag{3.19}$$

We shall show that the function $v(x)$ constructed is lower $\mathscr{C}_0$-semicontinuous. To this end we consider the arbitrary function $\varphi \in \mathbb{B}(A^-)$ and $\Phi(x) = M_x \varphi(x_t)$ where t is some fixed number from the interval $(0, \infty)$.

Since $\varphi(x)$ is an almost Borel function, $\Phi(x)$ will be an almost Borel function as well (see, for example, [12], chap. 1, prop. (5.8) and subsect. (10.21)). Further, let τ_n be the time of first entry into some compactum, with $P_x\{\tau_n \downarrow 0\} = 1$. Then by virtue of (1.40)

$$M_x \Phi(x_{\tau_n}) = M_x M_{x_{\tau_n}} \varphi(x_t) = M_x \theta_{\tau_n} \varphi(x_t) = M_x \varphi(x_{\tau_n + t}),$$

and by virtue of Fatou's lemma and the fact that the function $\varphi(x)$ is lower $\mathscr{C}_0$-semicontinuous, we have

$$\varliminf_n M_x \Phi(x_{\tau_n}) = \varliminf_n M_x \varphi(x_{\tau_n + t}) \geq M_x \varliminf_n \varphi(x_{\tau_n + t}) \geq M_x \varphi(x_t) = \Phi(x). \tag{3.20}$$

It is well-known ([31], Theor. 4.9) that an almost Borel function satisfying the inequality $\varliminf_n M_x \Phi(x_{\tau_n}) \geq \Phi(x)$ is lower $\mathscr{C}$-semicontinuous. This implies that each of the almost Borel functions

$$T_{2^{-n}} g(x), \qquad Q_n g(x) = \max\{g(x), T_{2^{-n}} g(x)\},$$
$$v_n(x) = \lim_N Q_n^N g(x), \qquad v(x) = \lim_n v_n(x)$$

is lower $\mathscr{C}_0$-semicontinuous.

We prove next the inequalities

$$v(x) \geq T_t v(x), \qquad t \geq 0. \tag{3.21}$$

Let us take the sequence of binary-rational numbers $r_i \downarrow t$, $i \to \infty$. By using successively (3.19), the fact that the trajectories of the process X are right continuous, the fact that the function $v(x)$ is lower $\mathscr{C}_0$-semicontinuous, the fact that $v(x)$ belongs to the class $\mathbb{B}(A^-)$, and, finally, Fatou's lemma, we get

$$v(x) \geq \varliminf_i T_{r_i} v(x) = \varliminf_i M_x v(x_{r_i}) \geq M_x \varliminf_i v(x_{r_i}) \geq M_x v(x_t) = T_t v(x),$$

which proves (3.21).

Let us assume that $u(x)$ is another excessive majorant of the function $g(x)$. Then it follows from $u(x) \geq g(x)$ that

$$u(x) = Q_n^N u(x) \geq Q_n^N g(x).$$

Hence $u(x) \geq v(x)$ and, therefore, $v(x)$ is the smallest excessive majorant of $g(x)$. This concludes the proof of the lemma. □

Remark 1. Let $g \in \mathbb{B}(A^-)$ and let

$$g^b(x) = \min(b, g(x)), \qquad b \geq 0.$$

Then for the smallest excessive majorant $v(x)$ of the function $g(x)$ we have the representations

$$v(x) = \lim_n \lim_b \lim_N Q_n^N g^b(x) = \lim_n \lim_N \lim_b Q_n^N g^b(x).$$

The proof follows from (2.51) and (3.17).

3.2.4

The lemmas in this subsection provide additional information about the structure of the smallest excessive majorant of the continuous function $g(x)$ in the case where the process X is a Feller process.

Lemma 4. *Let X be a Feller process and let $g(x) \geq C > -\infty$ be continuous. Then its smallest excessive majorant $v(x)$ is a lower semicontinuous function ($\varliminf_{y \to x} v(y) \geq v(x)$).*

PROOF. We can consider without loss of generality that the function $g(x)$ is nonnegative. Since the function $g(x)$ is continuous, each bounded function $g^m(x) = \min(m, g(x))$, $m \in N$, is also continuous. Since the process X is a Feller process the functions $T_t g^m(x)$, $t \geq 0$, are continuous. It follows from this (see the proof of the preceding lemma) that each of the functions $Q_n g^m(x)$, $Q_n^N g^m(x)$ is continuous. Hence the functions

$$v_n^m(x) = \lim_{N \to \infty} Q_n^N g^m(x) \quad \text{and} \quad v^m(x) = \lim_{n \to \infty} v_n^m(x)$$

are lower semicontinuous (as the limit of a monotone increasing sequence of continuous functions).

Since $v^{m+1}(x) \geq v^m(x)$, the function $\tilde{v}(x) = \lim_{m \to \infty} v^m(x)$ is lower semicontinuous (as the limit of the monotone increasing sequence of lower semicontinuous functions).

We need only to show that $\tilde{v}(x) = v(x)$, which can be proved as the analogous relation was in Lemma 3. □

Remark 2. Lemma 4 holds for continuous functions $g \in \mathbb{B}(A^-)$ as well so long as the functions $T_t g^m(x)$ are continuous for each $t \geq 0$ and $m \in N$.

In searching for the smallest excessive majorant of the nonnegative continuous functions $g(x)$ the following construction is useful.

Let

$$\tilde{Q}g(x) = \sup_{t \geq 0} T_t g(x),$$

$$\tilde{Q}^0 g(x) = g(x),$$

and

$$\tilde{Q}^N g(x) = \sup_{t \geq 0} T_t(\tilde{Q}^{N-1} g)(x),$$

where $\tilde{Q}^N$ is the Nth power of $\tilde{Q}$.

Lemma 5. *Let X be a Feller process and let the function $g(x) \geq C > -\infty$ be continuous. Then*

$$v(x) = \lim_{N \to \infty} \tilde{Q}^N g(x) \tag{3.22}$$

is lower semicontinuous and is the smallest excessive majorant of the function $g(x)$.

PROOF. Set $v_N(x) = \tilde{Q}^N g(x)$. Then

$$v_{N+1}(x) = \tilde{Q} v_N(x) = \sup_{t \geq 0} T_t v_N(x) \geq v_N(x) \geq g(x)$$

and for any $t \geq 0$

$$v_{N+1}(x) \geq T_t v_N(x).$$

Since $v_N(x) \uparrow v(x)$, $N \to \infty$,

$$v(x) \geq T_t v(x), \qquad t \geq 0, \tag{3.23}$$

and $v(x) \geq g(x)$.

We shall show that $v(x)$ is lower semicontinuous. Since $g(x)$ is continuous, $g^m(x) = \min(m, g(x))$, $m \in N$, is also continuous. Since the process X is a Feller process the function $T_t g^m(x)$ is continuous for any $t \geq 0$, $m \in N$. It follows from this (as in Lemma 4) that the functions $T_t g(x)$, $t \geq 0$, and $v_1(x) = \tilde{Q}g(x) = \sup_{t \geq 0} T_t g(x)$ are lower semicontinuous. We shall show by induction that each function $v_N(x)$, is also continuous.

Let the function $v_N(x)$ be lower semicontinuous for some $N \geq 1$. We shall prove that the function $v_{N+1}(x)$ is lower semicontinuous as well. To this end we construct the nondecreasing sequence $\{v_N^i(x)\}$, $i = 1, 2, \ldots$, of bounded continuous functions[4] such that

$$v_N^i(x) \uparrow v_N(x), \qquad i \to \infty.$$

Then the functions $T_t v_N^i(x)$ are continuous over x and it follows from the equations

$$v_{N+1}(x) = \sup_{t \geq 0} T_t v_N(x) = \sup_{t \geq 0} \lim_{i \to \infty} T_t v_N^i(x)$$

that $v_{N+1}(x)$ as well as $v(x) = \lim_{N \to \infty} v_N(x)$ are lower semicontinuous. Thus, $v(x) \geq g(x)$, $v(x) \geq T_t v(x)$ and, obviously, if $h(x)$ is some excessive majorant of $g(x)$, then $v(x) = \lim_{N \to \infty} \tilde{Q}^N g(x) \leq h(x)$. Thus, to complete the proof we need only to establish that $\lim_{t \downarrow 0} T_t v(x) = v(x)$. We have from (3.23) that $v(x) \geq \overline{\lim}_{t \downarrow 0} T_t v(x)$. On the other hand, since the function $v(x)$ is lower semicontinuous and the process X has right-continuous ((P_x-a.s.), $x \in E$) trajectories, we get by Fatou's lemma that

$$\underline{\lim}_{t \downarrow 0} T_t v(x) = \underline{\lim}_{t \downarrow 0} M_x v(x_t) \geq M_x \underline{\lim}_{t \downarrow 0} v(x_t) \geq v(x),$$

thus proving the lemma. □

3.2.5

In the case where the function $g \in \mathbb{B}(A^-, A^+)$, we can offer the following method for finding its smallest excessive majorant.

Let

$$\varphi(x) = M_x \left[\sup_{t \geq 0} g(x_t) \right], \qquad \varphi_n(x) = M_x \left[\sup_{k \in N} g(x_{k \cdot 2^{-n}}) \right].$$

If $f \in \mathbb{B}(A^-, A^+)$, we set (compare with Subsection 2.4.7)

$$G_n f(x) = \max\{g(x), T_{2^{-n}} f(x)\}, \qquad n \in N, \tag{3.24}$$

and let G_n^N be the Nth power of the operator G_n, $G_n^0 f = f$. We shall note that if $f(x) = g(x)$, then $G_n g(x) = Q_n g(x)$.

Lemma 6. *If the function $g \in \mathbb{B}(A^-, A^+)$, then its excessive majorant*

$$v(x) = \lim_{n \to \infty} \lim_{N \to \infty} G_n^N \varphi_n(x). \tag{3.25}$$

PROOF. Let $\tilde{v}_n(x) = \lim_{N \to \infty} G_n^N \varphi_n(x)$. According to Lemma 2.11, $\tilde{v}_n(x)$ coincides with the function $v_n(x)$ defined in (3.18). Applying Lemma 3 we shall obtain the required assertion, (3.25).

[4] For the proof that such a construction is possible see, for example, [47], chap. 7, Theorem 30, or [76], chap. 15, Theorem 10.

3.2.6

Lemma 7. *Let $f(x)$ be the excessive function satisfying the condition* A$^-$, *and let*

$$\sigma_B = \inf\{t > 0 : x_t \in B\}, \tag{3.26}$$

where B is an almost Borel set. Then the function

$$f_B(x) = M_x f(x_{\sigma_B}) \tag{3.27}$$

is excessive.

PROOF. Let $s \geq 0$ and let

$$\sigma_B^s = \inf\{t > s : x_t \in B\}. \tag{3.28}$$

It follows from Theorem 1.5 that the times σ_B^s are Markov. Further, as in the proof of Lemma 2.4, we can prove here that

$$T_t f_B(x) \leq f_B(x), \qquad x \in E, \qquad t \geq 0. \tag{3.29}$$

The function $f_B(x)$ is almost Borel (compare with the function $\Phi(x)$ in Lemma 3). Since $\sigma_B^s \downarrow \sigma_B$ for $s \downarrow 0$, Fatou's lemma and the fact that the process $f(x_t)$ is right continuous (see Property IV) yield

$$\varliminf_{t \downarrow 0} T_t f_B(x) = \varliminf_{t \downarrow 0} M_x f(x_{\sigma_B^t}) \geq M_x \varliminf_{t \downarrow 0} f(x_{\sigma_B^t}) = M_x f(x_{\sigma_B}) = f_B(x).$$

This, together with (3.29), leads us to the relation

$$\lim_{t \downarrow 0} T_t f_B(x) = f_B(x). \tag{3.30}$$

The lemma follows immediately from Property VI, and (3.29) and (3.30). □

Remark 3. For the times

$$\tau_B = \inf\{t \geq 0 : x_t \in B\}$$

the lemma is not, in general, true (for the times $\tau_B^s = \inf\{t \geq s : x_t \in B\}$ in the general case $\tau_B^s \nrightarrow \tau_B^0$, $s \downarrow 0$).

3.2.7

Let the function $g \in \mathbb{B}(\mathrm{A}^+)$ and let $v(x)$ be its smallest excessive majorant (known to exist if, in addition, $g \in \mathbb{B}(\mathrm{A}^-)$). For $\varepsilon \geq 0$ set

$$\tau_\varepsilon = \inf\{t \geq 0 : v(x_t) \leq g(x_t) + \varepsilon\}. \tag{3.31}$$

Lemma 8. *If the function $g \in \mathbb{B}(\mathrm{A}^+)$ and $v(x)$ is its smallest excessive majorant, then*

$$\varlimsup_{t \to \infty} v(x_t) = \varlimsup_{t \to \infty} g(x_t) \qquad (P_x\text{-a.s.}),\ x \in E \tag{3.32}$$

and for any $\varepsilon > 0$

$$P_x\{\tau_\varepsilon < \infty\} = 1, \qquad x \in E. \tag{3.33}$$

The proof is similar to that of Lemma 2.8.

3.3 Excessive characterization of the payoff and ε-optimal stopping times (under the condition A^-)

3.3.1

Theorem 1. *Let* $X = (x_t, \mathscr{F}_t, P_x)$, $t \geq 0$, *be a Markov process and let the function* $g \in \mathbb{B}(A^-)$. *Then*:

(1) *The payoff* $s(x)$ *is the smallest excessive majorant of the function* $g(x)$;
(2) $s(x) = \bar{s}(x)$.

PROOF. Let $v(x)$ be the smallest excessive majorant of the function $g(x)$, the existence of which is guaranteed by the assumption $g \in \mathbb{B}(A^-)$ and Lemma 3. By virtue of Lemma 1, for any $\tau \in \overline{\mathfrak{M}}$

$$M_x g(x_\tau) \leq M_x v(x_\tau) \leq v(x)$$

and, therefore,

$$s(x) \leq \bar{s}(x) \leq v(x). \tag{3.34}$$

Next, by making use of the notation of Lemma 3, we find that since $\mathfrak{N}(n) \subseteq \mathfrak{M}$,

$$v_n(x) \leq s(x).$$

But $v(x) = \lim_n v_n(x)$ and, therefore, $v(x) \leq s(x)$. This, together with (3.34), leads to the equations

$$s(x) = \bar{s}(x) = v(x),$$

thus proving the theorem. □

Corollary. *Let the time* $\bar{\tau} \in \overline{\mathfrak{M}}$ *be such that the associated gain* $f(x) = M_x g(x_{\bar{\tau}})$ *is an excessive function and* $f(x) \geq g(x)$. *Then* $f(x) = \bar{s}(x)$ *and the time* $\bar{\tau}$ *is* $(0, \bar{s})$*-optimal. If, in addition,* $\bar{\tau} \in \mathfrak{M}$, *this time will be an optimal stopping time.*

To illustrate this we consider the following example.[5] Let $W = (w_t, \mathscr{F}_t, P_x)$, $t \geq 0$, $x \in R$, be a Wiener process with $P_x\{w_0 = x\} = 1$, where

$$M_x[w_{t+s} - w_t] = \mu s, \qquad D_x[w_{t+s} - w_t] = s.$$

Take $g(x) = \max(0, x)$.

[5] Compare with Example 1 in Section 2.6.

It can be readily seen that for $\mu \geq 0$, $\bar{s}(x) \equiv +\infty$. The time $\bar{\tau}(\omega) \equiv +\infty$ is $(0, \bar{s})$-optimal.

Assume that $\mu < 0$. Let $\tau_\gamma = \inf\{t \geq 0 : w_t \in \Gamma_\gamma\}$, $\Gamma_\gamma = [\gamma, \infty)$. As in Example 1 in Section 2.6, we can show here that $\tau_\gamma \in \overline{\mathfrak{M}}$ and

$$f_\gamma(x) = M_x g(x_{\tau_\gamma}) = \begin{cases} \gamma e^{2\mu(\gamma - x)}, & x \leq \gamma, \\ x, & x > \gamma. \end{cases}$$

Setting $f^*(x) = \sup_\gamma f_\gamma(x)$ we find that $f^*(x) = f_{\gamma^*}(x)$ where $\gamma^* = -\frac{1}{2}\mu$. It is clear that $f_{\gamma^*}(x) \geq g(x)$, and it is immediate that $f(x) \geq T_t f(x)$ for $t \geq 0$. It follows from the Corollary to Theorem 1 that the time $\tau_{\gamma^*} = \inf\{t \geq 0 : w_t \in \Gamma_{\gamma^*}\}$ is $(0, \bar{s})$-optimal. It is interesting to note that $P_x\{\tau_{\gamma^*} = \infty\} > 0$ for all $x < \gamma^*$ so that the $(0, \bar{s})$-optimal time τ_{γ^*} is not a stopping time.

3.3.2

The disadvantage of the method used for proving Theorem 1 as compared with that used for proving Theorem 2.3 consists in the fact that it suggests no technique for constructing ε-optimal stopping times.

Let us recall the fact that the proof of Theorem 2.3 was based essentially on the relation $v(x) = M_x v(x_{\tau_\varepsilon})$ (see (2.71)) for the smallest excessive majorant $v(x)$ of the function $g(x)$. This enabled us to assert that the times τ_ε are (under assumption A^+) ε-optimal stopping times ($\varepsilon > 0$).

Hence it is natural to find the conditions under which this relation holds in the case of continuous time as well.

Theorem 2. *Let the function $g \in \mathbb{B}(A^-, A^+)$. Then for any $\varepsilon > 0$*

$$s(x) = M_x s(x_{\tau_\varepsilon}), \tag{3.35}$$

where

$$\tau_\varepsilon = \inf\{t \geq 0 : s(x_t) \leq g(x_t) + \varepsilon\}. \tag{3.36}$$

PROOF. We shall note first that by virtue of Lemma 8 the time τ_ε is (for $\varepsilon > 0$) a stopping time. Next, by virtue of Theorem 1 and Lemma 1

$$M_x s(x_{\tau_\varepsilon}) \leq s(x). \tag{3.37}$$

Hence it suffices to prove the inverse inequality. This follows from Lemmas 9 and 10 below. □

To formulate these lemmas we introduce some concepts and notations.

We shall say that the stopping time τ *belongs to the class* $\mathfrak{M}(x; \delta, \varepsilon)$ with $\delta \geq 0$ and $\varepsilon \geq 0$, if

$$P_x\{\tau < \tau_\varepsilon\} \leq \delta, \tag{3.38}$$

where τ_ε is the Markov time defined in (3.36).

Also, let $\mathfrak{M}(\delta, \varepsilon) = \bigcap_x \mathfrak{M}(x; \delta, \varepsilon)$.

Lemma 9. *If $g \in \mathbb{B}(A^-)$, then for any pair (δ, ε) with $\delta > 0$ and $\varepsilon > 0$*

$$s(x) = \sup_{\tau \in \mathfrak{M}(\delta, \varepsilon)} M_x g(x_\tau). \tag{3.39}$$

PROOF. On the set $\{\omega : \tau(\omega) < \tau_\varepsilon(\omega)\}$, $g(x_\tau) < s(x_\tau) - \varepsilon$. Hence, since $g(x_\tau) \le s(x_\tau)$, we have

$$\begin{aligned} M_x g(x_\tau) &= M_x I_{\{\tau < \tau_\varepsilon\}} g(x_\tau) + M_x I_{\{\tau \ge \tau_\varepsilon\}} g(x_\tau) \\ &= M_x I_{\{\tau < \tau_\varepsilon\}} [s(x_\tau) - \varepsilon] + M_x I_{\{\tau \ge \tau_\varepsilon\}} s(x_\tau) \\ &= M_x s(x_\tau) - \varepsilon P_x\{\tau < \tau_\varepsilon\}, \end{aligned}$$

which together with Lemma 1 leads to

$$M_x g(x_\tau) \le s(x) - \varepsilon P_x\{\tau < \tau_\varepsilon\}.$$

By the definition of the class $\mathfrak{M}(\delta, \varepsilon)$ we find from the above that

$$\sup_{\tau \in \mathfrak{M} - \mathfrak{M}(\delta, \varepsilon)} M_x g(x_\tau) \le s(x) - \varepsilon\delta,$$

which (with $\delta > 0$ and $\varepsilon > 0$) is equivalent to (3.39). □

Lemma 10. *Let $g \in \mathbb{B}(A^-, A^+)$. Then for any $\varepsilon > 0$*

$$s(x) = \sup_{\{\tau \in \mathfrak{M} : \tau \ge \tau_\varepsilon\}} M_x g(x_\tau) \tag{3.40}$$

and

$$s(x) \le M_x s(x_{\tau_\varepsilon}). \tag{3.41}$$

PROOF. Let us fix the point $x^0 \in E$ and set $\delta_n = 1/2^n$. By virtue of Lemma 9 it is possible to find the sequence of times $\tau_n \in \mathfrak{M}(x^0; \delta_n, \varepsilon)$, $n \in N$, such that the $M_{x^0} g(x_{\tau_n})$ do not decrease and

$$s(x^0) = \lim_{n \to \infty} M_{x^0} g(x_{\tau_n}). \tag{3.42}$$

Set $\sigma_n = \max(\tau_\varepsilon, \tau_n)$; by showing that

$$\underline{\lim}_{n \to \infty} M_{x^0} g(x_{\sigma_n}) \ge s(x^0), \tag{3.43}$$

(3.40) follows from the inequality $\sigma_n \ge \tau_\varepsilon$.

We have

$$\begin{aligned} \underline{\lim}_n M_{x^0} g(x_{\sigma_n}) &= \underline{\lim}_n M_{x^0} I_{\{\tau_n \ge \tau_\varepsilon\}} g(x_{\tau_n}) + \underline{\lim}_n M_{x^0} I_{\{\tau_n < \tau_\varepsilon\}} g(x_{\tau_\varepsilon}) \\ &= \underline{\lim}_n M_{x^0} g(x_{\tau_n}) - \overline{\lim}_n M_{x^0} I_{\{\tau_n < \tau_\varepsilon\}} g(x_{\tau_n}) \\ &\quad + \underline{\lim}_n M_{x^0} I_{\{\tau_n < \tau_\varepsilon\}} g(x_{\tau_n}) \\ &\ge s(x^0) - M_{x^0} \overline{\lim}_n I_{\{\tau_n < \tau_\varepsilon\}} g(x_{\tau_n}) + M_{x^0} \underline{\lim}_n I_{\{\tau_n < \tau_\varepsilon\}} g(x_{\tau_\varepsilon}). \end{aligned} \tag{3.44}$$

We have used here (3.42) and Fatou's lemma, whose applicability is guaranteed by the assumption $g \in \mathbb{B}(A^-, A^+)$. Since $P_{x^0}\{\tau_n < \tau_\varepsilon\} \le 2^{-n}$, by the Borel–Cantelli lemma we find that with P_{x^0}-probability 1

$$\lim_n I_{\{\tau_n < \tau_\varepsilon\}}(\omega) = 0.$$

Because $g \in \mathbb{B}(A^-, A^+)$ the two last terms in (3.44) vanish. This proves (3.43), from which in turn follows (3.40).

To prove (3.41) we note that by virtue of (3.40) and the obvious inequality $g(x_\tau) \le s(x_\tau)$

$$s(x) = \sup_{\{\tau \in \mathfrak{M}: \tau \ge \tau_\varepsilon\}} M_x g(x_\tau) \le \sup_{\{\tau \in \mathfrak{M}: \tau \ge \tau_\varepsilon\}} M_x s(x_\tau).$$

But $M_x s(x_\tau) \le M_x s(x_{\tau_\varepsilon})$ (Theorem 1 and Lemma 1), hence $s(x) \le M_x s(x_{\tau_\varepsilon})$. □

3.3.3

Theorem 3. *Let the function* $g \in \mathbb{B}_0(A^-, A^+)$.[6] *Then*:

(1) *For any* $\varepsilon > 0$ *the times* τ_ε *are* ε*-optimal stopping times;*
(2) *If the function* $g(x)$ *is upper semicontinuous, the time* τ_0 *is an optimal Markov time;*
(3) *If the function* $g(x)$ *is upper semicontinuous and* $P_x\{\tau_0 < \infty\} = 1$, $x \in E$, *the time* τ_0 *is an optimal stopping time;*
(4) *If in the class* $\overline{\mathfrak{M}}$ *(in* $\mathfrak{M}$*) there exists an optimal time* τ^*, *then* $P_x\{\tau_0 \le \tau^*\} = 1$, $x \in E$, *and the time* τ_0 *is optimal in the class* $\overline{\mathfrak{M}}$ *(in* $\mathfrak{M}$*).*

PROOF.

(1) The fact that the time τ_ε, $\varepsilon > 0$, belongs to the class $\mathfrak{M}$ follows from Lemma 8. Further, according to our assumption the function $g(x)$ is $\mathscr{C}_0$-continuous. The payoff $s(x)$ is also $\mathscr{C}_0$-continuous (by virtue of Theorem 1 and by Property V of excessive functions). Hence it follows from the fact that the trajectories of the process $\{x_t, t \ge 0\}$ are right continuous that the processes $\{s(x_t), t \ge 0\}$ and $\{g(x_t), t \ge 0\}$ are right continuous as well. This implies, in turn (see (3.36)), that

$$s(x_{\tau_\varepsilon}) \le g(x_{\tau_\varepsilon}) + \varepsilon \qquad (P_x\text{-a.s.}),\ x \in E. \tag{3.45}$$

Hence by virtue of Theorem 2

$$s(x) = M_x s(x_{\tau_\varepsilon}) \le M_x g(x_{\tau_\varepsilon}) + \varepsilon, \qquad x \in E, \tag{3.46}$$

thus proving (1).

(2) We note that if $\varepsilon_1 \ge \varepsilon_2 > 0$, then $P_x\{\tau_{\varepsilon_1} \le \tau_{\varepsilon_2}\} = 1$ and, therefore, $\lim_{\varepsilon \downarrow 0} \tau_\varepsilon$ exists (P_x-a.s.), $x \in E$. Set $\tau^* = \lim_{\varepsilon \downarrow 0} \tau_\varepsilon$.

[6] We recall the fact that the class $\mathbb{B}_0(A^-, A^+)$ consists of $\mathscr{C}_0$-continuous functions of the class $\mathbb{B}(A^-, A^+)$.

The time τ^* is Markov and, obviously, $\tau^* \leq \tau_0$ (P_x-a.s.), $x \in E$. By virtue of Theorem 1, (3.46), Fatou's lemma, the upper semicontinuity of $g(x)$, and the left quasicontinuity of the process $\{x_\tau, t \geq 0\}$

$$\begin{aligned}
\bar{s}(x) &= s(x) \\
&\leq \overline{\lim_{\varepsilon \downarrow 0}} M_x g(x_{\tau_\varepsilon}) \\
&\leq M_x \overline{\lim_{\varepsilon \downarrow 0}} g(x_{\tau_\varepsilon}) \\
&= M_x I_{\{\tau^* < \infty\}} \overline{\lim_{\varepsilon \downarrow 0}} g(x_{\tau_\varepsilon}) + M_x I_{\{\tau^* = \infty\}} \overline{\lim_{\varepsilon \downarrow 0}} g(x_{\tau_\varepsilon}) \\
&\leq M_x I_{\{\tau^* < \infty\}} g(x_{\tau^*}) + M_x I_{\{\tau^* = \infty\}} g(x_\infty) = M_x g(x_{\tau^*}),
\end{aligned} \tag{3.47}$$

i.e., the time τ^* is $(0, \bar{s})$-optimal. Therefore, it remains only to show that $\tau^* = \tau_0$ (P_x-a.s.), $x \in E$.

It is seen that $P_x\{\tau^* \leq \tau_0\} = 1, x \in E$. By the definition of the time τ_0, in order to prove $P_x\{\tau^* = \tau_0\} = 1, x \in E$, it suffices to show that $P_x\{s(x_{\tau^*}) = g(x_{\tau^*})\} = 1, x \in E$.

From the inequality $s(x_{\tau^*}) \geq g(x_{\tau^*})$, the inequality $s(x) \geq M_x s(x_{\tau^*})$ and the fact that the time τ^* is optimal (see (3.47)) we find

$$M_x s(x_{\tau^*}) \geq M_x g(x_{\tau^*}) = s(x) \geq M_x s(x_{\tau^*}). \tag{3.48}$$

$$s(x) = M_x s(x_{\tau^*}) = M_x g(x_{\tau^*}), \tag{3.49}$$

which together with the inequality $s(x_{\tau^*}) \geq g(x_{\tau^*})\}$ proves the required relation $P_x\{s(x_{\tau^*}) = g(x_{\tau^*})\} = 1, x \in E$.

(3) This assertion follows immediately from the previous one.

(4) Since $s(x_{\tau^*}) = g(x_{\tau^*})$, and since $s(x_{\tau_0}) = g(x_{\tau_0})$ (P_x-a.s.), $x \in E$, by (3.32) and the definition of τ_0, it follows that $P_x\{\tau_0 \leq \tau^*\} = 1, x \in E$. As can be easily seen, (3.48) holds for the optimal time τ^* and, therefore, by virtue of Theorem 1 and Lemma 1

$$\bar{s}(x) = s(x) = M_x g(x_{\tau^*}) = M_x s(x_{\tau^*}) \leq M_x s(x_{\tau_0}) = M_x g(x_{\tau_0}).$$

This proves the optimality of the time τ_0 (in the class $\overline{\mathfrak{M}}$ if $\tau^* \in \overline{\mathfrak{M}}$, and in the class $\mathfrak{M}$ if $\tau^* \in \mathfrak{M}$).

Corollary 1. *If $g \in \mathbb{B}_0(\mathrm{A}^-, \mathrm{A}^+)$ and is upper continuous,*

$$s(x) = M_x s(x_{\tau_0}). \tag{3.50}$$

The proof follows from (3.49) and the fact that the times τ^* and τ_0 coincide (see the proof in Subsection 3.2).

Corollary 2. *Let the function $g \in \mathbb{B}(\mathrm{A}^-, \mathrm{A}^+)$. Denote by $\hat{\mathfrak{C}} \subseteq \mathfrak{M}$ the class of stopping times $\tau_{\hat{C}} = \inf\{t \geq 0 : x_t \in \hat{C}\}$ where $\hat{C}$ is almost Borel. Then*

$$s(x) = \sup_{\tau \in \hat{\mathfrak{C}}} M_x g(x_\tau). \tag{3.51}$$

In other words, the value of the payoff $s(x)$ will not change if, instead of the class $\mathfrak{M}$, we consider only those stopping times which are times of first entry into almost Borel sets.

Corollary 3. *Let the function $g \in \mathbb{B}_0(A^-, A^+)$ be continuous, and let the payoff $s(x)$ be lower continuous (by Lemma 5 it is sufficient that the process X be a Feller process and that $g(x) \geq C > -\infty$). Denote by $\mathfrak{D} \subseteq \mathfrak{M}$ the class of stopping times of the form $\tau_D = \inf\{t \geq 0 : x_t \in D\}$ where D is closed. Then*

$$s(x) = \sup_{\tau \in \mathfrak{D}} M_x g(x_\tau). \tag{3.52}$$

For a proof we note that under our assumptions the sets $\{x : s(x) \leq g(x) + \varepsilon\}$, $\varepsilon > 0$, are closed.

3.4 Regular characterization of the payoff and ε-optimal stopping times (under the condition A^+)

3.4.1

If the function $g \in \mathbb{B}(A^-)$, the payoff $s(x)$ is the smallest excessive majorant of $g(x)$ (Theorem 1). Since in the case where the condition A^- is satisfied the classes of excessive and regular majorants of $g(x)$ coincide (Lemma 1), (1) of Theorem 1 can be reformulated as follows: If $g \in \mathbb{B}(A^-)$, the payoff $s(x)$ is the smallest regular majorant of $g(x)$. It turns out that the regular characterization of the payoff is preferable to the excessive characterization for the case of functions $g \in \mathbb{L}_0(A^+)$. In this regard it should be mentioned that the excessive functions are not, in general, regular if the condition A^- is dropped. It is not difficult to construct the pertinent example by generalizing Example 7 in Section 2.6 to the case of continuous time.

3.4.2

Recall that in considering problems of this type for the case of discrete time (Section 2.8) the following technique was used: first, we introduced along with the function $g(x)$ the functions $g_a(x) = \max\{a, g(x)\}$ where $a \leq 0$ ("cut" below), and the payoffs $s_a(x) = \sup_{\tau \in \mathfrak{M}} M_x g_a(x_\tau)$; second, we showed that $s_*(x) = \lim_{a \to -\infty} s_a(x)$ is regular and that $s(x) = \bar{s}(x) = s_*(x)$.

A similar method of investigating the structure of the payoff $s(x)$ will be used in the case of continuous time. The key point in proving the basic result of this subsection (Theorem 5) is:

Theorem 4. *Let $g \in \mathbb{L}_0(A^+)$. Then the function $s_*(x)$ is $\mathscr{C}_0$-continuous.*

3.4.3

Before proving this theorem we shall prove some useful lemmas.

Lemma 11. *Let $g \in \mathbb{B}(A^+)$. Then for each $a \leq 0$ and $x \in E$ the process $(s_a(x_t), \mathscr{F}_t, P_x)$, $t \geq 0$, is a right continuous uniformly integrable supermartingale.*[7]

PROOF. If $\tau \in \mathfrak{M}$, then

$$M_x g_a(x_\tau) \leq M_x \sup_{s \geq 0} g_a(x_s) \leq M_x \sup_{s \geq 0} g^+(x_s)$$

and

$$s_a(x) = \sup_{\tau \in \mathfrak{M}} M_x g_a(x_\tau) \leq M_x \sup_{s \geq 0} g^+(x_s).$$

We obtain from this, by virtue of the Markov property,

$$s_a(x_t) \leq M_{x_t} \sup_{s \geq 0} g^+(x_s) \leq M_x \left[\sup_{s \geq 0} g^+(x_s) | \mathscr{F}_t \right]. \tag{3.53}$$

By assumption, $g \in \mathbb{B}(A^+)$ and $Y = \sup_{s \geq 0} g^+(x_s)$ are integrable and the values $\{Y_t = M_x(Y | \mathscr{F}_t), t \geq 0\}$ are uniformly integrable. Since

$$a \leq s_a(x_t) \leq Y_t,$$

the family of random variables $\{s_a(x_t), t \geq 0\}$ is uniformly integrable.

Further, $g_a \in \mathbb{B}(A^-)$. By Theorem 1 the function $s_a(x)$ is the smallest excessive majorant of $g_a(x)$. Therefore, the function $s_a(x)$ is $\mathscr{C}_0$-continuous and the process $\{s_a(x_t), t \geq 0\}$ has right continuous trajectories. From the excessiveness of the function $s_a(x)$ and the Markovianness (P_x-a.s.), $x \in E$, we obtain

$$M_x[s_a(x_{t+u}) | \mathscr{F}_t] = M_{x_t} s_a(x_u) \leq s_a(x_t). \tag{3.54}$$

Thus, we have proved the lemma for all t, $0 \leq t < \infty$. The validity of the lemma for $0 \leq t \leq \infty$ follows from (3.54), the uniform integrability of the values $\{s_a(x_t), 0 \leq t < \infty\}$, and Theorem 1.9. □

Lemma 12. *Let $g \in \mathbb{B}(A^+)$. Then (P_x-a.s.), $x \in E$,*

$$\overline{\lim_t}\, g(x_t) = \overline{\lim_t}\, s(x_t) = \overline{\lim_t}\, s_*(x_t) = \lim_a \overline{\lim_t}\, s_a(x_t). \tag{3.55}$$

PROOF. By virtue of Lemma 8

$$\overline{\lim_t}\, s_a(x_t) = \overline{\lim_t}\, g_a(x_t).$$

Hence

$$\lim_a \overline{\lim_t}\, s_a(x_t) = \lim_a \overline{\lim_t}\, g_a(x_t) = \overline{\lim_t}\, g(x_t). \tag{3.56}$$

[7] The value $s_a(x_\infty)$ is understood to be $\lim_{t \to \infty} s_a(x_t)$, which exists by virtue of Theorems 1 and 1.9.

But

$$g(x) \leq s(x) \leq s_*(x) \leq s_a(x),$$

which together with (3.56) proves (3.55). □

Lemma 13. *Let $g \in \mathbb{B}(A^+)$. Then for any Markov times σ and τ with the property $P_x\{\sigma \leq \tau\} = 1$, $x \in E$, we have the inequality*

$$M_x[s_*(x_\tau)|\mathscr{F}_\sigma] \leq s_*(x_\sigma) \qquad (P_x\text{-a.s.}),\ x \in F. \tag{3.57}$$

PROOF. We note first that the function $s_*(x)$, being the limit of almost Borel functions $s_a(x)$, is almost Borel. The process $\{g(x_t), t \geq 0\}$ is separable and, therefore, by virtue of (3.55) the value $\overline{\lim}_t s_*(x_t)$ is measurable. It follows from the two above facts that for each Markov time τ the value $s_*(x_\tau)$ is measurable. Finally, by virtue of the condition $g \in \mathbb{B}(A^+)$ and the inequality

$$s_*(x_\tau) \leq s_a(x_\tau) \leq \sup_{t \geq 0} g^+(x_t) \tag{3.58}$$

it follows that the mathematical expectations $M_x s_*(x_\tau)$ and $M_x[s_*(x_\tau)|\mathscr{F}_\sigma]$ are defined.

According to Remark 2 to Lemma 2, to prove (3.57) it suffices to show that for any Markov times σ and τ with the property $P_x\{\sigma \leq \tau\} = 1$, $x \in E$,

$$M_x s_*(x_\tau) \leq M_x s_*(x_\sigma). \tag{3.59}$$

By virtue of Theorem 1 and Lemma 1

$$M_x s_a(x_\tau) \leq M_x s_a(x_\sigma). \tag{3.60}$$

By the Theorem on monotone convergence (whose validity follows from (3.58) and the assumption $g \in \mathbb{B}(A^+)$) (3.59) follows if we take into consideration the fact that by Lemma 12, $\lim_a \overline{\lim}_t s_a(x_t) = \overline{\lim}_t (s_*(x_t))$. □

Lemma 14. *Let $g \in \mathbb{B}(A^+)$. Then for the Markov time $\rho \in \overline{\mathfrak{M}}_g$ the system $(s_*(x_{\rho \wedge t}), \mathscr{F}_{\rho \wedge t}, P_x)$, $t \geq 0$, forms a supermartingale.*

PROOF. Let $s \leq t$. Set $\sigma = \rho \wedge s$, $\tau = \rho \wedge t$. By (3.57)

$$M_x[s_*(x_{\rho \wedge t})|\mathscr{F}_{\rho \wedge t}] \leq s_*(x_{\rho \wedge s}). \tag{3.61}$$

It is seen that

$$M_x s_*(x_{\rho \wedge t}) \leq M_x s_a(x_{\rho \wedge t}) \leq M_x \sup_{s \geq 0} g^+(x_s) < \infty,$$

and by (3.56) and the assumption $\rho \in \overline{\mathfrak{M}}_g$

$$-\infty < -M_x g^-(x_\rho) \leq M_x s_*(x_\rho) \leq M_x s_*(x_{\rho \wedge t}).$$

Therefore, $M_x|s_*(x_{\rho \wedge t})| < \infty$, $t \geq 0$. □

Lemma 15. *Let $g \in \mathbb{L}(A^+)$ and let S be a countable everywhere dense set in $(0, +\infty)$. Then for $r \downarrow 0$ the sequence $\{s_*(x_r)\}$, $r \in S$, converges (P_x-a.s.), $x \in E$, to the integrable random variable* $(\lim_{r \downarrow 0, r \in S} s_*(x_r))$.

PROOF. Let $\rho \in \overline{\mathfrak{M}}_g$. By virtue of [72], chap. 5, Theorem 21, $\lim_{r\downarrow 0, r\in S} s_*(x_{\rho\wedge r})$ exists for the supermartingale $(s_*(x_{\rho\wedge r}), \mathscr{F}_{\rho\wedge r}, P_x)$, $r \in S$, and

$$M_x\left|\lim_{r\downarrow 0, r\in S} s_*(x_{\rho\wedge t})\right| < \infty, \qquad x \in E. \tag{3.62}$$

By the hypothesis of the lemma, $g \in \mathbb{L}(\mathrm{A}^+)$ and, therefore, there exists a time $\rho \in \overline{\mathfrak{M}}_g$ such that $P_x\{\rho > 0\} = 1$, $x \in E$. Therefore, for this time

$$\lim_{r\downarrow 0, r\in S} s_*(x_{\rho\wedge\tau}) = \lim_{r\downarrow 0, r\in S} s_*(x_r) \qquad (P_x\text{-a.s.}), x \in E.$$

This, together with (3.62), proves the lemma. □

Lemma 16. *Let $g \in \mathbb{L}(\mathrm{A}^+)$ and let S be a countable everywhere dense set in $(0, +\infty)$. Then*

$$\lim_{r\downarrow 0, r\in S} s_*(x_r) = \lim_{t\downarrow 0} s_*(x_t) \qquad (P_x\text{-a.s.}), x \in E. \tag{3.63}$$

PROOF. Since the processes $\{s_a(x_t), t \geq 0\}$ are right continuous they are separable. Therefore, if I is an open interval, then $(P_x\text{-a.s.})$, $x \in E$,

$$\begin{aligned}\inf_{t\in I} s_*(x_t) &= \inf_{t\in I}\inf_a s_a(x_t)\\ &= \inf_a \inf_{t\in I} s_a(x_t)\\ &= \inf_a \inf_{t\in I\cap S} s_a(x_t) = \inf_{t\in I\cap S} s_*(x_t).\end{aligned}$$

Set $I = (0, 1/k)$. Then $(P_x\text{-a.s.})$, $x \in E$,

$$\underline{\lim}_{t\downarrow 0}\, s_*(x_t) = \lim_{k\to\infty}\inf_{0<t<1/k} s_*(x_t) = \lim_{k\to\infty}\inf_{\substack{0<r<1/k\\ r\in S}} s_*(x_r) = \underline{\lim}_{\substack{r\downarrow 0\\ r\in S}}\, s_*(x_r).$$

3.3.4

PROOF OF THEOREM 4. By Lemma 1 and Theorem 1, for $\rho \in \mathfrak{M}_g$

$$M_x[s_a(x_{\rho\wedge t})|\mathscr{F}_t] \leq s_a(x_t). \tag{3.64}$$

Hence, because $I_{\{\rho>t\}}$ is $\mathscr{F}_t$-measurable,

$$M_x I_{\{\rho>t\}} s_a(x_\rho) \leq M_x I_{\{\rho>t\}} s_a(x_t).$$

Therefore

$$\begin{aligned}M_x g_a(x_\rho) &= M_x I_{\{\rho\leq t\}} g_a(x_\rho) + M_x I_{\{\rho>t\}} g_a(x_\rho)\\ &\leq M_x I_{\{\rho\leq t\}} \sup_{0\leq s\leq t} g_a(x_s) + M_x I_{\{\rho<t\}} s_a(x_\rho)\\ &\leq M_x I_{\{\rho\leq t\}} \sup_{0\leq s\leq t} g_a(x_s) + M_x I_{\{\rho<t\}} s_a(x_t)\\ &\leq M_x\left[\left(\sup_{0\leq s\leq t} g_a(x_s)\right) \vee s_a(x_t)\right],\end{aligned}$$

and it follows that

$$s_a(x) \le M_x\left[\left(\sup_{0\le s\le t} g_a(x_s)\right) \vee s_a(x_t)\right].$$

By passing to the limit ($a \to -\infty$) we obtain (by the Theorem on monotone convergence)

$$s_*(x) \le M_x\left[\left(\sup_{0\le s\le t} g(x_s)\right) \vee s_*(x_t)\right]. \tag{3.65}$$

Let $t = r_n \downarrow 0$, $n \to \infty$, with $r_n \in S$, where S is countable and dense in $(0, \infty)$. It can be seen that the process $\{Y_t, t \ge 0\}$ with

$$Y_t = M_x\left[\sup_{s\ge 0} g^+(x_s) | \mathscr{F}_t\right]$$

is a uniformly integrable martingale and by virtue of [72], chap. 6, Theorem 3 the sequence Y_{r_n} converges (P_x-a.s.) to the integrable random variable Y_0.

We shall show that

$$\left(\sup_{0\le s\le r_n} g(x_s)\right) \vee s_*(x_{r_n}) \le Y_{r_n}. \tag{3.66}$$

Since $\sup_{0\le s\le r_n} g^+(x_s)$ is an $\mathscr{F}_{r_n}$-measurable variable, we have

$$\sup_{0\le s\le r_n} g(x_s) \le \sup_{0\le s\le r_n} g^+(x_s) = M_x\left[\sup_{0\le s\le r_n} g^+(x_s) | \mathscr{F}_{r_n}\right];$$

therefore, taking into account (3.53),

$$\begin{aligned}
&\left(\sup_{0\le s\le r_n} g(x_s)\right) \vee s_*(x_{r_n})\\
&\le M_x\left[\sup_{0\le s\le r_n} g^+(x_s) | \mathscr{F}_{r_n}\right] \vee M_x\left[\sup_{s\ge 0} g^+(x_s) | \mathscr{F}_{r_n}\right]\\
&\le M_x\left[\sup_{s\ge 0} g^+(x_s) | \mathscr{F}_{r_n}\right] = Y_{r_n}.
\end{aligned}$$

Since the variables $\{Y_{r_n}, n = 1, 2, \ldots\}$ are uniformly integrable, by virtue of (3.65), (3.66), and Remark 2 in Subsection 1.3,

$$\begin{aligned}
s_*(x) &\le \overline{\lim_{n\to\infty}}\, M_x\left[\left(\sup_{0\le s\le r_n} g(x_s)\right) \vee s_*(x_{r_n})\right]\\
&\le M_x\left[\overline{\lim_{n\to\infty}}\left(\sup_{0\le s\le r_n} g(x_s)\right) \vee s_*(x_{r_n})\right].
\end{aligned} \tag{3.67}$$

By virtue of the $\mathscr{C}_0$-continuity assumed of the function $g(x)$ and of the right continuity of the process $\{x_t, t \geq 0\}$, we get from (3.67) that

$$\begin{aligned} s_*(x) &\leq M_x\left[g(x_0) \vee \overline{\lim_{n\to\infty}}\, s_*(x_{r_n})\right] \\ &= M_x\left[g(x_0) \vee \lim_{n\to\infty} s_*(x_{r_n})\right] \\ &= M_x\left[g(x_0) \vee \lim_{r\downarrow 0, r\in S} s_*(x_r)\right], \end{aligned} \tag{3.68}$$

where by Lemma 15

$$\lim_{n\to\infty} s_*(x_{r_n}) = \lim_{r\downarrow 0, r\in S} s_*(x_r).$$

According to the Blumenthal 0–1 law ([31], p. 124), since $P_x\{x_0 = x\} = 1$, $x \in E$, we have (P_x-a.s.)

$$M_x\left[g(x_0) \vee \lim_{\substack{r\downarrow 0\\ r\in S}} s_*(x_r)\right] = g(x) \vee \lim_{\substack{r\downarrow 0\\ r\in S}} s_*(x_r). \tag{3.69}$$

We get from (3.68) and (3.69)

$$s_*(x) \leq g(x) \vee \lim_{\substack{r\downarrow 0\\ r\in S}} s_*(x_r) \qquad (P_x\text{-a.s.}),\ x \in E;$$

therefore, on the set $\{x : g(x) < s_*(x)\}$, (P_x-a.s.)

$$s_*(x) \leq \lim_{r\downarrow 0, r\in S} s_*(x_r).$$

Hence, by Lemma 16

$$\lim_{r\downarrow 0, r\in S} s_*(x_r) = \underline{\lim}_{r\downarrow 0, r\in S}\, s_*(x_r) = \underline{\lim}_{t\downarrow 0}\, s_*(x_t).$$

Hence, on the set $\{x : g(x) < s_*(x)\}$, (P_x-a.s.)

$$s_*(x) \leq \underline{\lim}_{t\downarrow 0}\, s_*(x_t), \tag{3.70}$$

i.e., on this set the function $s_*(x)$ is lower $\mathscr{C}_0$-continuous.

Since $s_*(x) \geq g(x)$ for all $x \in E$, $s_*(x) = g(x)$ on the set $E - \{x : s_*(x) > g(x)\}$; therefore, because $g(x)$ is $\mathscr{C}$-continuous, we have

$$s_*(x) = g(x) = \lim_{t\downarrow 0} g(x_t) \leq \underline{\lim}_{t\downarrow 0}\, s_*(x_t) \qquad (P_x\text{-a.s.}),$$
$$x \in E - \{x : s_*(x) > g(x)\}.$$

This, together with (3.70), demonstrates that the function $s_*(x)$ is lower $\mathscr{C}$-continuous (for all $x \in E$).

Further, since the functions $s_a(x)$ are $\mathscr{C}_0$-continuous and $s_a(x) \downarrow s_*(x)$, $a \to -\infty$, the function $s_*(x)$ is upper $\mathscr{C}_0$-continuous. Therefore, the function $s_*(x)$ is $\mathscr{C}_0$-continuous and Theorem 4 is proved. □

3.4.5

Theorem 5. *Let $X = (x_t, \mathscr{F}_t, P_x)$, $t \geq 0$, be a Markov process and let the function $g \in \mathbb{L}_0(A^+)$. Then:*

(1) *The payoff $s(x)$ is the smallest regular ($\mathscr{C}_0$-continuous) majorant of the function $g(x)$;*
(2) $s(x) = \bar{s}(x)$.

The proof of this theorem is based on some auxiliary assertions, many of which are similar to those used in proving Theorem 2.7 but now require finer methods of proof because the time parameter t is continuous.

For $\varepsilon \geq 0$ and $a \leq 0$ set

$$\begin{aligned} \sigma_\varepsilon^a &= \inf\{t \geq 0 : s_*(x_t) \leq g_a(x_t) + \varepsilon\}, \\ \tau_\varepsilon^a &= \inf\{t \geq 0 : s_a(x_t) \leq g_a(x_t) + \varepsilon\}, \\ \tau_\varepsilon^* &= \inf\{t \geq 0 : s_*(x_t) \leq g(x_t) + \varepsilon\}. \end{aligned} \tag{3.71}$$

Lemma 17. *Let $g \in \mathbb{L}(A^+)$. Then for $\varepsilon > 0$ and $a \leq 0$*

$$s_*(x) = M_x s_*(x_{\sigma_\varepsilon^a}), \qquad x \in E. \tag{3.72}$$

Proof. According to (3.35)

$$s_a(x) = M_x s_a(x_{\tau_\varepsilon^a}). \tag{3.73}$$

It is clear that $\sigma_\varepsilon^a \leq \tau_\varepsilon^a$ for $a \leq \alpha \leq 0$; hence, by virtue of Theorem 1 and Lemma 1,

$$M_x s_a(x_{\tau_\varepsilon^a}) \leq M_x s_a(x_{\sigma_\varepsilon^\alpha}),$$

which together with (3.73) yields the inequality

$$s_a(x) \leq M_x s_a(x_{\sigma_\varepsilon^\alpha}), \qquad a \leq \alpha \leq 0.$$

By Fatou's lemma we obtain

$$s_*(x) \leq \overline{\lim_{a \to -\infty}} M_x s_a(x_{\sigma_\varepsilon^\alpha}) \leq M_x \overline{\lim_{a \to -\infty}} s_a(x_{\sigma_\varepsilon^\alpha}) = M_x s_*(x_{\sigma_\varepsilon^\alpha}).$$

The inverse inequality $s_*(x) \geq M_x s_*(x_{\sigma_\varepsilon^\alpha})$ follows from Lemma 13.

Lemma 18. *Let $g \in \mathbb{L}_0(A^+)$. Then for $\varepsilon > 0$*

$$\lim_{\alpha \to -\infty} P_x\{g(x_{\sigma_\varepsilon^\alpha}) \leq \alpha\} = 0, \qquad x \in E, \tag{3.74}$$

$$P_x\{\tau_\varepsilon^* < \infty\} = 1. \tag{3.75}$$

Proof. Since the functions $g(x)$ and $s_*(x)$ are $\mathscr{C}_0$-continuous ($g(x)$ is by assumption, and $s_*(x)$ is by Theorem 4), the processes $\{g(x_t), t \geq 0\}$ and

$\{s_*(x_t), t \geq 0\}$ are right continuous (P_x-a.s.), $x \in E$, and by the definition of the times $\sigma_\varepsilon^\alpha$ [8]

$$s_*(x_{\sigma_\varepsilon^\alpha}) \leq g_\alpha(x_{\sigma_\varepsilon^\alpha}) + \varepsilon \qquad (P_x\text{-a.s.}),\ x \in E. \tag{3.76}$$

From this and (3.72) we find

$$\begin{aligned} -\infty &< g(x) \\ &\leq s_*(x) \\ &= M_x s_*(x_{\sigma_\varepsilon^\alpha}) \\ &\leq M_x g_\alpha(x_{\sigma_\varepsilon^\alpha}) + \varepsilon \\ &\leq \alpha P_x\{g(x_{\sigma_\varepsilon^\alpha}) \leq \alpha\} + M_x\left[\sup_{t \geq 0} g^+(x_t)\right] + \varepsilon. \end{aligned}$$

Hence ($\alpha < 0$)

$$P_x\{g(x_{\sigma_\varepsilon^\alpha}) \leq \alpha\} \leq -\frac{1}{\alpha}\left\{M_x\left[\sup_{t \geq 0} g^+(x_t) + \varepsilon - s_*(x)\right]\right\} \to 0, \qquad \alpha \to -\infty.$$

Finally, (3.75) can be proved in the same way as was the corresponding assertion in the case of discrete time (see the proof of Theorem 2.7). □

Lemma 19. *Let $g \in \mathbb{L}_0(A^+)$. Then for any $\varepsilon > 0$*

$$s_*(x) = M_x s_*(x_{\tau_\varepsilon^*}), \qquad x \in E. \tag{3.77}$$

Proof. The inequality $s_*(x) \geq M_x s_*(x_{\tau_\varepsilon^*})$ follows from (3.57). To prove the converse, we shall fix the point $x \in E$. It follows from (3.74) that there is a subsequence $\{\alpha_i\}$, $\alpha_i \to -\infty$, $i \to \infty$, such that

$$\lim_{i \to \infty} I_{\{g(x_{\sigma_\varepsilon^{\alpha_i}}) \leq \alpha_i\}}(\omega) = 0 \qquad (P_x\text{-a.s.}).$$

By taking advantage of (3.72), the fact that $\sigma_\varepsilon^{\alpha_i} = \tau_\varepsilon^*$ on the set

$$\{\omega : g(x_{\sigma_\varepsilon^{\alpha_i}}) \geq \alpha_i\},$$

and Fatou's lemma we find:

$$\begin{aligned} s_*(x) &= \overline{\lim_{i \to \infty}}\, M_x s_*(x_{\sigma_\varepsilon^{\alpha_i}}) \\ &\leq M_x \overline{\lim_{i \to \infty}}\, s_*(x_{\sigma_\varepsilon^{\alpha_i}}) \\ &= M_x \overline{\lim_{i \to \infty}}\, I_{\{g(x_{\sigma_\varepsilon^{\alpha_i}}) \geq \alpha_i\}} s_*(x_{\tau_\varepsilon^*}) + M_x \overline{\lim_{i \to \infty}}\, I_{\{g(x_{\sigma_\varepsilon^{\alpha_i}}) < \alpha_i\}} s_*(x_{\sigma_\varepsilon^{\alpha_i}}) \\ &= M_x s_*(x_{\tau_\varepsilon^*}). \end{aligned}$$

□

[8] Note that for $\varepsilon > 0$ the time $\sigma_\varepsilon^\alpha \in \mathfrak{M}$ since $\sigma_\varepsilon^\alpha \leq \tau_\varepsilon^\alpha$ and $P_x\{\tau_\varepsilon^\alpha < \infty\} = 1$ by virtue of Lemma 8.

3.4.6

Proof of Theorem 5. By Theorem 1, $\bar{s}_a(x) = s_a(x)$ and, since $\bar{s}(x) \le \bar{s}_a(x)$,

$$s(x) \le \bar{s}(x) \le s_*(x), \qquad x \in E. \tag{3.78}$$

We shall prove the converse: $s(x) \geqslant s_*(x)$.

By (3.75) the time $\tau_\varepsilon^* \in \mathfrak{M}$. As in the proof of (3.76), we can show that

$$s_*(x_{\tau_\varepsilon^*}) \le g(x_{\tau_\varepsilon^*}) + \varepsilon, \qquad (P_x\text{-a.s.}), x \in E. \tag{3.79}$$

From this and Lemma 19 we get

$$s_*(x) = M_x s_*(x_{\tau_\varepsilon^*}) \le M_x g(x_{\tau_\varepsilon^*}) + \varepsilon \le s(x) + \varepsilon. \tag{3.80}$$

This proves the required inequality $s_*(x) \le s(x)$ since $\varepsilon > 0$ is arbitrary.

Thus, $\bar{s}(x) = s(x) = s_*(x)$ and, by Theorem 4, each of these functions is $\mathscr{C}_0$-continuous.

Next let $v(x)$ be another regular majorant of the function $g(x)$. Then $v(x) \ge M_x v(x_\tau) \ge M_x g(x_\tau)$ and $v(x) \ge s(x)$. Consequently, $s(x)$ is the smallest regular ($\mathscr{C}_0$-continuous) majorant of the function $g(x)$ and the theorem is proved. □

3.4.7

Theorem 6. *Let the function $g \in \mathbb{L}_0(A^+)$. Then all the assertions of Theorem 3 hold true, i.e.:*

(1) *For any $\varepsilon > 0$ the times τ_ε are ε-optimal stopping times;*

(2) *If the function $g(x)$ is upper semicontinuous the time τ_0 is an optimal Markov time;*

(3) *If the function $g(x)$ is upper semicontinuous and $P_x\{\tau_0 < \infty\} = 1$, $x \in E$, the time τ_0 is an optimal stopping time;*

(4) *If in the class $\overline{\mathfrak{M}}$ (in $\mathfrak{M}$) there exists an optimal time τ^*, then $P_x\{\tau_0 \le \tau^*\} = 1$, $x \in E$, and the time τ_0 is optimal in the class $\overline{\mathfrak{M}}$ (in $\mathfrak{M}$).*

Proof. Note that it follows immediately from (3.80) that the stopping time $\tau_\varepsilon = \inf\{t \ge 0 : s(x_t) \le g(x_t) + \varepsilon\}$ is ε-optimal since $s(x) = s_*(x)$ and $\tau_\varepsilon^* = \tau_\varepsilon$. The remaining assertions can be proved as in Theorem 3.

Corollary. *If the function $g(x)$ is upper semicontinuous and belongs to the class $\mathbb{L}_0(A^+)$, and if $\lim_{t\to\infty} g(x_t) = -\infty$ (P_x-a.s.), $x \in E$, then τ_0 is an optimal stopping time.*

Proof. For some $x_0 \in E$, $P_x\{\tau_0 = \infty\} > 0$. Then $\bar{s}(x_0) = -\infty$, which fact contradicts the inequalities $\bar{s}(x_0) \ge g(x_0) > -\infty$. Therefore, $\tau_0 \in \mathfrak{M}$ and the Corollary follows from (2) of Theorem 6.

3.5 Regular characterization of the payoff (the general case)

3.5.1

Theorem 7. *Let the function $g \in \mathbb{L}_0$. Then:*

(1) *The payoff $s(x)$ is the smallest $\mathfrak{M}_g$-regular majorant of the function $g(x)$;*
(2) $s(x) = \bar{s}(x)$.

PROOF. (1) For $b \geq 0$ set

$$g^b(x) = \min\{b, g(x)\},$$

$$s^b(x) = \sup_{\tau \in \mathfrak{M}_g} M_x g^b(x_\tau),$$

$$s^*(x) = \lim_{b \to \infty} s^b(x).$$

If $\sigma, \tau \in \mathfrak{M}_g$ and $P_x\{\sigma \leq \tau\} = 1$, $x \in E$, then by Theorem 5

$$-\infty < M_x g^-(x_\tau) \leq M_x g^b(x_\tau) \leq M_x s^b(x_\tau) \leq M_x s^b(x_\sigma).$$

By the Theorem on monotone convergence we have from this

$$M_x s^*(x_\tau) \leq M_x s^*(x_\sigma). \tag{3.81}$$

The function $s^*(x)$, being the limit of the (nondecreasing) sequence of $\mathscr{C}_0$-continuous almost Borel functions $s^b(x)$, is lower $\mathscr{C}_0$-continuous and almost Borel. This fact together with (3.81) proves that $s^*(x)$ is $\mathfrak{M}_g$-regular.

(2) Let us prove $s^*(x) = s(x) = \bar{s}(x)$.

Since $g^b(x) \leq g(x)$,

$$s^b(x) = \sup_{\tau \in \mathfrak{M}_g} M_x g^b(x_\tau) \leq \sup_{\tau \in \mathfrak{M}_g} M_x g(x_\tau) = s(x),$$

and, therefore, $s^*(x) \leq s(x)$.

Further, if $\tau \in \overline{\mathfrak{M}}_g$, then

$$M_x g^b(x_\tau) \leq \bar{s}^b(x) = s^b(x) \leq s^*(x),$$

and by the Theorem on monotone convergence

$$M_x g(x_\tau) \leq s^*(x).$$

Therefore, $\bar{s}(x) \leq s^*(x)$ which, together with the inequality $s^*(x) \leq s(x)$, proves the required relation $s^*(x) = s(x) = \bar{s}(x)$.

Next, if $v(x)$ is also the $\mathfrak{M}_g$-regular majorant of $g(x)$, then

$$v(x) \geq M_x v(x_\tau) \geq M_x g(x_\tau)$$

and, therefore, $v(x) \geq s(x)$, i.e., $s(x)$ is the smallest $\mathfrak{M}_g$-regular majorant of the function $g(x)$. □

3.5.2

As follows from the theorem proved above, for any two Markov times σ and τ from $\mathfrak{M}_g$ with the property $P_x\{\sigma \le \tau\} = 1, x \in E$, we have the inequality

$$M_x\, s(x_\tau) \le M_x s(x_\sigma). \tag{3.82}$$

From the proof given above it is easily seen that this inequality is also true for any times σ and τ from $\mathfrak{M}$ for which there is a time $\rho \in \mathfrak{M}_g$ such that $P_x\{\sigma \le \tau \le \rho\} = 1$, $x \in E$. (In fact, $-\infty < -M_x g^-(x_\rho) \le M_x s^b(x_\rho) \le M_x s^b(x_\tau) \le M_x s^b(x_\sigma)$ and (3.81) holds true again by the Theorem on monotone convergence.)

The theorem which follows shows the form in which this inequality extends to the times from the classes $\overline{\mathfrak{M}}_g$ and $\overline{\mathfrak{M}}$.

Theorem 8. *Let the function $g \in \mathbb{L}_0$, and let*

$$\hat{s}_\tau(\omega) = \begin{cases} s(x_t(\omega)), & \omega \in \{\omega : \tau(\omega) = t\}, \\ \overline{\lim}_{t\to\infty}\, g(x_t(\omega)), & \omega \in \{\omega : \tau(\omega) = +\infty\}. \end{cases}$$

Then for any two Markov times σ and τ from $\overline{\mathfrak{M}}$ such that there exists $\rho \in \overline{\mathfrak{M}}_g$ with $P_x\{\sigma \le \tau \le \rho\} = 1, x \in E$,

$$M_x(\hat{s}_\tau | \mathscr{F}_\sigma) \le \hat{s}_\sigma \qquad (P_x\text{-a.s.}),\ x \in E, \tag{3.83}$$

and, in particular,

$$M_x \hat{s}_\tau \le M_x \hat{s}_\sigma, \qquad x \in E. \tag{3.84}$$

Proof. From (3.55) we have

$$\overline{\lim_t}\, s^b(x_t) = \overline{\lim_t}\, g^b(x_t) \qquad (P_x\text{-a.s.}),\ x \in E, \tag{3.85}$$

and by Theorem 5

$$-\infty < -M_x g^-(x_\rho) \le M_x s^b(x_\rho) \le M_x s^b(x_\tau) \le M_x s^b(x_\sigma). \tag{3.86}$$

By the Theorem on monotone convergence and (3.85) we obtain for the times σ and τ

$$M_x I_{\{\tau<\infty\}} s(x_\tau) + M_x I_{\{\tau=\infty\}} \lim_b \overline{\lim_t}\, g^b(x_t)$$

$$\le M_x I_{\{\sigma<\infty\}} s(x_\sigma) + M_x I_{\{\sigma=\infty\}} \lim_b \overline{\lim_t}\, g^b(x_t). \tag{3.87}$$

But $\lim_b \overline{\lim}_t\, g^b(x_t) = \overline{\lim}_t\, g(x_t)$ which together with (3.87) leads to (3.84), from which (3.83) can be deduced as in Lemma 2. □

3.5.3

The theorem given in this subsection contains additional information about the structure of the (finite) payoff $s(x)$.

Theorem 9. *Let the function $g \in \mathbb{L}_0$ and let $s(x) < \infty$, $x \in E$. Then the function $s(x)$ is $\mathscr{C}_0$-continuous.*

Proof. Since $s(x) = s^*(x) = \lim_b s^b(x)$ where $s^b(x)$ is a $\mathscr{C}_0$-continuous function, $s^b(x) \leq s^{b+1}(x)$, the payoff $s(x)$ is lower $\mathscr{C}_0$-semicontinuous. The proof of the upper semicontinuity of the payoff $s(x)$ proceeds in a way similar to the proof of the lower $\mathscr{C}_0$-semicontinuity of the function $s_*(x)$ (see Theorem 4).

In fact, let S be a countable, everywhere dense set in $(0, +\infty)$. From Theorem 7 for $\rho \in \overline{\mathfrak{M}}_g$, $r \in S$, we get

$$M_x s(x_{\rho \wedge r}) \leq s(x). \tag{3.88}$$

Since $s(x) < \infty$ by assumption, and

$$M_x s(x_{\rho \wedge r}) \geq M_x \hat{s}_\rho \geq -M_x g^-(x_\rho) > -\infty$$

by Theorem 8, the sequence $\{s(x_{\rho \wedge r}), r \in S\}$ is uniformly integrable for each $\rho \in \overline{\mathfrak{M}}_g$. It follows from (3.83) that this sequence forms a supermartingale. The arguments used in Lemma 15 show that the sequence $\{s(x_{\rho \wedge r}), r \in S\}$ converges to the integrable random variable $(\lim_{r \downarrow 0, r \in S} s(x_{\rho \wedge r}))$ for $r \downarrow 0$ and that

$$\lim_{r \downarrow 0, r \in S} s(x_{\rho \wedge r}) = \lim_{r \downarrow 0, r \in S} s(x_r) \qquad (P_x\text{-a.s.}),\ x \in E. \tag{3.89}$$

It follows from Theorem 1.4 that

$$\lim_{r \downarrow 0, r \in S} M_x s(x_{\rho \wedge r}) = M_x \lim_{r \downarrow 0, r \in S} s(x_{\rho \wedge r}). \tag{3.90}$$

From (3.88)–(3.90) it follows that

$$M_x \lim_{r \downarrow 0, r \in S} s(x_r) \leq s(x). \tag{3.91}$$

Further, as in Lemma 16, we can show here that

$$\overline{\lim_{r \downarrow 0, r \in S}}\ s(x_r) = \overline{\lim_{t \downarrow 0}}\ s(x_t). \tag{3.92}$$

Finally (compare with (3.69)),

$$M_x \overline{\lim_{r \downarrow 0, r \in S}}\ s(x_r) = \overline{\lim_{r \downarrow 0, r \in S}}\ s(x_r) \qquad (P_x\text{-a.s.}),\ x \in E,$$

which together with (3.91) and (3.92) leads to

$$\overline{\lim_{t \downarrow 0}}\ s(x_t) \leq s(x) \qquad (P_x\text{-a.s.}),\ x \in E. \qquad \square$$

3.5.4

It follows from Theorem 6 (compare also with Theorem 3) that if $g \in \mathbb{L}_0(A^+)$ and if in the class $\overline{\mathfrak{M}}$ there exists an optimal time τ^*, then the time $\tau_0 = \inf\{t \geq 0 : s(x_t) = g(x_t)\}$ is also optimal and $P_x\{\tau_0 \leq \tau^*\} = 1$, $x \in E$.

The theorem which follows extends this result to the case where $g \in \mathbb{L}_0$ and $|s(x)| < \infty$, $x \in E$.

Theorem 10. *Let $g \in \mathbb{L}_0$ and let $|s(x)| < \infty$, $x \in E$. Then, if in the class $\overline{\mathfrak{M}}$ (in $\mathfrak{M}$) there exists an optimal Markov time (stopping time) τ^*, the time τ_0 is an optimal time in the class $\overline{\mathfrak{M}}$ (in $\mathfrak{M}$) and $P_x\{\tau_0 \leq \tau^*\} = 1$, $x \in E$.*

Proof. It is clear that the time $\tau^* \in \overline{\mathfrak{M}}_g$. By virtue of the optimality of the time τ^*, the inequality $g(x) \leq s(x)$, and Theorem 8 (with $\tau = \tau^*$ and $\sigma = 0$),

$$\begin{aligned} s(x) &= M_x g(x_{\tau^*}) \\ &= M_x I_{\{\tau^* < \infty\}} g(x_{\tau^*}) + M_x I_{\{\tau^* = \infty\}} \overline{\lim_t}\, g(x_t) \\ &\leq M_x I_{\{\tau^* < \infty\}} s(x_{\tau^*}) + M_x I_{\{\tau^* = \infty\}} \overline{\lim_t}\, g(t) \leq s(x), \end{aligned}$$

from which we find

$$M_x g(x_{\tau^*}) = M_x I_{\{\tau^* < \infty\}} s(x_{\tau^*}) + M_x I_{\{\tau^* = \infty\}} \overline{\lim_t}\, g(x_t). \tag{3.93}$$

It is obvious that

$$g(x_{\tau^*}) \leq I_{\{\tau^* < \infty\}} s(x_{\tau^*}) + I_{\{\tau^* = \infty\}} \overline{\lim_t}\, g(x_t) \qquad (P_x\text{-a.s.}),\ x \in E. \tag{3.94}$$

Because $|s(x)| < \infty$, we conclude from (3.93) and (3.94) that

$$g(x_{\tau^*}) = I_{\{\tau^* < \infty\}} s(x_{\tau^*}) + I_{\{\tau^* = \infty\}} \overline{\lim_t}\, g(x_t) \qquad (P_x\text{-a.s.}),\ x \in E,$$

and, therefore, on the set $\{\omega : \tau^* < \infty\}$

$$g(x_{\tau^*}) = s(x_{\tau^*}).$$

From this, by the definition of the time τ_0, we conclude that on the set $\{\omega : \tau^* < \infty\}$

$$\tau_0 \leq \tau_* \qquad (P_x\text{-a.s.}),\ x \in E.$$

On the set $\{\omega : \tau^* = \infty\}$ this inequality is obvious. Therefore, $\tau_0 \leq \tau^*$ (P_x-a.s.), $x \in E$.

We shall show next that the time τ_0 is optimal.

The functions $g(x)$ and $s(x)$ are $\mathscr{C}_0$-continuous (see Theorem 9) and, therefore, the processes $\{g(x_t), t \geq 0\}$ and $\{s(x_t), t \geq 0\}$ are right continuous. Hence on the set $\{\omega : \tau_0 < \infty\}$

$$g(x_{\tau_0}) = s(x_{\tau_0}). \tag{3.95}$$

This relation together with (3.93) and (3.84) (with $\tau = \tau^*$ and $\sigma = \tau_0$) leads to the inequalities

$$\begin{aligned} s(x) &= M_x g(x_{\tau^*}) \\ &= M_x I_{\{\tau^* < \infty\}} s(x_{\tau^*}) + M_x I_{\{\tau^* = \infty\}} \overline{\lim_t}\, g(x_t) \\ &\leq M_x I_{\{\tau_0 < \infty\}} s(x_{\tau_0}) + M_x I_{\{\tau_0 = \infty\}} \overline{\lim_t}\, g(x_t) \\ &= M_x I_{\{\tau_0 < \infty\}} g(x_{\tau_0}) + M_x I_{\{\tau_0 = \infty\}} \overline{\lim_t}\, g(x_t) = M_x g(x_{\tau_0}), \end{aligned}$$

from which it follows that the time τ_0 is optimal. □

3.6 The construction of regular majorants

3.6.1

If the function $g \in \mathbb{B}(A^-)$, the payoff $s(x) = \sup_{\tau \in \mathfrak{M}} M_x g(x_\tau)$ is the smallest excessive majorant of $g(x)$ (Theorem 1). By Lemma 3, this can be constructed by means of:

$$s(x) = \lim_n \lim_N Q_n^N g(x). \tag{3.96}$$

At the same time, if the condition A^- is violated the payoff $s(x)$ will be (say, in the case $g \in \mathbb{L}_0(A^+)$) the smallest regular majorant of the function $g(x)$. This will not, in general, coincide with the smallest excessive majorant of $g(x)$. The question naturally arises as to the techniques for constructing the smallest regular majorants.

The main result of this section (compare with Theorem 2.6) is that if $g \in \mathbb{L}(a^-)$, then the payoff $s(x)$, being the smallest $\mathfrak{M}_g$-regular majorant of $g(x)$, will coincide with the smallest excessive majorant of a function $G(x)$ $(\geq g(x))$ constructed in a particular way. In constructing the payoff for $g \in \mathbb{L}_0(a^-)$ we use the relation

$$s(x) = \lim_n \lim_N Q_n^N G(x). \tag{3.97}$$

3.6.2

Before formulating the above result in precise terms we present some necessary concepts and auxiliary assertions.

Let the function $h \in \mathbb{B}$ satisfy the condition

$$h(x) = M_x h(x_\infty), \tag{3.98}$$

where, as usual, $h(x_\infty) = \overline{\lim}_t\, h(x_t)$.

An example of such a function is

$$h(x) = -M_x g^-(x_\infty), \tag{3.99}$$

where $M_x g^-(x_\infty) < \infty$ (see the proof of Theorem 12).

Set

$$G(x) = \max\{g(x), h(x)\},$$
$$Q_n G(x) = \max\{G(x), T_{2^{-n}} G(x)\},$$
$$V_n(x) = \lim_{N\to\infty} Q_n^N G(x),$$
$$V(x) = \lim_n V_n(x) = \lim_n \lim_N Q_n^N G(x). \tag{3.100}$$

Remark 1. It can be readily seen that

$$\begin{aligned} T_{2^{-n}} G(x) &= M_x G(x_{2^{-n}}) \\ &\geq M_x h(x_{2^{-n}}) \\ &= M_x M_{x_{2^{-n}}} h(x_\infty) \\ &= M_x M_x[h(x_\infty) | \mathscr{F}_{2^{-n}}] = M_x h(x_\infty) = h(x) > -\infty. \end{aligned}$$

Hence the existence of $\lim_N Q_n^N G(x)$ follows from Lemma 2.6. This limit, $V_n(x)$, is (by the same lemma) the smallest excessive majorant of the function $G(x)$ (with respect to the operator $T_{2^{-n}}$). Further, since $V_{n+1}(x)$ is excessive with respect to the operator $T_{2^{-(n+1)}}$,

$$\begin{aligned} T_{2^{-n}} V_{n+1} &= T_{2^{-(n+1)}} T_{2^{-(n+1)}} V_{n+1}(x) \\ &= T_{2^{-(n+1)}} M_x V_{n+1}(x_{2^{-(n+1)}}) \\ &= M_x M_{x_{2^{-(n+1)}}} V_{n+1}(x_{2^{-(n+1)}}) \\ &\leq M_x V_{n+1}(x_{2^{-(n+1)}}) = T_{2^{-(n+1)}} V_{n+1}(x) \leq V_{n+1}(x), \end{aligned}$$

and, therefore, the function $V_{n+1}(x)$ is also excessive with respect to the operator $T_{2^{-n}}$. Hence $V_n(x) \leq V_{n+1}(x)$, and, therefore, the limit $\lim_n V_n(x)$ exists.

3.6.3

Theorem 11. *Let the function $g \in \mathbb{L}_0$ and let the function $h \in \mathbb{B}$ be such that:*

(a) $h(x) = M_x h(x_\infty)$, $x \in E$;
(b) $h(x) \leq s(x)$, $x \in E$.

Then:

(1) *$s(x)$ is the smallest excessive majorant of the function*

$$G(x) = \max\{g(x), h(x)\};$$

(2) *The payoff $s(x)$ and the payoff $S_G(x) = \sup_{\tau\in\mathfrak{M}_G} M_x G(x_\tau)$ coincide;*
(3) $s(x) = \lim_n \lim_N Q_n^N G(x)$.

To prove the above we shall need:

Lemma 20. *Let the functions $g(x)$ and $h(x)$ belong to the class $\mathbb{B}$, $h(x)$ satisfying relation* (3.98). *Then:*

(a) *The function $V(x)$ belongs to the class $\mathbb{B}$;*
(b) $T_t V(x) \leq V(x)$, $t \geq 0$, $x \in E$;

(c) *The function $V(x)$ is upper $\mathscr{C}_0$-continuous;*
(d) *$V(x)$ is the smallest excessive majorant of $G(x)$.*

PROOF.
(a) In the case where the functions $g(x)$ and $h(x) \in \mathbb{B}(A^-)$ the corresponding proof is contained in Lemma 3. This proof will be valid for the present case if only we justify the inequality (see (3.20))

$$M_x \lim_n \varphi(x_{t+\tau_n}) \leq \lim_n M_x \varphi(x_{t+\tau_n}), \tag{3.101}$$

the validity of which was guaranteed under the conditions of Lemma 3 by Fatou's lemma.

Thus, let the function $\varphi \in \mathbb{B}$, $\varphi \geq G$. Let us show that in this case for any $t \geq 0$ the function $\Phi(x) = M_x \varphi(x_t)$ belongs to the class $\mathbb{B}$. Fix the point $x \in E$ and let $\{\tau_n\}$ be a sequence of Markov times that are the times of first entry into compact sets and such that for all[9] $\omega \in \Omega$, $\tau_n(\omega) \downarrow 0$, $n \to \infty$.

The function $\Phi(x)$ is almost Borel (see the proof of Lemma 3) and from the strong Markov property we have:

$$\begin{aligned} \underline{\lim}_n M_x \Phi(x_{\tau_n}) &= \underline{\lim}_n M_x T_t \varphi(x_{\tau_n}) \\ &= \underline{\lim}_n M_x M x_{\tau_n} \varphi(x_t) = \underline{\lim}_n M_x \varphi(x_{t+\tau_n}). \end{aligned} \tag{3.102}$$

Set

$$\eta_n = M_x[-h^-(x_\infty) | \mathscr{F}_{t+\tau_n}], \qquad n = 1, 2, \ldots.$$

Then (P_x-a.s.)

$$\eta_n = M_{x_{t+\tau_n}}[-h^-(x_\infty)] \leq M_{x_{t+\tau_n}} h(x_\infty) = h(x_{t+\tau_n}) \leq \varphi(x_{t+\tau_n}), \tag{3.103}$$

since $h \leq G \leq \varphi$.

The function $-h^-(x_\infty)$ is integrable, and hence the family of random variables $\{\eta_n, n = 1, 2, \ldots\}$ is uniformly integrable and $\lim_n \eta_n$ exists with P_x-probability 1. Therefore, by (3.102), (3.103), the remark to Theorems 1.2–1.4, and the lower $\mathscr{C}_0$-semicontinuity of $\varphi(x)$,

$$\begin{aligned} \underline{\lim}_n M_x \Phi(x_{\tau_n}) &= \underline{\lim}_n M_x \varphi(x_{t+\tau_n}) \\ &= M_x \underline{\lim}_n \varphi(x_{t+\tau_n}) \geq M_x \varphi(x_t) = \Phi(x). \end{aligned}$$

It follows from the above ([31], theor. 4.9) that the almost Borel function $\Phi(x)$ is lower $\mathscr{C}_0$-semicontinuous. As in Lemma 3, we can deduce that the function $V(x)$ belongs to the class $\mathbb{B}$, i.e., it is almost Borel, lower $\mathscr{C}_0$-semicontinuous, and satisfies the inequality $V(x) > -\infty$.

[9] We may assume that $\tau_n(\omega) \downarrow 0$, $n \to \infty$, for all $\omega \in \Omega$, without loss of generality because we can always use, if necessary,

$$\tilde{\tau}_n(\omega) = \begin{cases} \tau_n(\omega), & \omega \in \{\omega : \tau_n(\omega) \downarrow 0\}, \\ 0, & \omega \notin \{\omega : \tau_n(\omega) \downarrow 0\}. \end{cases}$$

(b) Let us consider the function $V_n(x) = \lim_N Q_n^N G(x)$. As in Lemma 3, we can show here that for all $l, k \in N$,

$$T_{l \cdot 2^{-k}} V_n(x) \leq V_n(x),$$

from which we have

$$T_{l \cdot 2^{-k}} V(x) \leq V(x).$$

Let $t \in [0, \infty)$. We shall choose the sequence of binary-rational numbers $\{r_i\}$ such that $r_i \downarrow t$, $i \to \infty$, and set

$$\eta_i = M_x[-h^-(x_\infty) | \mathscr{F}_{r_i}].$$

The sequence $\{\eta_i\}$ converges (P_x-a.s.), $i \to \infty$,

$$\eta_i \leq M_{x_{r_i}} h(x_\infty) \leq V(x_{r_i});$$

further, by virtue of (a) above and the Remark to Theorems 1.2–1.4

$$\begin{aligned} V(x) &\geq \varliminf_i T_{r_i} V(x) \\ &= \varliminf_i M_x V(x_{r_i}) \\ &\geq M_x \varliminf_i V(x_{r_i}) \\ &\geq M_x \varliminf_{h \downarrow 0} V(x_{t+h}) \geq M_x V(x_t) = T_t V(x). \end{aligned} \tag{3.104}$$

(c) We shall prove that the function $V(x)$ is upper $\mathscr{C}_0$-continuous. To this end it suffices to show that for any sequence of Markov times $\{\tau_n\}$ that are the times of first entry into compact sets contained in the open neighborhood of the point x and such that $P_x\{\tau_n \downarrow 0\} = 1$, the inequality

$$\varlimsup_n M_x V(x_{\tau_n}) \leq V(x)$$

is satisfied.

Let the Markov time τ take on values in the set $\{l \cdot 2^{-n}, l \in N\}$. We prove the inequality

$$M_x V(x_\tau) \leq V(x).$$

To this end we shall show first that

$$(V^b(x_{l \cdot 2^{-n}}), \mathscr{F}_{l \cdot 2^{-n}}, P_x), \qquad l \in N,$$

where $V^b(x) = \min(b, V(x))$, $b \geq 0$, is a uniformly integrable supermartingale. Since $V^b(x_{l \cdot 2^{-n}}) \leq b$ and

$$\begin{aligned} V^b(x_{l \cdot 2^{-n}}) &\geq -h^-(x_{l \cdot 2^{-n}}) \\ &= M_{x_{l \cdot 2^{-n}}}(-h^-(x_\infty)) \\ &= M_x(-h^-(x_\infty) | \mathscr{F}_{l \cdot 2^{-n}}) = \eta_{l \cdot 2^{-n}}, \end{aligned}$$

where the sequence $\{\eta_{l \cdot 2^{-n}}\}$, $l \in N$, is uniformly integrable, the sequence $\{V^b(x_{l \cdot 2^{-n}}\}$, $l \in N$, is also uniformly integrable. By virtue of (3.104)

$$\begin{aligned} M_x[V^b(x_{l \cdot 2^{-n}+t}) | \mathscr{F}_{l \cdot 2^{-n}}] &= T_t V^b(x_{l \cdot 2^{-n}}) \\ &= M_{x_{l \cdot 2^{-n}}} \min[b, V(x_t)] \\ &\le \min[b, M_{x_{l \cdot 2^{-n}}} V(x_t)] \\ &= \min[b, T_t V(x_{l \cdot 2^{-n}})] \le \min[b, V(x_{l \cdot 2^{-n}})] \\ &= V^b(x_{l \cdot 2^{-n}}). \end{aligned}$$

This implies that the sequence

$$(V^b(x_{l \cdot 2^{-n}}), \mathscr{F}_{l \cdot 2^{-n}}, P_x) \qquad l \in N,$$

is a (uniformly integrable) supermartingale, from the known properties of which (Theorem 1.9) it follows that with P_x-probability 1

$$\lim_{l \to \infty} V^b(x_{l \cdot 2^{-n}})$$

exists, $M_x |\lim_{l \to \infty} V^b(x_{l \cdot 2^{-n}})| < \infty$ (P_x-a.s.) and for all $m \in N$

$$M_x\left[\lim_{l \to \infty} V^b(x_{l \cdot 2^{-n}}) | \mathscr{F}_{m \cdot 2^{-n}}\right] \le V^b(x_{m \cdot 2^{-n}}) \qquad (P_x\text{-a.s.}).$$

It follows from Theorem 1.11 (see (1.28)) that

$$M_x[V^b(x_\tau) | \mathscr{F}_\sigma] \le V^b(x_\sigma) \qquad (P_x\text{-a.s.}) \tag{3.105}$$

where σ and τ ($\sigma \le \tau$) take on values in the set $\{l \cdot 2^{-n}, l \in N\}$. From this, by setting $\sigma = 0$ and $b \to \infty$, we get

$$M_x V(x_\tau) \le V(x). \tag{3.106}$$

We shall prove that this inequality holds for any $\tau \in \overline{\mathfrak{M}}$ as well. Set

$$\tau_n = \begin{cases} l \cdot 2^{-n} & \text{if } (l-1)2^{-2} < \tau \le l \cdot 2^{-n}, \\ +\infty & \text{if } \tau = +\infty, \end{cases}$$

It is clear that $\tau_n \downarrow \tau$. Since we have the inequality

$$\varliminf_{h \downarrow 0} V(x_{\tau+h}) \ge V(x_\tau) \qquad (P_x\text{-a.s.}), x \in E,$$

for the lower $\mathscr{C}_0$-continuous functions $V(x)$ (compare with [31], cor. 3 to theor. 4.9), it follows that (P_x-a.s.), $x \in E$,

$$\begin{aligned} \varliminf_n V(x_{\tau_n}) &= \lim_n \inf_{s \ge n} V(x_{\tau_s}) \\ &\ge \lim_n \inf_{t \le 2^{-n}} V(x_{\tau+t}) \\ &\ge \varliminf_{t \downarrow 0} V(x_{\tau+t}) \ge V(x_\tau). \end{aligned}$$

Since $V(x_{\tau_n}) \geq M_x[-h^-(x_\infty)|\mathscr{F}_{\tau_n}] = \eta_n$, by (3.106) and the Remark to Theorems 1.2–1.4

$$M_x V(x_\tau) \leq M_x \varliminf_n V(x_{\tau_n}) \leq \varliminf_n M_x V(x_{\tau_n}) \leq V(x) \qquad (3.107)$$

(for any $\tau \in \overline{\mathfrak{M}}$).

Let $\{\tau_n\}$ be a sequence of Markov times with the property $P_x\{\tau_n \downarrow 0\} = 1$. Then, as shown above,

$$M_x V(x_{\tau_n}) \leq V(x)$$

and, therefore, $\overline{\lim}_n M_x V(x_{\tau_n}) \leq V(x)$. This is sufficient (see (c) of the proof) for the function $V(x)$ to be upper $\mathscr{C}_0$-continuous.

(d) It follows from (a) and (c) that the function $V(x)$ is $\mathscr{C}_0$-continuous and from (b) that $V(x)$ is excessive (see Definition 2 in Section 2). It can also be seen that $V(x) \geq G(x)$, i.e., $V(x)$ is the excessive majorant of $G(x)$. We need only to show that the function $V(x)$ is the smallest excessive majorant of $G(x)$.

Let $\varphi(x)$ be the excessive majorant of $G(x)$. Then, in particular, $T_{2^{-n}}\varphi(x) \leq \varphi(x)$ and therefore

$$Q_n^N \varphi(x) = Q_n^{N-1} \max\{\varphi(x), T_{2^{-n}}\varphi(x)\} = Q_n^{N-1}\varphi(x) = \cdots = \varphi(x).$$

Since $\varphi(x) \geq G(x)$, $\varphi(x) = Q_n^N\varphi(x) \geq Q_n^N G(x)$. Hence

$$\varphi(x) \geq \lim_n \lim_N Q_n^N G(x) = V(x),$$

thus proving the lemma.

3.6.4

Proof of Theorem 11. By virtue of Lemma 20 we need only to show that

$$V(x) = s(x) = S_G(x).$$

Let us prove that $S_G(x) \leq V(x)$. Since $G(x_\tau) \leq V(x_\tau)$, $\tau \in \overline{\mathfrak{M}}$, we find from (3.107) that

$$M_x G(x_\tau) \leq M_x V(x_\tau) \leq V(x),$$

and, therefore,

$$S_G(x) = \sup_{\tau \in \overline{\mathfrak{M}}} M_x G(x_\tau) \leq V(x).$$

Let us now prove the inequality $S_G(x) \geq V(x)$.

For $\varepsilon > 0$ let

$$\tau_\varepsilon^n = \inf\{k \cdot 2^{-n}: V_n(x_{k \cdot 2^{-n}}) \leq G(x_{k \cdot 2^{-n}}) + \varepsilon\}, \qquad k \in N.$$

Assume first that the function $G(x) \leq b < \infty$. The function $G(x)$ is $\mathscr{C}_0$-continuous and $M_x G(x_{2^{-n}}) > -\infty$. Hence $V_n(x)$ is the smallest excessive majorant of $G(x)$ with respect to $T_{2^{-n}}$ (Lemma 2.6).[10] It can also be seen that $G \in \mathbb{L}(A^+)$ and, therefore (Lemma 1.8),

$$\varlimsup_{k\to\infty} V_n(x_{k\cdot 2^{-n}}) = \varlimsup_{k\to\infty} G(x_{k\cdot 2^{-n}}) \qquad (P_x\text{-a.s.}),\ x \in E,$$

and

$$P_x(\tau_\varepsilon^n < \infty\} = 1. \tag{3.108}$$

By virtue of Lemma 1.7

$$V_n(x) = M_x I_{\{\tau_\varepsilon^n < k\cdot 2^{-n}\}} V_n(x_{\tau_\varepsilon^n}) + M_x I_{\{\tau_\varepsilon^n \geq k\cdot 2^{-n}\}} V_n(x_{k\cdot 2^{-n}});$$

since $V_n(x) \leq b < \infty$, by Fatou's lemma

$$V_n(x) \leq M_x \varlimsup_{k\to\infty} I_{\{\tau_\varepsilon^n < k\cdot 2^{-n}\}} V_n(x_{\tau_\varepsilon^n}) + M_x \varlimsup_{k\to\infty} I_{\{\tau_\varepsilon^n \geq k\cdot 2^{-n}\}} V_n(x_{k\cdot 2^{-n}}).$$

Taking into account (3.108) we have from the above the inequality

$$V_n(x) \leq M_x V_n(x_{\tau_\varepsilon^n}), \qquad \varepsilon > 0. \tag{3.109}$$

To get rid of the assumption $G \leq b < \infty$, we shall introduce the variables

$$G^b(x) = \min\{b, G(x)\},\ V_n^b(x) = \lim_N Q_n^N G^b(x)$$

and prove that $\lim_{b\to\infty} V_n^b(x) = V_n(x)$.

It is seen that the values $V_n^b(x)$ do not decrease as $b \to \infty$. Let us set $\bar{V}_n(x) = \lim_{b\to\infty} V_n^b(x)$ and prove that the function $\bar{V}_n(x)$ is the smallest excessive majorant of the function $G(x)$ with respect to $T_{2^{-n}}$.

The function $V_n^b(x)$ is the smallest excessive majorant of the function $G^b(x)$ with respect to $T_{2^{-n}}$. Therefore,

$$\begin{aligned} T_{2^{-n}} \bar{V}_n(x) &= T_{2^{-n}} \lim_{b\to\infty} V_n^b(x) \\ &= \lim_{b\to\infty} T_{2^{-n}} V_n^b(x) \\ &\geq \lim_{b\to\infty} V_n^b(x) = \bar{V}_n(x), \end{aligned}$$

i.e., $\bar{V}_n(x)$ is excessive with respect to $T_{2^{-n}}$. Since $V_n^b(x) \geq G^b(x)$ and $G^b(x) \uparrow G(x)$, $\bar{V}_n(x) \geq G(x)$, i.e., $\bar{V}_n(x)$ is the excessive majorant of the function $G(x)$.

If now $\varphi(x)$ is also the excessive majorant of $G^b(x)$, then $\varphi(x) \geq G^b(x)$, and, since $V_n^b(x)$ is the smallest excessive majorant, $V_n^b(x) \leq \varphi(x)$. Therefore $\bar{V}_n(x) = \lim_{b\to\infty} V_n^b(x) \leq \varphi(x)$.

Thus, $\bar{V}_n(x)$ is the smallest excessive majorant of the function $G(x)$ with respect to $T_{2^{-n}}$. Therefore, $\bar{V}_n(x) = V_n(x)$ and $\lim_{b\to\infty} V_n^b(x) = V_n(x)$.

[10] I.e., $V_n \in \mathbb{B}$ and satisfies (3.7) for $t = k \cdot 2^{-n}$, $k \in N$.

It follows from (3.109) (satisfied by $V_n^b(x)$) that

$$\begin{aligned} V_n^b(x) &= M_x V_n^b(x_{\tau_\varepsilon^n}) \\ &= M_x G^b(x_{\tau_\varepsilon^n}) + \varepsilon \\ &\leq M_x G(x_{\tau_\varepsilon^n}) + \varepsilon \\ &= \sup_{\tau \in \overline{\mathfrak{M}}} M_x G(x_\tau) + \varepsilon = S_G(x) + \varepsilon, \end{aligned}$$

and, therefore,

$$V(x) = \lim_n V_n(x) = \lim_n \lim_{b \to \infty} V_n^b(x) \leq S_G(x).$$

Thus, $V(x) \leq S_G(x)$.

Further, the inequality

$$s(x) \leq S_G(x) \tag{3.110}$$

is obvious. To complete the proof we need only to establish that

$$S_G(x) \leq s(x). \tag{3.111}$$

We note that we have not used the assumption that the function $g \in \mathbb{L}_0$ (we used only the assumption $g \in \mathbb{L}$), nor that $h(x) \leq s(x)$. Let us take advantage of these assumptions.

Set

$$G^b(x) = \min[b, G(x)],$$

$$S_G^b(x) = \sup_{\tau \in \overline{\mathfrak{M}}} M_x G^b(x_\tau), \qquad b \geq 0.$$

The function $g^b(x) = \min[b, g(x)]$ satisfies the condition A^+; from Theorem 5 (proved for the functions $g \in \mathbb{L}_0(A^+)$) we have

$$M_x s^b(x_\tau) \leq s^b(x), \qquad \tau \in \overline{\mathfrak{M}}, \qquad x \in E, \tag{3.112}$$

where $s^b(x)$ is the payoff for $g^b(x)$. Further, since $h(x) \leq s(x)$,

$$\begin{aligned} M_x G^b(x_\tau) &= M_x \min[\max(g(x_\tau), h(x_\tau)), b] \\ &= M_x \max[g^b(x_\tau), h^b(x_\tau)] \\ &\leq M_x \max[s^b(x_\tau), h^b(x_\tau)] = M_x s^b(x_\tau). \end{aligned}$$

Therefore, by virtue of (3.112)

$$S_{G^b}(x) = \sup_{\tau \in \overline{\mathfrak{M}}} M_x G^b(x_\tau) \leq \sup_{\tau \in \overline{\mathfrak{M}}} M_x s^b(x_\tau) \leq s^b(x) \leq s(x).$$

Letting $b \to \infty$ we get

$$S_G(x) \leq s(x),$$

which proves (3.111).

Thus, $S_G(x) = s(x) = V(x)$ and $V(x)$ is the smallest excessive majorant of $G(x)$, and the theorem is completely proved.

3.6.5

The following theorem deals with the special class of functions $h = h(x)$ that enables us to weaken the assumptions about the function $g(x)$ made in Theorem 11.

Theorem 12. *Let the function* $g \in \mathbb{B}(a^-)$. *Then*:

(1) $s(x)$ *is the smallest excessive majorant of the function* $G(x) = \max[g(x), -M_x g^-(x_\infty)]$;

(2) $s(x) = S_G(x)$;

(3) $s(x) = \lim_n \lim_N Q_n^N G(x)$.

PROOF. Set

$$h(x) = -M_x g^-(x_\infty).$$

We shall show that the function $h \in \mathbb{B}$.

It is clear that $h(x) > -\infty$, $x \in E$. To prove that the function $h(x)$ is almost Borel we need only to prove ([31], theor. 5.13) that this function is $\overline{\mathscr{B}}$-measurable (recall that $\overline{\mathscr{B}}$ is the completion of the σ-algebra $\mathscr{B}$ with respect to the family of all probability measures on $\mathscr{B}$), and that $\lim_{t \downarrow 0} T_t h(x) = h(x)$ for each $x \in E$. The last fact follows immediately from the Markov property:

$$T_t h(x) = M_x[-M_{x_t} g^-(x_\infty)] = -M_x M_x[g^-(x_\infty) | \mathscr{F}_t] = h(x).$$

Further, the random variable $g(x_t)$ is $\overline{\mathscr{F}}_t^X$-measurable[11] for each $t \geq 0$ and, due to the separability of the process $\{g(x_t), t \geq 0\}$, the variable $g(x_\infty)$ is $\overline{\mathscr{F}}^X$-measurable. From this and [31] (theor. 3.1) it follows that the function $h(x)$ is $\overline{\mathscr{B}}$-measurable. This fact together with the property $\lim_{t \downarrow 0} T_t h(x) = h(x)$ proves that the function $h(x)$ is almost Borel.

Let $\{\tau_n\}$ be a sequence of Markov times with $P_x\{\tau_n \downarrow 0\} = 1$. Then it follows from the strong Markov property that

$$\begin{aligned}\lim_n M_x h(x_{\tau_n}) &= \lim_n M_x\{-M_{x_{\tau_n}} g^-(x_\infty)\} \\ &= -\lim_n M_x M_x[g^-(x_\infty) | \mathscr{F}_{\tau_n}] \\ &= -M_x g^-(x_\infty) = h(x),\end{aligned}$$

and it follows from [31] (theor. 4.9) that the function $h(x)$ is $\mathscr{C}_0$-continuous.

Thus, the function $h(x)$ is almost Borel and $\mathscr{C}_0$-continuous. It follows from [31] theor. 4.11 that the random function $\{h(x_t), t \geq 0\}$ is right continuous so that, therefore, the process $\{h(x_t), t \geq 0\}$ is separable. Thus, the function $h \in \mathbb{B}$.

We shall show that $h(x) = M_x h(x_\infty)$. By virtue of the fact that the process $\{h(x_t), t \geq 0\}$ is separable for any countable set $S \in [0, \infty)$

$$\overline{\lim_{t \to \infty}}\, h(x_t) = \overline{\lim_{s \to \infty, s \in S}}\, h(x_s).$$

[11] $\overline{\mathscr{F}}_t^X$ denotes completion of the σ-algebra $\mathscr{F}_t^X = \sigma\{\omega : x_s, s \leq t\}$ over the system of all probability measures $P_\mu(\cdot) = \int_E P_x(\cdot)\mu(dx)$; $\overline{\mathscr{F}}^X = \sigma(\bigcup_t \overline{\mathscr{F}}_t^X)$.

Hence

$$
\begin{aligned}
M_x h(x_\infty) &= M_x \varlimsup_{s\to\infty, s\in S} h(x_s) \\
&= -M_x \varlimsup_{s\to\infty, s\in S} M_{x_s} g^-(x_\infty) \\
&= -M_x \varlimsup_{s\to\infty, s\in S} M_x[g^-(x_\infty)|\mathscr{F}_s] \\
&= -M_x M_x[g^-(x_\infty)|\mathscr{F}_\infty] \\
&= -M_x g^-(x_\infty) = h(x). \qquad (3.113)
\end{aligned}
$$

Let us note that in proving Theorem 11 we used the condition $g \in \mathbb{L}_0$ to prove only the inequality $S_G(x) \le s(x)$. Hence to prove the theorem we can make use of Theorem 11 if we show that the inequality indicated holds under the assumption $g \in \mathbb{B}(a^-)$.

If $s(x) = \infty$, then $S_G(x) = \infty$; hence we may consider only the case $s(x) < \infty$.

For $\tau \in \overline{\mathfrak{M}}$ set

$$
\begin{aligned}
A &= \{\omega : \tau(\omega) < \infty, -M_{x_\tau} g^-(x_\infty) \le g(x_\tau)\}, \\
B &= \{\omega : \tau(\omega) < \infty, -M_{x_\tau} g^-(x_\infty) > g(x_\tau)\},
\end{aligned}
$$

$$
\sigma_\tau(\omega) = \begin{cases} \tau(\omega), & \omega \in A, \\ +\infty, & \omega \notin A, \end{cases}
$$

and let us show that

$$
M_x G(x_\tau) \le M_x g(x_{\sigma_\tau}). \qquad (3.114)
$$

The set $B \in \mathscr{F}_\tau$. Hence

$$
\begin{aligned}
M_x I_{\{\tau<\infty\}} G(x_\tau) &= M_x I_{\{\tau<\infty\}} \max[g(x_\tau), -M_{x_\tau} g^-(x_\infty)] \\
&= M_x I_A g(x_\tau) - M_x I_B M_x[g^-(x_\infty)|\mathscr{F}_\tau] \\
&\le M_x I_A g(x_\tau) + M_x I_B g(x_\infty) = M_x I_{\{\tau<\infty\}} g(x_{\sigma_\tau}). \qquad (3.115)
\end{aligned}
$$

It follows from (3.113) that $h(x_\infty) = -g^-(x_\infty)$ (P_x-a.s.) and, therefore,

$$
G(x_\infty) = \max\{g(x_\infty), -g^-(x_\infty)\} = g(x_\infty),
$$

which together with (3.115) proves (3.114).

We shall show also that for any $\tau \in \overline{\mathfrak{M}}$ the time $\sigma_\tau \in \overline{\mathfrak{M}}_g$. From (3.115) and the fact that the function $I_{\{\tau<\infty\}}(\omega)$ is $\mathscr{F}_\tau$-measurable we get

$$
\begin{aligned}
M_x g(x_{\sigma_\tau}) &= M_x I_{\{\tau<\infty\}} g(x_{\sigma_\tau}) + M_x I_{\{\tau=\infty\}} g(x_{\sigma_\tau}) \\
&\ge M_x I_{\{\tau<\infty\}} G(x_\tau) + M_x I_{\{\tau=\infty\}} g(x_\infty) \\
&\ge -M_x I_{\{\tau<\infty\}} M_x[g^-(x_\infty)|\mathscr{F}_\tau] - M_x I_{\{\tau=\infty\}} g^-(x_\infty) \\
&= -M_x I_{\{\tau<\infty\}} g^-(x_\infty) - M_x I_{\{\tau=\infty\}} g^-(x_\infty) \\
&= -M_x g^-(x_\infty) = h(x) > -\infty.
\end{aligned}
$$

Therefore,

$$\sigma_\tau \in \overline{\mathfrak{M}}_g.$$

Finally, we find from (3.114) that

$$\begin{aligned} S_G(x) &= \sup_{\tau \in \overline{\mathfrak{M}}} M_x G(x_\tau) \\ &\leq \sup_{\tau \in \overline{\mathfrak{M}}} M_x g(x_{\sigma_\tau}) \\ &\leq \sup_{\sigma \in \overline{\mathfrak{M}}_g} M_x g(x_\sigma) = s(x), \end{aligned}$$

which fact, as noted above, proves the validity of all the assertions of Theorem 11. Thus, we have proved the theorem. □

3.6.6

It follows from Theorem 7 that if $g \in \mathbb{L}_0$, then for any two Markov times σ and $\tau \in \mathfrak{M}_g$ with the property $P_x\{\sigma \leq \tau\} = 1, x \in E$, the payoff $s(x)$ satisfies the inequality

$$M_x s(x_\tau) \leq M_x s(x_\sigma). \tag{3.116}$$

The methods discussed above in this section enable us to extend the result to the times from the class $\overline{\mathfrak{M}}$.

Theorem 13. *Let either condition be satisfied*:

(a) $g \in \mathbb{L}_0$ *and there exists a function* $h \in \mathbb{B}$ *such that* $h(x) = M_x h(x_\infty)$ *and* $h(x) \leq s(x)$;
(b) $g \in \mathbb{B}(a^-)$.

Then for any Markov times σ *and* $\tau \in \overline{\mathfrak{M}}$ *such that* $P_x(\sigma \leq \tau\} = 1, x \in E$, *(3.116) is satisfied and, therefore, the payoff* $s(x)$ *is the smallest regular majorant of the function* $g(x)$.

PROOF. Let us consider again the functions $v(x)$ and $v^b(x) = \min\{b, v(x)\}$. By Lemma 20 they are excessive.

It follows from [31] (theorem 4.11) that the process $\{v^b(x_t), t \geq 0\}$ is right continuous. Next, as in Lemma 20, we can show here that the process $\{v^b(x_t), t \geq 0\}$ is a uniformly integrable supermartingale. Hence it follows from Theorem 1.9 that there exists an integrable limit $\lim_{t\to\infty} v^b(x_t)$ such that

$$M_x\left[\lim_t v^b(x_t) | \mathscr{F}_s\right] \leq v^b(x_s) \qquad (P_x\text{-a.s.}), x \in E.$$

From Theorem 1.11, for any Markov times σ and $\tau \in \overline{\mathfrak{M}}$ with $P_x\{\sigma \leq \tau\} = 1$ we have

$$M_x[v^b(x_\tau) | \mathscr{F}_\sigma] \leq v^b(x_\sigma) \qquad (P_x\text{-a.s.}), x \in E,$$

and, therefore, $M_x v^b(x_\tau) \leq M_x v^b(x_\sigma)$. From this, assuming $b \to \infty$, we find that

$$M_x v(x_\tau) \leq M_x v(x).$$

The assertions of the theorem follow immediately from Theorems 11 and 12, in which we proved that the payoff $s(x) = v(x)$ under the conditions (a) and (b), respectively.

3.7 $\varepsilon(x)$-optimal Markov times

3.7.1

If the function $g \in \mathbb{L}_0(A^+)$, then (by virtue of Theorem 6) for any $\varepsilon > 0$ the times

$$\tau_\varepsilon = \inf\{t \geq 0 : s(x_t) \leq g(x_t) + \varepsilon\} \tag{3.117}$$

are ε-optimal, i.e.,

$$s(x) \leq M_x g(x_{\tau_\varepsilon}) + \varepsilon. \tag{3.118}$$

However, if the condition A^+ is violated the times τ_ε need not, in general, be ε-optimal (see Example 6 in Section 2.6).

To investigate the question of ε-optimality in the case where the condition A^+ is violated we shall introduce the following:

Definition. The Markov time τ is called $\varepsilon(x)$-*optimal on the set* E_0 ($E_0 \subseteq E$) if for all $x \in E_0$ the mathematical expectations $M_x g(x_\tau)$ are defined and

$$s(x) \leq M_x g(x_\tau) + \varepsilon(x), \qquad x \in E_0. \tag{3.119}$$

If $\varepsilon(x) \equiv \varepsilon$, $E_0 = E$, this definition becomes the definition given in Section 1 of the ε-optimality of the time τ. However, as noted above, there are, in general, no ε-optimal times. We shall consider here the cases in which we can assert that there exist $\varepsilon(x)$-optimal Markov times for a function $\varepsilon(x) = \varepsilon s(x)$.

Theorem 14. *Let the following be satisfied: the function* $g \in \mathbb{L}_0$; $\varepsilon(x) = \varepsilon s(x)$; *for all* $x \in E_0 = \{x : s(x) < \infty\}$

$$P_x\left\{\overline{\lim_t}\, g(x_t) \geq 0\right\} = 1.$$

Then for any $\varepsilon > 0$ *the times*

$$\sigma_\varepsilon^* = \inf\{t \geq 0 : s(x_t) \leq g(x_t) + \varepsilon s(x_t)\}$$

are $\varepsilon \cdot s(x)$-*optimal Markov times on the set* E_0, *i.e.*,

$$M_x g(x_{\sigma_\varepsilon^*}) \geq s(x)[1 - \varepsilon], \qquad x \in E_0. \tag{3.120}$$

3.7.2

Before proving the theorem we shall prove two lemmas; we shall say that the Markov time $\tau \in \mathfrak{M}_g$ *belongs only to the class* $\mathfrak{R}(x; \delta, \varepsilon)$, where $x \in E$, $\delta \geq 0$, and $\varepsilon \geq 0$, if

$$M_x I_{\{\tau < \sigma_\varepsilon^*\}} s(x_\tau) \leq \delta \tag{3.121}$$

(compare with the definition of the class $\mathfrak{M}(x; \delta, \varepsilon)$ in Subsection 3.2).

Lemma 21. *Let* $g \in \mathbb{L}_0$. *Then for* $x \in E_0 = \{x: s(x) < \infty\}$, $\delta > 0$, $\varepsilon > 0$, *the class* $\mathfrak{R}(x; \delta, \varepsilon)$ *is the sufficient class of Markov times for* $x \in E_0$:

$$s(x) = \sup_{\tau \in \mathfrak{R}(x;\, \delta,\, \varepsilon)} M_x g(x_\tau), \qquad x \in E_0.$$

Proof. By the definition of the time σ_ε^* on the set $\{\omega: \tau < \sigma_\varepsilon^*\}$ we have the inequality

$$g(x_\tau) < s(x_\tau) - \varepsilon s(x_\tau).$$

From this together with the obvious inequality $g(x_\tau) \leq s(x_\tau)$ we get for $\tau \in \mathfrak{M}_g$

$$\begin{aligned} M_x g(x_\tau) &= M_x I_{\{\tau < \sigma_\varepsilon^*\}} g(x_\tau) + M_x I_{\{\tau \geq \sigma_\varepsilon^*\}} g(x_\tau) \\ &\leq M_x I_{\{\tau < \sigma_\varepsilon^*\}} [s(x_\tau) - \varepsilon s(x_\tau)] + M_x I_{\{\tau \geq \tau_\varepsilon^*\}} s(x_\tau) \\ &= M_x s(x_\tau) - \varepsilon M_x I_{\{\tau < \sigma_\varepsilon^*\}} s(x_\tau). \end{aligned} \tag{3.122}$$

For all $\sigma, \tau \in \mathfrak{M}_g$ with the property $P_x\{\sigma \leq \tau\} = 1$

$$M_x s(x_\tau) \leq M_x s(x_\sigma) \leq s(x)$$

by virtue of (3.82).

It follows from (3.122) and the definition of the class $\mathfrak{R}(x; \delta, \varepsilon)$ that for all $\tau \in \mathfrak{M}_g - \mathfrak{R}(x; \delta, \varepsilon)$

$$M_x g(x_\tau) \leq s(x) - \varepsilon\delta,$$

and, therefore,

$$\sup_{\tau \in \mathfrak{M}_g - \mathfrak{R}(x;\, \delta,\, \varepsilon)} M_x g(x_\tau) \leq s(x) - \varepsilon\delta.$$

But $s(x) = \sup_{\tau \in \mathfrak{M}_g} M_x g(x_\tau)$. Hence, due to the fact that $\varepsilon > 0$ and $\delta > 0$ we find that if $x \in E_0$, then

$$s(x) = \sup_{\tau \in \mathfrak{R}(x;\, \delta,\, \varepsilon)} M_x g(x_\tau). \qquad \square$$

Lemma 22. *Let* $g \in \mathbb{L}_0$ *and* $P_x\{\overline{\lim}_t\, g(x_t) \geq 0\} = 1$, $x \in E_0$. *Then for all* $\varepsilon > 0$ *and* $x \in E_0$

$$s(x) = M_x I_{\{\sigma_\varepsilon^* < \infty\}} s(x_{\sigma_\varepsilon^*}) + M_x I_{\{\sigma_\varepsilon^* = \infty\}} \overline{\lim_t}\, g(x_t). \tag{3.123}$$

PROOF. Since $P_x\{\overline{\lim}_t g(x_t) \geq 0\} = 1$, $x \in E_0$, $M_x g(x_\infty) > -\infty$, $x \in E_0$. We obtain from Theorem 8

$$M_x I_{\{\sigma_\varepsilon^* < \infty\}} s(x_{\sigma_\varepsilon^*}) + M_x I_{\{\sigma_\varepsilon^* = \infty\}} \overline{\lim_t}\, g(x_t) \leq s(x), \qquad x \in E_0 \quad (3.124)$$

Let us prove now the converse. Let $\delta_n = 2^{-n}$ and $x \in E_0$. By virtue of Lemma 21 one can find a sequence $\tau_n \in \mathfrak{R}(x; \delta, \varepsilon)$ such that

$$s(x) = \lim_n M_x g(x_{\tau_n}).$$

We shall make use of (3.84) with $\tau = \tau_n$ and $\sigma = \min\{\tau_n, \sigma_\varepsilon^*\}$.

Taking into account that $s(x) = \bar{s}(x) \geq M_x \overline{\lim}_t g(x_t) \geq 0$, we have

$$\begin{aligned} s(x) &= \lim_n M_x g(x_{\tau_n}) \\ &\leq \overline{\lim_n}\, M_x I_{\{\tau_n < \infty\}} s(x_{\tau_n}) \\ &\leq \overline{\lim_n}\, M_x I_{\{\tau_n \wedge \tau_\varepsilon^* < \infty\}} s(x_{\tau_n \wedge \sigma_\varepsilon^*}) \\ &\leq \overline{\lim_n}\, [M_x I_{\{\tau_n \wedge \sigma_\varepsilon^* < \infty\} \cap \{\tau_n < \sigma_\varepsilon^*\}} s(x_{\tau_n}) + M_x I_{\{\tau_n \wedge \sigma_\varepsilon^* < \infty\} \cap \{\tau_n \geq \sigma_\varepsilon^*\}} s(x_{\sigma_\varepsilon^*})] \\ &\leq \overline{\lim_n}\, M_x[I_{\{\tau_n < \sigma_\varepsilon^*\}} s(x_{\tau_n})] + \overline{\lim_n}\, M_x[I_{\{\sigma_\varepsilon^* < \infty\} \cap \{\tau_n \geq \sigma_\varepsilon^*\}} s(x_{\sigma_\varepsilon^*})]. \end{aligned} \quad (3.125)$$

Since $\tau_n \in \mathfrak{R}(x; \delta, \varepsilon)$,

$$\lim_n M_x I_{\{\tau_n < \sigma_\varepsilon^*\}} s(x_{\tau_n}) \leq \lim_n \delta_n = 0. \quad (3.126)$$

Note that

$$\{\sigma_\varepsilon^* < \infty\} \cap \{\tau_n \geq \sigma_\varepsilon^*\} \subseteq \{\sigma_\varepsilon^* < \infty\}. \quad (3.127)$$

By virtue of (3.125)–(3.127) and the condition $P_x\{\overline{\lim}_t g(x_t) \geq 0\} = 1$

$$s(x) \leq M_x I_{\{\sigma_\varepsilon^* < \infty\}} s(x_{\sigma_\varepsilon^*}) + M_x I_{\{\sigma_\varepsilon^* > \infty\}} \overline{\lim_t}\, g(x_t),$$

which together with (3.124) proves (3.123). □

3.7.3

PROOF OF THEOREM 14. Let us show that the integral $M_x g(x_{\sigma_\varepsilon^*})$ exists for all $x \in E_0$.

On the set E_0 the function $s(x)$ is $\mathscr{C}_0$-continuous (Theorem 9) and the processes $\{g(x_t), t \geq 0\}$ and $\{s(x_t), t \geq 0\}$ are right continuous.

It follows from the definition of the time σ_ε^* that

$$\begin{aligned} g(x_{\sigma_\varepsilon^*}) &= I_{\{\sigma_\varepsilon^* < \infty\}} g(x_{\sigma_\varepsilon^*}) + I_{\{\sigma_\varepsilon^* = \infty\}} \overline{\lim_t}\, g(x_t) \\ &\geq (1 - \varepsilon) I_{\{\sigma_\varepsilon^* < \infty\}} s(x_{\sigma_\varepsilon^*}) + I_{\{\sigma_\varepsilon^* = \infty\}} \overline{\lim_t}\, g(x_t). \end{aligned}$$

It can be seen that for $0 < \varepsilon \leq 1$

$$M_x g(x_{\sigma_\varepsilon^*}) \geq 0.$$

If $\varepsilon > 1$, then

$$g(x_{\sigma_\varepsilon^*}) \geq (\varepsilon - 1)\{-I_{\{\sigma_\varepsilon^* < \infty\}} s(x_{\sigma_\varepsilon^*}) - I_{\{\sigma_\varepsilon^* = \infty\}} \overline{\lim_t}\, g(x_t)\},$$

and we find from (3.84) (since $s(x) < \infty$, $x \in E_0$) that

$$M_x g(x_{\sigma_\varepsilon^*}) \geq (\varepsilon - 1)(-s(x)) > -\infty.$$

Therefore, $M_x g(x_{\sigma_\varepsilon^*})$ is defined for all $x \in E_0$.

Let us prove (3.120). By virtue of Lemma 22 and the fact that the processes $\{g(x_t), t \geq 0\}$ and $\{s(x_t), t \geq 0\}$ are right continuous, we have

$$\begin{aligned} s(x) &= M_x I_{\{\sigma_\varepsilon^* < \infty\}} s(x_{\sigma_\varepsilon^*}) + M_x I_{\{\sigma_\varepsilon^* = \infty\}} \overline{\lim_t}\, g(x_t) \\ &\leq M_x I_{\{\sigma_\varepsilon^* < \infty\}} [g(x_{\sigma_\varepsilon^*}) - \varepsilon s(x_{\sigma_\varepsilon^*})] + M_x I_{\{\sigma_\varepsilon^* = \infty\}} \overline{\lim_t}\, g(x_t) \\ &\leq M_x g(x_{\sigma_\varepsilon^*}) + \varepsilon M_x I_{\{\sigma_\varepsilon^* < \infty\}} s(x_{\sigma_\varepsilon^*}). \end{aligned} \tag{3.128}$$

But $P_x\{\overline{\lim_t}\, g(x_t) \geq 0\} = 1$, $x \in E_0$. Hence

$$0 \leq M_x I_{\{\sigma_\varepsilon^* = \infty\}} \overline{\lim_t}\, g(x_t),$$

and it follows from Lemma 22 that

$$\varepsilon M_x I_{\{\sigma_\varepsilon^* < \infty\}} s(x_{\sigma_\varepsilon^*}) = \varepsilon[s(x) - M_x I_{\{\sigma_\varepsilon^* = \infty\}} \overline{\lim_t}\, g(x_t)] \leq \varepsilon s(x).$$

Therefore, we obtain from (3.128)

$$s(x) \leq M_x g(x_{\sigma_\varepsilon^*}) + \varepsilon s(x),$$

i.e.,

$$s(x)[1 - \varepsilon] \leq M_x g(x_{\sigma_\varepsilon^*}).$$

The theorem is proved. □

Remark 1. The condition $P_x\{\overline{\lim}_t\, g(x_t) \geq 0\} = 1$, $x \in E_0$, in the formulation of Theorem 14 can be replaced by the condition $-\infty < M_x \overline{\lim}_t\, g(x_t)$, $x \in E_0$. In this case we should take instead of the function $\varepsilon(x) = \varepsilon s(x)$ the function

$$\varepsilon(x) = \varepsilon[s(x) - M_x \overline{\lim_t}\, g(x_t)].$$

Remark 2. Theorem 14 will hold true in the case of discrete time (under the assumption $g \in \mathbb{L}$ and $P_x\{\overline{\lim}_n\, g(x_n) \geq 0\} = 1$, $x \in E_0 = \{x : s(x) < \infty\}$). As applied to Example 6 in Section 2.6 it follows from the above that the time of first entry into the set $\Gamma = \{0\} \cup \{n : n \geq (1 - \varepsilon)/\varepsilon\}$ is $\varepsilon(x) = \varepsilon s(x)$-optimal ($\varepsilon > 0$).

3.8 Equations for the payoff and generalized Stefan problem: the conditions for "smooth pasting"

3.8.1

In the case of discrete time $n = 0, 1, 2, \ldots$ the payoff $s(x)$ satisfies the recursive equation

$$s(x) = \max\{g(x), Ts(x)\}, \tag{3.129}$$

which is to be used in seeking the payoff. It is naturally desirable to obtain the analog of this equation for the case of continuous time.

If the function $g \in \mathbb{B}(A^-)$, $s(x) \geq g(x)$ and $s(x) \geq T_t s(x)$, $t \geq 0$, by virtue of Theorem 1. Consequently,

$$s(x) \geq \max\{g(x), T_t s(x)\}. \tag{3.130}$$

By analogy with the case of discrete time one would expect that at least for some $t > 0$, equality holds in (3.130). However, this is not true, generally speaking. Nevertheless, if (in 3.130) we replace time t by a suitable (nontrivial) Markov time τ, we can obtain a relation analogous to (3.129).

To this end we set

$$\Gamma_0 = \{x : g(x) = s(x)\}, C_0 = E - \Gamma_0.$$

We know from Theorem 3 that if the function $g \in \mathbb{B}_0(A^-, A^+)$ is upper semicontinuous, the time τ_0, being the time of first entry into the set Γ_0, is an optimal Markov time. Therefore, knowledge about the payoff $s(x)$ enables us to describe the optimal stopping rule in this case. It is clear that the payoff $s(x)$ coincides with the function $g(x)$ in the domain Γ_0 of *stopping of observations.*

The theorem which follows shows the property of the payoff $s(x)$ in the domain C_0 of *continuing observations.*

Theorem 15. *Let the function $g \in \mathbb{L}_0$. Then for each point $x \in C_0 \cap \{x : s(x) < \infty\}$ the payoff $s(x)$ belongs to the domain of definition of the characteristic operator $\mathfrak{U}$ (in topology $\mathscr{C}_0$) and is a solution of the generalized Stefan problem:*

$$\begin{aligned} \mathfrak{U}s(x) &= 0, \qquad x \in C_0 \cap (x : s(x) < \infty\}, \\ s(x) &= g(x), \qquad x \in \Gamma_0 = E - C_0. \end{aligned} \tag{3.131}$$

PROOF. By virtue of the definition of the operator $\mathfrak{U}$ it suffices to show that for each point $x \in C_0 \cap \{x : s(x) < \infty\}$ there exists a neighborhood $\mathfrak{U} \in \mathscr{C}_0 \cap \mathscr{B}$ of the point x such that for any neighborhood $V \in \mathscr{C}_0 \cap \mathscr{B}$, $V \subseteq U$, the equality

$$s(x) = M_x s(x_{\sigma(V)}) \tag{3.132}$$

is satisfied, where $\sigma(V)$ is the time of first departure of V.

Let $x \in C_0$ and $s(x) \lessgtr \infty$. Then there will be $a \leq 0$ and $\varepsilon' > 0$ such that

$$x \in \tilde{U} = \{x : g_a(x) + \varepsilon' < s^{|a|}(x)\},$$

where $g_a(x) = \max\{a, g(x)\}$, $s^{|a|}(x) = \sup_{\tau \in \mathfrak{M}} M_x g^{|a|}(x_\tau)$, and $g^{|a|}(x) = \min\{|a|, g(x)\}$, with $s^{|a|}(x) \uparrow s(x)$, $a \downarrow -\infty$ (see the proof of Theorem 7). It follows from the $\mathscr{C}_0$-continuity of the functions $g_a(x)$ and $s^{|a|}(x)$ (Theorem 5) that $\tilde{U} \in \mathscr{C}_0$. By virtue of [31] (corollary to Theorem 4.8) $\mathscr{C}_0 \cap \mathscr{B}$ is the basis of topology $\mathscr{C}_0$. Therefore, there exists a neighborhood $U \in \mathscr{C}_0 \cap \mathscr{B}$ such that $x \in U \subseteq \tilde{U}$.

For $\varepsilon \geq 0$ and $b \geq 0$ set

$$\tau_\varepsilon^b = \inf\{t \geq 0 : s^b(x_t) \leq g^b(x_t) + \varepsilon\}.$$

By virtue of Lemma 19 and Theorem 5, for $\varepsilon > 0$ and $b \geq 0$

$$s^b(x) = M_x s^b(x_{\tau_\varepsilon^b}). \tag{3.133}$$

It is seen that $\sigma(U) \leq \tau_{\varepsilon'}^b$ for all $b \geq |a|$, $P_x\{\tau_{\varepsilon'}^b < \infty\} = 1$, $\varepsilon' > 0$. Hence it follows from the fact that the payoff $s^b(x)$ is $\mathfrak{M}_g$-regular (Theorem 5) that

$$s^b(x) \leq M_x s^b(x_{\sigma(V)}) \leq M_x s^b(x_{\sigma(U)}) \leq M_x s^b(x_{\tau_{\varepsilon'}^b}),$$

where $V \subseteq U$, $V \in \mathscr{C}_0 \cap \mathscr{B}$. This together with (3.134) yields the relation

$$s^b(x) = M_x s^b(x_{\sigma(V)}) = M_x s^b(x_{\sigma(U)}), \tag{3.134}$$

from which we obtain the required equality, (3.132), by passing to the limit $(b \to \infty)$.

It follows from the definition of the operator $\mathfrak{U}$ and (3.132) that

$$\mathfrak{U}s(x) = 0, \qquad x \in \mathscr{C}_0 \cap \{x : s(x) < \infty\}.$$

It is clear also that

$$s(x) = g(x), \qquad x \in \Gamma_0 = E - C_0,$$

which fact proves the theorem. □

Remark 1. If the process $X = (x_t, \mathscr{F}_t, P_x)$ considered is an l-dimensional Wiener process and it has been known *a priori* that $s(x)$ is twice continuously differentiable, the operator $\mathfrak{U}$ becomes the Laplace differential operator $\frac{1}{2}\sum_{i=1}^{l} \partial^2/\partial x_i^2$ (see [31]), and (assuming $g \in \mathbb{L}_0$) the payoff $s(x)$ is one of the solutions of the Stefan (differential) problem:

$$\sum_{i=1}^{l} \frac{\partial^2}{\partial x_i} s(x) = 0, \qquad x \in C_0 \cap \{x : s(x) < \infty\},$$

$$s(x) = g(x), \qquad x \in \Gamma_0 = E - C_0. \tag{3.135}$$

This justifies the fact that problem (3.131) was referred to as a *generalized Stefan problem.*

Remark 2. It should be emphasized that unlike ordinary boundary problems, in the Stefan problems under consideration the domain $\mathscr{C}_0$ of continuing observations (where the "equation" $\mathfrak{A}s(x) = 0$ holds true) is unknown and is therefore to be found at the same time as the function $s(x)$. In this connection the Stefan problems are referred to as problems with *free* (or *unknown*) *boundaries*.

Remark 3. We have considered in this chapter only the problems in which the cost of the observations is zero. By analogy with Section 2.14 we can investigate here the case where the cost can be defined, for example, as

$$s(x) = \sup M_x\left[g(x_\tau) - \int_0^\tau c(x_s)ds\right],$$

where the cost of observations $c(x) \geq 0$, $c(x)$ and $g(x) \in \mathbb{L}_0$, and sup is taken over the stopping times for which the mathematical expectations

$$M_x\left[g(x_\tau) - \int_0^\tau c(x_s)ds\right], \qquad x \in E,$$

are defined.

By using the methods suggested in the present chapter and the methods described in Section 2.14 (see Theorem 2.24 and Remark 2 to this theorem), it is easy to prove that under the assumptions made about the functions $c(x)$ and $g(x)$ the payoff $s(x)$ belongs to the domain of definition of the characteristic operator (in topology $\mathscr{C}_0$) and is a solution of the generalized Stefan problem:

$$\begin{aligned} \mathfrak{U}s(x) &= -c(x), \qquad x \in \{x : g(x) < s(x) < \infty\}, \\ s(x) &= g(x), \qquad x \in \{x : g(x) = s(x)\}. \end{aligned}$$

3.8.2

Very simple examples[12] illustrate that the Stefan problem (3.131) has, as a rule, a nonunique solution. Hence to distinguish the unique solution (coinciding with the payoff $s(x)$) we need to find additional conditions which $s(x)$, the function sought, must necessarily satisfy.

We shall consider below the case where we can find additional conditions which the function $s(x)$ satisfies on the boundary $\partial\Gamma_0$ of the domain of stopping of observations $\Gamma_0 = \{x : s(x) = g(x)\}$. These conditions may also be insufficient for finding the payoff $s(x)$. However, in the problems to be considered in Chapter 4 these conditions enable us to define completely the payoff $s(x)$ and the structure of the optimal stopping rule.

Let $X = (x_t, \mathscr{F}_t, P_x)$, $t \geq 0$, be a one-dimensional continuous (standard) Markov process with state space $(E, \mathscr{B})$, where $E \subseteq R$ and the function $g \in \mathbb{L}_0(A^-, A^+)$. Set $s(x) = \sup_{\tau \in \mathfrak{M}} M_x g(x_\tau)$, $\Gamma_0 = \{x \in E : s(x) = g(x)\}$, $C_0 = E - \Gamma_0$, and let $\partial\Gamma_0$ be the boundary of the set Γ_0. We shall assume that for each point $y \in \partial\Gamma_0$ and sufficiently small $\rho > 0$ the set $V_\rho^-(y) = \{x : y - \rho < x < y\} \subseteq C_0$ and $V_\rho^+(y) = \{x : y + \rho > x \geq y\} \subseteq \Gamma_0$. It is

[12] See, for example, [83].

seen that $s(y) = g(y)$ if $y \in \partial\Gamma_0$ and $V_\rho^-(y) \cup V_\rho^+(y) = V_\rho(y)$ where $V_\rho(y) = \{x : |x - y| < \rho\}$.

Let

$$\sigma_\rho(y) = \inf\{t \geq 0 : x_t \in E - V_\rho(y)\}.$$

Theorem 16. *Let $X = (x_t, \mathscr{F}_t, P_x)$, $t \geq 0$, be a one-dimensional continuous (standard) Markov process with state space $E \subseteq R$ and $g \in \mathbb{L}_0(A^-, A^+)$. Assume that)*[13]

(A_1) $g(y) = T_{\sigma_\rho(y)} g(y) + o(\rho), \qquad y \in \partial\Gamma_0$;

(A_2) *In some neighborhood $V_\rho^-(y) \cup \{y\}$ of the point $y \in \partial\Gamma_0$, the left derivatives*

$$\frac{d^- g(x)}{dx} \quad \text{and} \quad \frac{d^- s(x)}{dx}$$

exist and are continuous;

(A_3) *For small $\rho > 0$, $P_y\{x_{\sigma_\rho(y)} = y - \rho\} \geq c > 0$.*

Then at the point $y \in \partial\Gamma_0$ we have the condition of "smooth pasting":

$$\frac{d^- s(x)}{dx} = \frac{d^- g(x)}{dx}\bigg|_{x = y \in \partial\Gamma_0}. \tag{3.136}$$

PROOF. Let $f(x) = s(x) - g(x)$ and let us show that

$$T_{\sigma_\rho(y)} f(y) = o(\rho). \tag{3.137}$$

In fact, by virtue of Theorem 1 and Lemma 1, $s(y) \geq T_{\sigma_\rho(y)} s(y)$. By using the condition (A_1), we find from the above that

$$g(y) = T_{\sigma_\rho(y)} g(y) + o(\rho) = s(y) \geq T_{\sigma_\rho(y)} s(y);$$

therefore

$$o(\rho) = T_{\sigma_\rho(y)} s(y) - T_{\sigma_\rho(y)} g(y) = T_{\sigma_\rho(y)} f(y) \geq 0,$$

which proves (3.137).

By virtue of (3.32) and Theorem 1

$$\overline{\lim_t}\, s(x_t) = \overline{\lim_t}\, g(x_t) \qquad (P_x\text{-a.s}),\ x \in E.$$

Hence

$$T_{\sigma_\rho(y)} f(y) = \int_{\{\omega : \sigma_\rho(y) < \infty\}} f(x_{\sigma_\rho(y)}) dP_y.$$

[13] $T_{\sigma_\rho(y)} g(y) = M_y g(x_{\sigma_\rho(y)}) = \int_{\{\omega: \sigma_\rho(y) < \infty\}} g(x_{\sigma_\rho(y)}) dP_y + \int_{\{\omega: \sigma_\rho(y) = \infty\}} \overline{\lim_t}\, g(x_t) dP_y$.

By expanding the function $f(x)$ in the neighborhood of the point y in a Taylor series and making use of the fact that the process X is continuous we find that for sufficiently small $\rho > 0$

$$\begin{aligned}
T_{\sigma_\rho(y)} f(y) &= \int_{\{\omega: \sigma_\rho(y) < \infty\}} f(x_{\sigma_\rho(y)}) dP_y \\
&= \int_{\{\omega: x_{\sigma_\rho(y)} = y - \rho\}} f(x_{\sigma_\rho(y)}) dP_y \\
&= \frac{d^- f(x)}{dx}\bigg|_{x=y} \cdot \int_{\{\omega: x_{\sigma_\rho(y)} = y - \rho\}} [x_{\sigma_\rho(y)} - y] dP_y + R_1(\rho) \\
&= -\rho \frac{d^- f(x)}{dx}\bigg|_{x=y} \cdot P_y\{x_{\sigma_\rho(y)} = y - \rho\} + R_1(\rho),
\end{aligned} \tag{3.138}$$

where $R_1(\rho) = o(\rho)$ by virtue of condition (A_2).

According to condition (A_3), for sufficiently small ρ the probability $P_y\{x_{\sigma_\rho(y)} = y - \rho\} \geq c > 0$. Hence the required relation, (3.136), follows immediately from (3.137) and (3.138). □

Remark 3. If $V_\rho^+(y) \subseteq C_0$, relation (3.136) is to be replaced by the equation of the right derivatives:

$$\frac{d^+ s(x)}{dx} = \frac{d^+ g(x)}{dx}\bigg|_{x = y \in \partial\Gamma_0}. \tag{3.139}$$

(In this case condition (A_2) needs to be modified in an obvious manner.)

Remark 4. The deduction described of "smooth pasting" conditions (3.136) and (3.139) extends to the case of l-dimensional processes (for details, see [49]).

The result given below follows immediately from Theorems 15 and 16.

Theorem 17. *Let the conditions made in Theorem 16 be satisfied. Then the payoff $s(x)$ is the solution of the following generalized Stefan problem:*

$$\begin{aligned}
\mathfrak{U} s(x) = 0, &\qquad x \in C_0 = E - \Gamma_0, \\
s(x) = g(x), &\qquad x \in \Gamma_0,
\end{aligned} \tag{3.140}$$

$$\frac{d^- s(x)}{dx} = \frac{d^- g(x)}{dx}\bigg|_{x=y}, \qquad y \in \partial\Gamma_0,\ V_0^-(y) \subseteq C_0,$$

$$\frac{d^+ s(x)}{dx} = \frac{d^+ g(x)}{dx}\bigg|_{x=y}, \qquad y \in \partial\Gamma_0,\ V_0^+(y) \subseteq C_0.$$

Notes to Chapter 3

3.1–2. As in the case of discrete time, the theory of optimal stopping rules for continuous time parameter develops in a martingale direction and in a Markov direction. For the martingale direction, see, for instance, Fakejev [41], Dochviri [27], and Thompson [101].

The results of the present chapter are related to the Markov direction which was further developed in Zvonkin [112] for sequentially controlled Markov processes. The definitions and properties of excessive functions can be found in Hunt [55], Dynkin [31], Blumenthal and Getoor [12], and Meyer [72]. Lemma 2 is due to A. Engelbert [37]. The technique of constructing the smallest excessive majorant of the function $g(x)$ given in Lemma 3 was described by Grigelionis and Shiryayev [49]. Another technique was presented by Dynkin [32].

3.3. In the case $g(x) \geq 0$ the assertion of Theorem 1 that the payoff $s(x)$ is the smallest excessive majorant of $g(x)$ was proved by Dynkin [32]. The payoff $\bar{s}(x)$ for Markov processes with continuous time has been examined here for the first time. The example given in Section 1 can be found in Taylor [100]. Theorems 2 and 3 were proved by G. Yu. Engelbert [38]. The assertions of these theorems made under more restrictive assumptions are given in Dynkin [32] and Shiryayev [94].

3.4. Theorems 4, 5, and auxiliary assertions of this section are due to A. Engelbert [37]. $\mathscr{R}$-regularity, assumed in investigating the structure of the payoffs $s(x)$ and $\bar{s}(x)$ for Markov processes with continuous time, was introduced by Shiryayev [94]. For the assertions of Theorem 6, see Shiryayev [94], chap. 3, §4, and also see G. Yu. Engelbert [39] and A. Engelbert [37].

3.5. Theorems 7–10 are given in A. Engelbert [37].

3.6. The discussion presented in this section follows A. Engelbert [36].

3.7. $\varepsilon(x)$—optimal times were investigated by P. Katyshev and by A. Engelbert [35].

3.8. The discussion presented in this section follows Grigelionis and Shiryayev [49], Shiryayev [94], G. Yu. Engelbert [39], and A. Engelbert [37]. The "smooth pasting" condition was used in solving specific problems in Mikhalevich [74], Chernoff [16], Lindley [67], Bather [7], Shiryayev [91], Whittle [110], and Stratanovich [98]. Theorem 16 is due to Grigelionis and Shiryayev [49]. For the deduction of the "smooth pasting" conditions on the boundary of a stopping domain see also Grigelionis [50] and Krylov [61].

Some applications to problems of mathematical statistics

4

4.1 The sequential testing of two simple hypotheses (discrete time)

4.1.1

The objective of the present chapter is to show how the methods for finding optimal stopping rules are used for solving problems of statistical sequential analysis.

The problem of sequential testing of two simple hypotheses is discussed in Sections 1 and 2. The problem of the earliest detection of the time at which the probabilistic characteristics change in the observable process (the *disorder* (*disruption*) *problem*) is investigated in Sections 3 and 4.

4.1.2

We now formulate the general problem of testing two statistical hypotheses.

In a measure space $(\Omega, \mathscr{F})$ we are given two probability measures P^0, P^1 and a sequence of random variables $\xi_1, \xi_2, \ldots$, whose joint distribution is P^θ where θ is an unknown parameter taking on values 0 and 1. Our problem is to estimate the true value of parameter θ from the observations $\xi_1, \xi_2, \ldots$ with minimum loss.

We shall consider the case where $\xi_1, \xi_2, \ldots$ form a sequence of independent uniformly distributed random variables with probability density[1] (with respect to measure μ) $p_\theta(x)$ with respect to each measure P^θ, $\theta = 0, 1$.

We shall investigate the Bayes and fixed error probability formulation of this problem, taking into consideration the assumptions on the structure of the unknown parameter.

[1] This assumption is not restrictive since we can always take, for example, $\mu = \frac{1}{2}(P^0 + P^1)$.

4.1.3

For the Bayes formulation, in a measure space $(\Omega, \mathscr{F})$, let random variables $\theta, \xi_1, \xi_2, \ldots$ and a probability measure P^π be given such that

(1°)
$$P^\pi\{\theta = 1\} = \pi, \qquad P^\pi\{\theta = 0\} = 1 - \pi \tag{4.1}$$

(π is a fixed number from $[0, 1]$);

(2°) For each set $A \in \mathscr{F}^\xi = \sigma\{\omega : \xi_1, \xi_2, \ldots\}$

$$P^\pi\{A \cap [\theta = j]\} = \pi P^j(A), \qquad j = 0, 1, \tag{4.2}$$

where P^1 and P^0 are two probability measures on $(\Omega, \mathscr{F}^\xi)$ independent of π and possessing the property that for any $n \geq 1$

(3°)
$$P^j\{\xi_1 \leq x_1, \ldots, \xi_n \leq x_n\} = \prod_{k=1}^{n} P^j\{\xi_k \leq x_k\}. \tag{4.3}$$

Let us note that it follows from (1°) and (2°) that for each $A \in \mathscr{F}^\xi$ we have

$$P^\pi(A) = \pi P^1(A) + (1 - \pi)P^0(A).$$

In other words, (compare with (2.206) and (2.207)) we assume that θ takes on the two values 1 and 0 with *a priori* probabilities π and $1 - \pi$. We also assume that $\xi_1, \xi_2, \ldots$ are (conditionally) independent and uniformly distributed (with distribution functions $F^j(x) = P^j\{\xi_1 \leq x\}$ independent of π) if the random variable θ is equal to j.

Let $\mathfrak{M}^\xi = \{\tau\}$ be a set of Markov times $\tau = \tau(\omega)$ with values in $N = \{0, 1, \ldots\}$ (with respect to the system of σ-algebras $\{\mathscr{F}_n^\xi\}$, $n \in N$, where $\mathscr{F}_n^\xi = \sigma\{\omega : \xi_1, \ldots, \xi_n\}$ for $n \geq 1$, and $\mathscr{F}_0^\xi = \{\varnothing, \Omega\}$), and let $\mathscr{D}_\tau^\xi = \{d\}$ be a set of $\mathscr{F}_\tau^\xi$-measurable functions $d = d(\omega)$ taking on values 0 and 1.

As in the problem of testing two hypotheses $\langle\!\langle H_0 : \theta = 0 \rangle\!\rangle$ and $\langle\!\langle H_1 : \theta = 1 \rangle\!\rangle$, the choice of time τ implies the choice of a rule for terminating the observation of $\xi_1, \xi_2, \ldots$; the *terminal decision function* $d = d(\omega)$ indicates that either the hypothesis H_0 or the hypothesis H_1 should be accepted. We shall accept the hypothesis H_1 if $d(\omega) = 1$, and we shall accept the hypothesis H_0 if $d(\omega) = 0$.

Definition 1. We call a pair of functions $\delta = (\tau, d)$ in which $\tau \in \mathfrak{M}^\xi$ and $d \in \mathscr{D}_\tau^\xi$ a *decision function* or a *decision rule with a set of rules* $\Delta = \{\delta\}$.

Each decision rule $\delta = (\tau, d)$ naturally implies losses of two kinds: the loss due to the cost of the observations; and the loss due to a terminal decision error.

We shall assume the cost of one observation to be $c > 0$. Then the average loss of the first kind can be naturally characterized by $cM^\pi\tau$ where $M^\pi\tau$ is the mathematical expectation of the observation of duration τ. To describe the loss of the second kind we need the function

$$W(\theta, d) = \begin{cases} a & (\theta = 1, d = 0), \\ b & (\theta = 0, d = 1), \\ 0 & (\theta = i, d = i; i = 0; 1), \end{cases}$$

where $a > 0$, $b > 0$.

The average loss of the terminal decision $d = d(\omega)$ is

$$\begin{aligned} M^\pi W(\theta(\omega), d(\omega)) &= aP^\pi\{\theta = 1, d = 0\} + bP^\pi\{\theta = 0, d = 1\} \\ &= a\pi P^\pi\{d = 0 \mid \theta = 1\} + b(1 - \pi)P^\pi\{d = 1 \mid \theta = 0\}. \end{aligned} \tag{4.4}$$

The probability $P^\pi\{d = 0 \mid \theta = 1\} = P^1\{d = 0\}$ is called an *error probability of the first kind* and is denoted by $\alpha^\pi(\delta)$. Similarly, $\beta^\pi(\delta) = P^\pi\{d = 1 \mid \theta = 0\} = P^0\{d = 1\}$ is called an *error probability of the second kind.*

Finally, let

$$\rho^\pi(\delta) = cM^\pi\tau + M^\pi W(\theta, d) \tag{4.5}$$

be the total loss (or *risk*) of the decision rule $\delta = (\tau, d)$.

Definition 2. Let π be a fixed number from $[0, 1]$. The decision rule $\delta_\pi^* = (\tau_\pi^*, d_\pi^*)$ is called a *π-Bayes rule* if

$$\rho(\delta_\pi^*) = \inf_{\delta \in \Delta} \rho^\pi(\delta). \tag{4.6}$$

We shall show that the search for π-Bayes decision rules can be reduced to solving a problem of optimal stopping for a Markov random function constructed in a specific manner.

Let $\pi_n^\pi = P^\pi\{\theta = 1 \mid \mathscr{F}_n^\xi\}$ be the *a posteriori* probability of the hypothesis $H_1 : \theta = 1$. Since $\mathscr{F}_0^\xi = \{\varnothing, \Omega\}$, $\mathrm{P}^\pi\{\pi_0^\pi = \pi\} = 1$. Using Bayes' formula we can easily show that for all $n \geq 1$ (P^π-a.s.)

$$\pi_n^\pi = \frac{\pi p_1(\xi_1) \cdots p_1(\xi_n)}{\pi p_1(\xi_1) \cdots p_1(\xi_n) + (1 - \pi)p_0(\xi_1) \cdots p_0(\xi_n)}. \tag{4.7}$$

From this we have

$$\pi_{n+1}^\pi = \frac{\pi_n^\pi p_1(\xi_{n+1})}{\pi_n^\pi p_1(\xi_{n+1}) + (1 - \pi_n^\pi)p_0(\xi_{n+1})}. \tag{4.8}$$

Thus, the system of statistics $\{\pi_n^\pi\}$, $n = 0, 1, \ldots$, is transitive.

In the case above, (2.186) is satisfied (see the corresponding verification in (2.215)). Hence the system $\Pi^\pi = (\pi_n^\pi, \mathscr{F}_n^s, P^\pi)$, $n = 0, 1, \ldots$, forms (for a given π) a Markov random function.

Let

$$\rho^\pi = \inf_{\delta \in \Delta} \rho^\pi(\delta). \tag{4.9}$$

Lemma 1. *Let $\delta = (\tau, d)$ be a decision rule and let $\tilde{\delta} = (\tau, \tilde{d})$ be another decision rule having the same time τ and such that*

$$\tilde{d} = \begin{cases} 1, & a\pi_\tau^\pi \geq b(1 - \pi_\tau^\pi), \\ 0, & a\pi_\tau^\pi < b(1 - \pi_\tau^\pi). \end{cases} \tag{4.10}$$

Then

$$\rho^\pi(\tilde{\delta}) \leq \rho^\pi(\delta), \tag{4.11}$$

$$\rho^\pi(\tilde{\delta}) = M^\pi[c\tau + g(\pi_\tau^\pi)], \tag{4.12}$$

$$\rho^\pi = \inf_{\tau \in \mathfrak{M}^\xi} M^\pi[c\tau + g(\pi_\tau^\pi)] \tag{4.13}$$

where $g(\pi) = \min\{a\pi, b(1 - \pi)\}$.

PROOF. If $\pi = 0$ or $\pi = 1$, then the assertions of the lemma are obvious. Hence we assume that $0 < \pi < 1$.

Let us consider measures P^0, P^1, and P^π on $(\Omega, \mathscr{F}^\xi)$. Since for $A \in \mathscr{F}^\xi$

$$P^\pi(A) = \pi P^1(A) + (1 - \pi)P^0(A),$$

it follows that for $0 < \pi < 1$ the measures P^1 and P^0 are absolutely continuous with respect to measure P^π. Let

$$\frac{dP^i}{dP^\pi}(\omega)$$

be the Radon–Nikodym derivative of measure P^i with respect to measure $P^\pi (i = 0, 1)$, and let

$$\frac{dP^i}{dP^\pi}(\mathscr{F}_n^\xi)(\omega)$$

be the Radon–Nikodym derivative of the restriction of measures P^i and P^π to a σ-algebra $\mathscr{F}_n^\xi$.

It can be readily seen that (P^π-a.s.)

$$\pi_n^\pi(\omega) = \pi \frac{dP^1}{dP^\pi}(\mathscr{F}_n^\xi)(\omega) = \pi M^\pi\left[\frac{dP^1}{dP^\pi}(\omega) | \mathscr{F}_n^\xi\right].$$

It follows that for any stopping time τ (with respect to the system $\mathscr{F}^\xi = \{\mathscr{F}_n^\xi\}, n \geq 0$)

$$\pi_\tau^\pi(\omega) = \pi M^\pi\left[\frac{dP^1}{dP^\pi}(\omega) | \mathscr{F}_\tau^\xi\right] \qquad (P^\pi\text{-a.s.})$$

since[2]

$$\pi_\tau^\pi(\omega) = \pi_n^\pi(\omega) \qquad (\{\tau = n\}; (P^\pi\text{-a.s.}))$$

[2] We recall that the random variables ξ and η given on $(\Omega, \mathscr{F}, P)$ are said to coincide (P-a.s.) on the set A (written: $\xi = \eta(A; P\text{-a.s.})$) if $P\{A \cap [\xi \neq \eta]\} = 0$.

(by definition, $\pi_\tau^\pi = \sum_{n=0}^\infty \pi_n^\pi I_{\{\tau=n\}}$) and

$$M^\pi\left[\frac{dP^1}{dP^\pi}(\omega)|\mathscr{F}_\tau^\xi\right] = M^\pi\left[\frac{dP^1}{dP^\pi}(\omega)|\mathscr{F}_n^\xi\right] \qquad (\{\tau = n\}; \qquad (P^\pi\text{-a.s.}))$$

(see [69], lem. 1.9).

Thus we have

$$\begin{aligned} P^\pi\{\theta = 1, d = 0\} &= \pi P^1\{d = 0\} \\ &= \pi M^1[1 - d(\omega)] \\ &= \pi M^\pi\left[(1 - d(\omega))\frac{dP^1}{dP^\pi}(\omega)\right] \\ &= M^\pi\left[(1 - d(\omega))M^\pi\left(\pi\frac{dP^1}{dP^\pi}(\omega)|\mathscr{F}_\tau^\xi\right)\right] \\ &= M^\pi[(1 - d(\omega))\pi_\tau^\pi(\omega)], \end{aligned}$$

and, similarly,

$$P^\pi\{\theta = 0, d = 1\} = M^\pi[d(\omega)(1 - \pi_\tau^\pi(\omega))].$$

By virtue of (4.4) and (4.5) we have

$$\begin{aligned} \rho^\pi(\delta) &= cM^\pi\tau + M_\tau^\pi W(\theta, d) \\ &= M^\pi(c\tau + a\pi_\tau^\pi[1 - d(\omega)] + b[1 - \pi_\tau^\pi]d(\omega)\} \\ &\geq M^\pi\{c\tau + \min[a\pi_\tau^\pi, b(1 - \pi_\tau^\pi)]\} \\ &= M^\pi\{c\tau + g(\pi_\tau^\pi)\} = \rho^\pi(\tilde{\delta}), \end{aligned}$$

thus proving the lemma. □

The lemma implies, in particular, that in seeking π-Bayes decision rules it is sufficient to consider only the rules of the type $\tilde{\delta} = (\tau, d)$ in which $\tau \in \mathfrak{M}^\xi$ and

$$d(\omega) = \begin{cases} 1, & \pi_\tau^\pi(\omega) \geq \dfrac{b}{a + b}, \\ 0, & \pi_\tau^\pi(\omega) < \dfrac{b}{a + b}. \end{cases} \tag{4.14}$$

It also follows from the lemma that the problem of finding a π-Bayes decision rule is reduced to the problem of finding (for a given π) a time τ_π^* such that

$$M^\pi[c\tau_\pi^* + g(\pi_{\tau_\pi^*}^\pi)] = \inf_{\tau \in \mathfrak{M}^\xi} M^\pi[c\tau + g(\pi_\tau^\pi)]. \tag{4.15}$$

We have seen (Section 2.15) that this problem can be solved by solving the pertinent problem for the Markov process (constructed in a specific manner)

$$\Pi = (\pi_n, \mathscr{F}_n, P_\pi), \qquad n \geq 0,$$

with values in a state space $(E, \mathscr{B}) = ([0, 1], \mathscr{B}([0, 1]))$ and defined on the extended space $(\tilde{\Omega}, \tilde{\mathscr{F}}) = (\Omega \times [0, 1], \mathscr{F} \times \mathscr{B}([0, 1]))$ as follows:

$$\pi_0(\tilde{\omega}) = \pi, \qquad \pi_n(\tilde{\omega}) = \pi_n^{\pi}(\omega),$$

for $\tilde{\omega} = (\omega, \pi)$ with $\omega \in \Omega$, $\pi \in [0, 1]$;

$$\mathscr{F}_n = \sigma\{\tilde{\omega} : \tilde{\xi}_0(\tilde{\omega}), \tilde{\xi}_1(\tilde{\omega}), \ldots, \tilde{\xi}_n(\tilde{\omega})\}$$

where $\tilde{\xi}_0(\tilde{\omega}) = \pi$, $\tilde{\xi}_n(\tilde{\omega}) = \xi_n(\omega)$, $n = 1, 2, \ldots$;

$$P_\pi(A) = P^\pi(A_\pi) I_{A^\omega}(\pi)$$

for the set $A \in \mathscr{F}_\infty = \sigma(\bigcup \mathscr{F}_n)$, where

$$A_\pi = \{\omega : (\omega, \pi) \in A\}, \; A^\omega = \{\pi : (\omega, \pi) \in A\}.$$

Let $\Pi = (\pi_n, \mathscr{F}_n, P_\pi)$, $n \geq 0$, $\pi \in [0, 1]$, be the Markov process constructed and let

$$\rho(\pi) = \inf_{\tau \in \mathfrak{M}[F]} M_\pi[c\tau + g(\pi_\tau)]. \tag{4.16}$$

It can easily be shown[3] that it is sufficient to take inf in (4.16) only over the class of stopping times $\mathfrak{M}^1[F] = \{\tau \in \mathfrak{M}[F] : M_\pi \tau < \infty, 0 \leq \pi \leq 1\}$. Then by virtue of Theorem 2.23 we have[4] that the risk $\rho(\pi)$ satisfies the recursive equation

$$\rho(\pi) = \min\{g(\pi), c + T\rho(\pi)\} \tag{4.17}$$

and the time

$$\tau_0 = \inf\{n \geq 0 : \rho(\pi_n) = g(\pi_n)\} \tag{4.18}$$

is an optimal stopping time (in the class $\mathfrak{M}^1[F^\pi] \subseteq \mathfrak{M}^1[F]$).

Let us investigate the structure of the function $\rho(\pi)$. Let

$$Q_{(1,c)} g(\pi) = \min\{g(\pi), c + Tg(\pi)\}.$$

It can easily be seen from (4.7) and the fact that the function $g(\pi)$ is convex upward that the function

$$\begin{aligned} Tg(\pi) &= M_\pi g(\pi_1) \\ &= \int_{-\infty}^{\infty} g\left(\frac{\pi p_1(x)}{\pi p_1(x) + (1 - \pi) p_0(x)}\right) [\pi p_1(x) + (1 - \pi) p_0(x)] d\mu(x) \end{aligned}$$

is also convex upward. This fact implies that the functions $Q_{(1,c)} g(\pi)$, $Q^2_{(1,c)} g(\pi), \ldots$ also have the property given above. Therefore, the function $\rho(\pi) = \lim_n Q^n_{(1,c)} g(\pi)$ (Theorem 2.23) is convex upward. It follows from the convexity (upward) of the functions $\rho(\pi)$ and $T\rho(\pi)$ that they are continuous

[3] Compare with Section 2.1.

[4] It is obviously not difficult to formulate the results of Theorem 2.23 for a risk case.

on the interval (0, 1) ([31], theor. 0.8). Hence, by virtue of (4.18), the domain of continuing observations $C_0^* = \{\pi : \rho(\pi) < g(\pi)\}$ is

$$C_0^* = \{\pi : A^* < \pi < B^*\}$$

where $0 \leq A^* \leq B^* \leq 1$ (Figure 6).

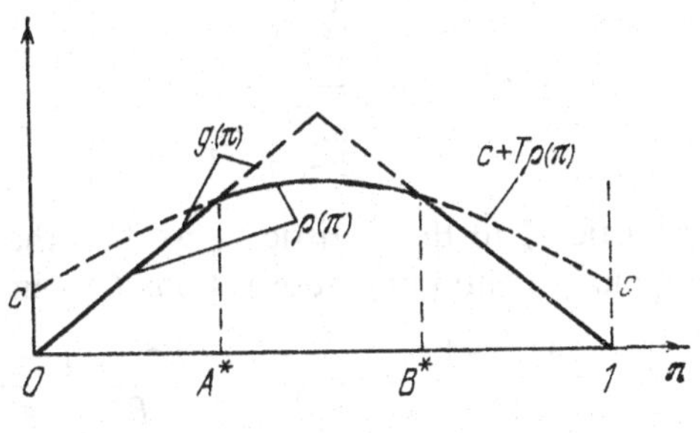

Figure 6

It can easily be seen that if $\rho(\pi) \equiv g(\pi)$, then $A^* = B^* = b/(a + b)$ and, therefore, the domain of *continuing observations* $C_0^* = \varnothing$. If $\rho(\pi) < g(\pi)$ at least at one point, then $A^* < B^*$.

Thus, we have found the structure of the optimal stopping rule for "Problem (4.15)." According to Section 2.15 the knowledge of this optimal rule enables us to describe the optimal stopping rule in "Problem 4.13" for any given π, $0 \leq \pi \leq 1$. Namely, the optimal time τ_π^* (see (4.15)) is the following:

$$\tau_\pi^* = \min\{n \geq 0 : \pi_n^\pi \in \Gamma_0^*\},$$

where

$$\Gamma_0^* = [0, 1] - (A^*, B^*) = [0, A^*] \cup [B^*, 1].$$

This, together with (4.14), proves:

Theorem 1. *In the problem of sequential testing of two hypotheses the π-Bayes rule $\delta_\pi^* = (\tau_\pi^*, d_\pi^*)$ has the following structure*:

$$\tau_\pi^* = \min\{n \geq 0 : \pi_n^\pi \notin (A^*, B^*)\}, \tag{4.19}$$

$$d_\pi^* = \begin{cases} 1, & a\pi_{\tau_\pi^*}^\pi \geq b(1 - \pi_{\tau_\pi^*}^\pi), \\ 0, & a\pi_{\tau_\pi^*}^\pi < b(1 - \pi_{\tau_\pi^*}^\pi), \end{cases} \tag{4.20}$$

where A^* and B^* are constants, $0 \leq A^* \leq B^* \leq 1$.

Remark 1. The terminal decision d_π^* can be written

$$d_\pi^* = \begin{cases} 1, & \pi_{\tau_\pi^*}^\pi \geq B^*, \\ 0, & \pi_{\tau_\pi^*}^\pi \leq A^*. \end{cases} \tag{4.21}$$

Remark 2. Let

$$\varphi_n = \frac{p_1(\xi_1)\cdots p_1(\xi_n)}{p_0(\xi_1)\cdots p_0(\xi_n)}, \qquad \varphi_0 \equiv 1.$$

Then for all $\pi < 1$ and $n \geq 0$ we have

$$\pi_n^\pi = \frac{\dfrac{\pi}{1-\pi}\varphi_n}{1 + \dfrac{\pi}{1-\pi}\varphi_n}.$$

Going from the statistic π_n^π to the statistic φ_n we find that (for a given π, $0 < \pi < 1$) the domain of continuing observations C_0^* can be written

$$C_0^* = \left\{\varphi : \frac{A^*}{1-A^*}\cdot\frac{1-\pi}{\pi} < \varphi < \frac{B^*}{1-B^*}\cdot\frac{1-\pi}{\pi}\right\}.$$

To describe the π-Bayes rule δ_π^* completely we need to define the unknown constants A^* and B^* in (4.19). However, the problem of determining them generally presents difficulties. We shall consider a similar problem, the sequential testing of two simple hypotheses for a Wiener process, in Section 2. By using the results given in Chapter 3 we can find the system of equations by which A^* and B^* are defined uniquely.

4.1.4

The fixed error probability formulation involves no probabilistic assumptions on the unknown parameter θ. On a measure space $(\Omega, \mathscr{F})$ let two probability measures P_0, P_1 and a sequence of independent (with respect to each measure P_i, $i = 0, 1$) random variables $\xi_1, \xi_2, \ldots$ be given. Let

$$\mathscr{F}_0^\xi = \{\varnothing, \Omega\},$$
$$\mathscr{F}_n^\xi = \sigma\{\omega : \xi_1, \ldots, \xi_n\}, \qquad n \geq 1,$$

let $\mathfrak{M}^\xi = \{\tau\}$ be the class of stopping times $\tau = \tau(\omega)$ (with respect to $F^\xi = \{\mathscr{F}_n^\xi\}, n \geq 0$, and let $\mathscr{D}_\tau^\xi = \{d\}$ be the set of $\mathscr{F}_\tau^\xi$-measurable functions $d = d(\omega)$ taking on values 0 and 1. Next, let $\Delta^\xi(\alpha, \beta)$ be the class of decision rules $\delta = (\tau, d)$ such that: $M_0\tau < \infty$ and $M_1\tau < \infty$; the error probabilities $\alpha(\delta) = P_1\{d(\omega) = 0\} \leq \alpha$ and $\beta(\delta) = P_0\{d(\omega) = 0\} \leq \beta$, where the nonnegative numbers α and β are subject to the condition $\alpha + \beta < 1$.

Following A. Wald we formulate the problem of testing two simple hypotheses as follows (see [106]).

Let two nonnegative numbers α and β be given, $\alpha + \beta < 1$. It is required to find in the class $\Delta^\xi(\alpha, \beta)$ a rule $\tilde{\delta} = (\tilde{\tau}, \tilde{d})$ such that

$$\mathrm{M}_0\tilde{\tau} \leq M_0\tau \quad \text{and} \quad M_1\tilde{\tau} \leq M_1\tau \tag{4.22}$$

simultaneously for all $\delta = (\tau, d) \in \Delta^\xi(\alpha, \beta)$.

It should be noted that the existence of such a rule minimizing two mathematical expectations simultaneously is not obvious at all; it was A. Wald's profound accomplishment to conjecture the existence of such a rule and determine its structure.

Let us consider the statistics $\varphi_0 \equiv 1$ and

$$\varphi_n = \frac{p_1(\xi_1) \cdots p_1(\xi_n)}{p_0(\xi_1 \cdots p_0(\xi_n)}, \qquad n \geq 1,$$

and let

$$\lambda_0 = \ln \varphi_0 \equiv 0,$$

$$\lambda_n = \ln \varphi_n = \sum_{k=1}^{n} \ln \frac{p_1(\xi_k)}{p_0(\xi_k)}.$$

It is clear that the domain of continuing observations can be written in terms of the above statistics in the π-Bayes formulation of the problem as follows:

$$C_0^* = \left\{\lambda : \ln\left(\frac{A^*}{1 - A^*} \cdot \frac{1 - \pi}{\pi}\right) < \lambda < \ln\left(\frac{B^*}{1 - B^*} \cdot \frac{1 - \pi}{\pi}\right)\right\}$$

or

$$C_0^* = \{\lambda : A^*(\pi) < \lambda < B^*(\pi)\},$$

where

$$A^*(\pi) = \ln\left(\frac{A^*}{1 - A^*} \cdot \frac{1 - \pi}{\pi}\right), \qquad B^*(\pi) = \ln\left(\frac{B^*}{1 - B^*} \cdot \frac{1 - \pi}{\pi}\right).$$

It turns out that the domain structure remains unchanged for the optimal decision rule in the fixed error probability formulation as well.

Theorem 2. *Let nonnegative numbers α and β be such that $\alpha + \beta < 1$, and let there be numbers $\tilde{A}$ and $\tilde{B}$, $\tilde{A} < 0 < \tilde{B}$, having the property that for the rule $\tilde{\delta} = (\tilde{\tau}, \tilde{\delta})$ with*

$$\tilde{\tau} = \inf\{n \geq 0 : \lambda_n \notin (\tilde{A}, \tilde{B})\},$$

$$\tilde{d} = \begin{cases} 1, & \lambda_{\tilde{\tau}} \geq \tilde{B}, \\ 0, & \lambda_{\tilde{\tau}} \leq \tilde{A}, \end{cases} \tag{4.23}$$

the error probabilities $\alpha(\tilde{\delta})$ and $\beta(\tilde{\delta})$ are equal to α and β respectively.

Then the rule[5] *$\tilde{\delta} = (\tilde{\tau}, \tilde{d})$ in the class $\Delta^{\xi}(\alpha, \beta)$ is optimal in the sense that for any $\delta = (\tau, d) \in \Delta^{\xi}(\alpha, \beta)$*

$$M_0 \tilde{\tau} \leq M_0 \tau, \qquad M_1 \tilde{\tau} \leq M_1 \tau. \tag{4.24}$$

[5] If we go from λ_n to the statistic $\varphi_n = e^{\lambda_n}$, we obtain a rule called the *sequential probability ratio test*.

A proof of optimality of the rule $\tilde{\delta} = (\tilde{\tau}, \tilde{d})$, essentially based on the properties of π-Bayes rules $\delta_\pi^* = (\tau_\pi^*, d_\pi^*)$ (see Theorem 1), is given in [66] (chap. 3, §12) and hence will be omitted here.

Remark 1. The case where $\alpha + \beta \geq 1$ is not interesting for the following reason. Let us consider a randomized decision rule in which the hypothesis H_0 is accepted with probability $1 - \alpha$ and the hypothesis H_1 with probability α without making observations; more precisely, let $(\tilde{\Omega}, \tilde{\mathscr{F}}, \tilde{P})$ be an auxiliary probability space and let $\eta = \eta(\tilde{\omega})$, $\tilde{\omega} \in \tilde{\Omega}$, be a random variable taking on values 0 and 1 with probabilities $1 - \alpha$ and α, respectively. Then, if $\eta = 0$, we shall accept hypothesis H_0 and if $\eta = 1$, hypothesis H_1. For this randomized decision rule the duration of observations is zero and the error probabilities satisfy the constraints given.

Remark 2. Theorem 2 yields the optimality of the rule $\tilde{\delta} = (\tilde{\tau}, \tilde{d})$ in the class of the rules $\delta = (\tau, d) \in \Delta^\xi(\alpha, \beta)$ in which $M_0\tau < \infty$, $M_1\tau < \infty$. In fact, we can show (see [15]) that the rule $\tilde{\delta}$ is optimal in a wider class of rules $\delta = (\tau, d)$ for which $M_0\tau$ and $M_1\tau$ may be infinite.

Remark 3. It may occur that for given α and β the error probabilities of the first and second kind are not equal to α and β, respectively, for any choice of thresholds A and B. In this case Theorem 2 does not guarantee that there is an optimal rule among the rules $\delta_{(A,B)} = (\tau_{(A,B)}, d_{(A,B)}) \in \Delta^\xi(\alpha, \beta)$ such that

$$\tau_{(A,B)} = \inf\{n \geq 0 : \lambda_n \notin (A, B)\},$$

$$d_{(A,B)} = \begin{cases} 1, & \lambda_{\tau(A,B)} \geq B, \\ 0, & \lambda_{\tau(A,B)} \leq A. \end{cases} \tag{4.25}$$

We give now an example illustrating the fact that the rules of the type (4.25) are not optimal.

Let the densities (with respect to Lebesgue measure) $p_0(x)$ and $p_1(x)$ be given by

$$p_0(x) = \begin{cases} 1, & x \in [0, 1], \\ 0, & x \notin [0, 1], \end{cases}$$

$$p_1(x) = \begin{cases} 1, & x \in [a, a + 1], \\ 0, & x \notin [a, a + 1], \end{cases}$$

where $0 < a < 1$. Then for all x, $0 \leq x \leq a + 1$,

$$\ln \frac{p_1(x)}{p_0(x)} \begin{cases} \infty, & x \in [0, a), \\ 0, & x \in [a, 1], \\ -\infty, & x \in (1, 1 + a]. \end{cases} \tag{4.26}$$

It can be seen from (4.26) that for any choice of thresholds $A < 0 < B$ the error probabilities are given by

$$P_0\{d_{(A,B)} = 1\} = 0, \qquad P_1\{_{(A,B)} = 0\} = 0,$$

with

$$M_0 \tau_{(A,B)} = M_1 \tau_{(A,B)} = \frac{1}{a}.$$

Hence, if $\alpha > 0$ and $\beta > 0$, we cannot have error probabilities of first and second kind equal to $\alpha > 0$ and $\beta > 0$, respectively, for any choice of constants A and B, $A < 0 < B$.

At the same time, rules other than $\delta_{(A,B)}$ for which the error probabilities are equal to the given values of $\alpha > 0$ and $\beta > 0$ exist, and the mathematical expectations of observation times (for each of the hypotheses H_0 and H_1) are less than a^{-1}.

Thus, for instance, let $\delta_h(\tau_h, d_h)$ be such that

$$\tau_h = \inf\{n \geq 1 : \xi_n \notin (a + h, 1 - h)\},$$

$$d_h = \begin{cases} 1, & \xi_{\tau_h} \in [1 - h, a + 1], \\ 0, & \xi_{\tau_h} \in [0, a + h], \end{cases}$$

where $0 < h < 1 - (a/2)$. Then

$$\alpha(\delta_h) = P_1\{d_h = 1\} = h, \qquad \beta(\delta_h) = P_0\{d_h = 0\} = h,$$

and

$$M_0 \tau_h = M_1 \tau_h = \frac{1}{a + 2h} < \frac{1}{a} = M_0 \tau_{(A,B)} = M_1 \tau_{(A,B)}.$$

4.1.5

In connection with Theorem 2 the question naturally arises as to how to determine the thresholds $\tilde{A}$ and $\tilde{B}$ and the mathematical expectations of the observation durations $M_0 \tilde{\tau}$ and $M_1 \tilde{\tau}$ from the numbers α and β. As in the π-Bayes problem, it is rather difficult to evaluate these. However, one can find estimates which are adequate for applications.

Let $\delta_{(A,B)} = (\tau_{(A,B)}, d_{(A,B)})$ be a decision rule such that

$$\tau_{(A,B)} = \inf\{n \geq 1 : \lambda_n \notin (A, B)\},$$

$$d_{(A,B)} = \begin{cases} 1, & \lambda_{\tau_{(A,B)}} \geq B, \\ 0, & \lambda_{\tau_{(A,B)}} \leq A, \end{cases} \tag{4.27}$$

where

$$\lambda_n = \ln \varphi_n = \sum_{k=1}^{n} \ln \frac{p_1(\xi_k)}{p_0(\xi_k)}.$$

It follows from Theorem 2 that the optimal rule $\tilde{\delta}$ is a rule of the type given by (4.27).

Let

$$\alpha(A, B) = \alpha(\delta_{(A,B)}) = P_1\{d_{(A,B)} = 0\}$$

and

$$\beta(A, B) = \beta(\delta_{(A, B)}) = P_0\{d_{(A, B)} = 1\}.$$

Theorem 3. *If for given constants A and B we have*

$$P_i\{\tau_{(A, B)} < \infty\} = 1, \qquad i = 0, 1, \tag{4.28}$$

and $\alpha(A, B) < 1$, $\beta(A, B) < 1$, then

$$\ln \frac{\alpha(A, B)}{1 - \beta(A, B)} \leq A, \qquad B \leq \ln \frac{1 - \alpha(A, B)}{\beta(A, B)}. \tag{4.29}$$

PROOF. For given A and B the probability is

$$\begin{aligned}
\alpha(A, B) &= P_1\{\lambda_{\tau(A, B)} \leq A\} \\
&= \int_{\{\lambda_{\tau(A, B)} \leq A\}} P_1(d\omega) \\
&= \int_{\{\lambda_{\tau(A, B)} \leq A\}} \frac{p_1(\xi_1) \cdots p_1(\xi_{\tau(A, B)})}{p_0(\xi_1) \cdots p_0(\xi_{\tau(A, B)})} P_0(d\omega) \\
&= \int_{\{\lambda_{\tau(A, B)} \leq A\}} e^{\lambda_{\tau(A, B)}} P_0(d\omega) \leq e^A P_0\{\lambda_{\tau(A, B)} \leq A\} \\
&= e^A[1 - \beta(A, B)],
\end{aligned}$$

from which the first inequality in (4.29) follows. We can prove the second inequality in (4.29) in a similar way. □

The lemma which follows contains the conditions which, in particular, guarantee that the requirements given by (4.28) in Theorem 3 are satisfied.

Lemma 2. *Let $-\infty < A \leq 0 \leq B < \infty$ and let*

$$P_i\left\{\left|\ln \frac{p_1(\xi_1)}{p_0(\xi_1)}\right| > 0\right\} > 0, \quad i = 0, 1.$$

Then $P_i\{\tau_{(A, B)} < \infty\} = 1$ and there exists a $t_0 > 0$ such that for all $t \leq t_0$

$$M_i e^{t\tau(A, B)} < \infty, \qquad i = 0, 1. \tag{4.30}$$

PROOF. Let

$$z_k = \ln \frac{p_1(\xi_k)}{p_0(\xi_k)}, \qquad s_k = z_1 + \cdots + z_k,$$

and let $C = B - A$. Assume first that $P_i\{|z_1| \leq C\} = p_i < 1$. Then

$$\begin{aligned}
\{\omega : \tau_{(A, B)} \geq n\} &= \{\omega : A < s_k < B, 1 \leq k \leq n - 1\} \\
&\subseteq \{\omega : |z_k| \leq C, 1 \leq k \leq n - 1\},
\end{aligned}$$

and, therefore,

$$P_i\{\tau_{(A,B)} \geq n\} \leq p_i^{n-1}, \tag{4.31}$$

from which

$$\begin{aligned} M_i e^{t\tau_{(A,B)}} &= \sum_{k=1}^{\infty} e^{tk} P_i\{\tau_{(A,B)} = k\} \\ &\leq \sum_{k=1}^{\infty} e^{tk} P_i\{\tau_{(A,B)} \geq k\} \leq e^t \sum_{k=0}^{\infty} (e^t p_i)^k < \infty, \end{aligned}$$

if $e^t p_i < 1$.

Next, let $P_i\{|z_1| \leq C\} = 1$. Then there will be a finite $m \geq 1$ such that

$$0 < P_i\{|z_1 + \cdots + z_m| \leq C\} = p_i < 1,$$

and hence

$$\begin{aligned} &P_i\{\tau_{(A,B)} \geq mk\} \leq p_i^{k-1}, \\ &P_i\{\tau_{(A,B)} \geq n\} \leq p_i^{[n/m]-1} \leq p_i^{-2} p_i^{n/m}. \end{aligned} \tag{4.32}$$

Therefore,

$$M_i e^{t\tau_{(A,B)}} \leq p_i^{-2} \sum_{k=1}^{\infty} (e^t p_i^{1/m})^k < \infty,$$

if $e^t p_i^{1/m} < 1$.

It follows from (4.30) that $\mathrm{P}_i\{\tau_{(A,B)} < \infty\} = 1$ and that $M_i \tau_{(A,B)}^n < \infty$ for all $n \geq 1$.

4.1.6

In establishing (lower) bounds for the average number of necessary observations we rely basically on a general result known as Wald's identity.

Let $(\Omega, \mathscr{F}, P)$ be a probability space and let $\xi, \xi_1, \xi_2, \ldots$ be a sequence of independent identically distributed random variables. Let $\mathscr{F}_n = \sigma\{\omega : \xi_1, \ldots, \xi_n\}$, $s_n = \xi_1 + \cdots + \xi_n$, and let τ be a Markov time (with respect to the system $F = \{\mathscr{F}_n\}$, $n \geq 1$) taking on values $1, 2, \ldots$.

Lemma 3. (Wald's identity). *If $M|\xi| < \infty$, $M\tau < \infty$, then*

$$Ms_\tau = M\xi \cdot M\tau. \tag{4.33}$$

If, in addition, $M\xi^2 < \infty$, then

$$M[s_\tau - Ms_\tau]^2 = D\xi \cdot M\tau \tag{4.34}$$

where $D\xi = M[\xi - M\xi]^2$.

Proof. Let $\tau_N = \min(\tau, N)$ where $N < \infty$. Let $\eta_n = s_n - nM\xi$. It is easily seen that the elements $(\eta_n, \mathscr{F}_n, P)$, $n \geq 1$, form a martingale.

It is obvious that

$$M|\eta_{\tau_N}| < \infty, \lim_{n\to\infty} \int_{\{\tau_N > n\}} |\eta_n| dP = 0. \tag{4.35}$$

Hence we can apply Theorem 1.12, by which

$$M_{\eta_{\tau_N}} = M_{\eta_1} = 0,$$

and, therefore,

$$Ms_{\tau_N} = M\xi \cdot M\tau_N. \tag{4.36}$$

This proves Wald's identity for Markov times bounded with probability 1.

Next let us consider the general case.

We have from (4.33) that

$$M\{|\xi_1| + \cdots + |\xi_{\tau_N}|\} = M|\xi| \cdot M\tau_N \le M|\xi| \cdot M\tau < \infty. \tag{4.37}$$

Since with probability 1

$$\tau_N \uparrow \tau, \qquad \sum_{i=1}^{\tau_N} |\xi_i| \uparrow \sum_{i=1}^{\tau} |\xi_i|, \qquad N \to \infty,$$

we have from (4.37) that

$$M\{|\xi_1| + \cdots + |\xi_\tau|\} = \lim_{N\to\infty} M\{|\xi_1| + \cdots + |\xi\tau_N|\} \le M|\xi| \cdot M\tau < \infty.$$

Therefore,

$$\begin{aligned} M|\eta_\tau| &\le M|s_\tau| + M\tau \cdot M|\xi| \\ &\le M\{|\xi_1| + \cdots + |\xi_\tau|\} + M\tau \cdot M|\xi| < \infty. \end{aligned} \tag{4.38}$$

We shall show that

$$\lim_{n\to\infty} \int_{\{\tau > n\}} |\eta_n| dP = 0. \tag{4.39}$$

We have

$$\begin{aligned} |\eta_n| &\le |\xi_1 - M\xi_1| + \cdots + |\xi_n - M\xi_n| \\ &\le |\xi_1| + \cdots + |\xi_n| + nM|\xi|. \end{aligned}$$

On the set $\{\omega : \tau > n\}$

$$|\eta_n| \le |\xi_1| + \cdots + |\xi_\tau| + \tau M|\xi|.$$

Since $M(|\xi_1| + \cdots + |\xi_\tau|) < \infty$ and $M\tau < \infty$, we have

$$\int_{\{\tau > n\}} |\eta_n| dP \le \int_{\{\tau > n\}} \{|\xi_1| + \cdots + |\xi_\tau|\} dP + M|\xi| \cdot \int_{\tau > n\}} \tau dP \to 0,$$

thus proving (4.39).

By virtue of (4.38) and (4.39) Theorem 1.12 can be applied to the martingale $(\eta_n, \mathscr{F}_n, P)$, $n \geq 1$. Hence $M\eta_\tau = 0$ or $M(s_\tau - \tau M\xi) = 0$. Since $M|s_\tau| < \infty$, $M\tau < \infty$, and $M|\xi| < \infty$, it follows that $Ms_\tau = M\tau \times M\xi$, which proves the first part of the lemma.

Noting that the process $(\eta_n^2 - nD\xi, \mathscr{F}_n, P)$, $n \geq 1$, is a martingale, we can prove (4.34) in a similar fashion.

The considerations above which rely on Theorem 1.12 enable us to obtain relations (similar to (4.33), (4.34)) involving moments of higher orders. The easiest way to obtain these relations for bounded Markov times τ $(P(\tau \leq N) = 1, N < \infty)$ is from the fundamental identity of sequential analysis:

$$Me^{\lambda s_\tau}[\varphi(\lambda)]^{-\tau} = 1, \tag{4.40}$$

where the complex value λ is such that $\varphi(\lambda) = Me^{\lambda\xi}$ exists and does not vanish.

(4.40) follows immediately from Theorem 1.11 if we note that the elements $(e^{\lambda s_n}[\varphi(\lambda)]^{-n}, \mathscr{F}_n, P)$, $n \geq 1$, form a martingale and

$$Me^{\lambda s_1}[\varphi(\lambda)]^{-1} = 1. \tag{4.41}$$

By passing to the limit from the "truncated" times $\tau_N = \min(\tau, N)$ we are able sometimes to prove (4.40) for Markov times τ belonging to the class $\mathfrak{M}$.

4.1.7

We shall next give bounds for the average number of necessary observations in problems of testing N competing hypotheses, where we assume, in general, that $N \geq 2$. Since we have so far considered the case of two hypotheses only, we need to introduce some additional notation.

On a measure space $(\Omega, \mathscr{F})$ let measures P_θ, $\theta = 0, 1, \ldots, N-1$, and a sequence of independent identically distributed (with respect to each measure P_θ, $\theta = 0, 1, \ldots, N-1$) random variables $\xi_1, \xi_2, \ldots$ be given. We may assume without loss of generality that the probability distributions $P_\theta(x) = P_\theta\{\omega : \xi_1 \leq x\}$ have densities $p_\theta(x)$ with respect to some (σ-finite) measure μ.

Let us set $\mathscr{F}_n = \sigma\{\omega : \xi_1, \ldots, \xi_n\}$, $n \geq 1$, and let $\tau = \tau(\omega)$ be a Markov time (with respect to the system $F = \{\mathscr{F}_n\}$) such that $P_\theta\{\tau < \infty\} = 1$ for all $\theta = 0, 1, \ldots, N-1$. Also let $d = d(\omega)$ be an $\mathscr{F}_\tau$-measurable function taking on N values $d_0, \ldots, d_{N-1}$. The value $d = d_i$ will be interpreted as implying that the hypothesis $H_i : \theta = i$ is accepted.

Let

$$\alpha_{ij} = P_i\{d(\omega) = d_j\}, \qquad 0 \leq i, j \leq N-1,$$

be the probability that the hypothesis H_j is accepted when $\theta = i$ and the decision rule $\delta = (\tau, d)$ is used.

Theorem 4. *Let $i(0 \le i \le N-1)$ be fixed and let $\alpha_{ik} = 0$ when the probability $\alpha_{jk} = 0$ for $j \neq i$. Also let $\mu\{x : p_i(x) \neq p_j(x)\} > 0$, $i \neq j$. Then*

$$M_i\tau \ge \max_{j \neq i} \frac{\sum_{k=0}^{N-1} \alpha_{ik} \ln \dfrac{\alpha_{ik}}{\alpha_{jk}}}{M_i \ln \dfrac{p_i(\xi_1)}{p_j(\xi_1)}}, \tag{4.42}$$

where the expressions of the form $0 \cdot \ln(0/c)$ are zero for any $c \ge 0$.

PROOF. If $M_i\tau = \infty$, (4.42) is obvious. Hence from now on we shall assume $M_i\tau < \infty$.

Fix $j \neq i$ as well as i. Then by virtue of the Jensen inequality and the assumption $\mu\{x : p_i(x) \neq p_j(x)\} > 0$

$$0 < M_i \ln \frac{p_i(\xi_1)}{p_j(\xi_1)} = \int_{\{x : p_i(x) > 0\}} \ln \frac{p_i(x)}{p_j(x)} p_i(x) d\mu(x) \tag{4.43}$$

(see [64], pp. 25–26). If $M_i \ln(p_i(\xi_1)/(p_j(\xi_1)) = \infty$, then (4.42) becomes trivial.

Let $M_i \ln(p_i(\xi_1)/p_j(\xi_1)) < \infty$. Then by virtue of Lemma 3

$$M_i \ln \frac{p_i(\xi_1) \cdots p_i(\xi_\tau)}{p_i(\xi_1) \cdots p_j(\xi_\tau)} = M_i\tau \cdot M_i \ln \frac{p_i(\xi_1)}{p_j(\xi_1)}.$$

Let $\eta = (p_j(\xi_1) \cdots p_j(\xi_\tau)/p_i(\xi_1) \cdots p_i(\xi_\tau))$ and let $D_i = \{0 \le k \le N-1 : \alpha_{ik} \neq 0\}$.

By virtue of the Jensen inequality we have

$$\begin{aligned}
M_i \ln \frac{p_i(\xi_1) \cdots p_i(\xi_\tau)}{p_j(\xi_1) \cdots p_j(\xi_\tau)} &= -M_i \ln \eta \\
&= -\sum_{k=0}^{N-1} \int_{\{d(\omega) = d_k\}} \ln \eta P_i(d\omega) \\
&= -\sum_{k \in D_i} \int_{\{d(\omega) = d_k\}} \ln \eta P_i(d\omega) \\
&= -\sum_{k \in D_i} P_i\{d(\omega) = d_k\} \int_\Omega \ln \eta P_i(d\omega \mid d(\omega) = d_k) \\
&\ge -\sum_{k \in D_i} P_i\{d(\omega) = d_k\} \ln \int_\Omega \eta P_i(d\omega \mid d(\omega) = d_k) \\
&= -\sum_{k \in D_i} P_i\{d(\omega) = d_k\} \\
&\quad \times \ln \left\{ \sum_{n=1}^{\infty} \int_{A_{kn}} \frac{p_j(\xi_1) \cdots p_j(\xi_n)}{p_i(\xi_1) \cdots p_i(\xi_n)} \cdot \frac{P_i(d\omega)}{P_i\{d(\omega) = d_k\}} \right\},
\end{aligned} \tag{4.44}$$

with

$$A_{kn} = \{\tau = n\} \cap \{p_i(\xi_1) \cdots p_i(\xi_n) > 0\} \cap \{d(\omega) = d_k\}.$$

But

$$\int_{A_{kn}} \frac{p_j(\xi_1) \cdots p_j(\xi_n)}{p_i(\xi_1) \cdots p_i(\xi_n)} \cdot \frac{P_i(d\omega)}{P_i\{d(\omega) = d_k\}}$$
$$= \frac{1}{a_{ik}} \int_{A_{kn}} P_j(d\omega) \leq \frac{1}{\alpha_{ik}} \int_{\{\tau = n\} \cap \{d = d_k\}} P_j(d\omega). \tag{4.45}$$

Hence it follows from (4.44) and (4.45) that

$$M_i \ln \frac{p_i(\xi_1) \cdots p_i(\xi_\tau)}{p_j(\xi_1) \cdots p_j(\xi_\tau)} \geq - \sum_{k \in D_i} \alpha_{ik} \ln \left\{ \frac{1}{\alpha_{ik}} \int_{\{d = d_k\}} P_j(d\omega) \right\}$$
$$= - \sum_{k \in D_i} \alpha_{ik} \ln \frac{\alpha_{jk}}{\alpha_{ik}}$$
$$= \sum_{k \in D_i} \alpha_{ik} \ln \frac{\alpha_{ik}}{\alpha_{jk}} = \sum_{k=0}^{N-1} \alpha_{ik} \ln \frac{\alpha_{ik}}{\alpha_{jk}}. \tag{4.46}$$

Comparing this inequality with (4.43) we obtain the estimate

$$M_i \tau \geq \frac{\sum_{k=0}^{N} \alpha_{ik} \ln \frac{\alpha_{ik}}{a_{jk}}}{M_i \ln \frac{p_i(\xi_1)}{p_i(\xi_1)}}, \tag{4.47}$$

from which (4.42) follows immediately.

Corollary 1. *Let $N = 2$. Assuming $\alpha = \alpha_{10}$ and $\beta = \alpha_{01}$, we find from* (4.42) *that*

$$M_0 \tau \geq \frac{\omega(\beta, \alpha)}{M_0 \ln \frac{p_0(\xi_1)}{p_1(\xi_1)}}, \qquad M_1 \tau \geq \frac{\omega(\alpha, \beta)}{M_1 \ln \frac{p_1(\xi_1)}{p_0(\xi_1)}}, \tag{4.48}$$

where the function

$$\omega(x, y) = (1 - x) \ln \frac{1 - x}{y} + x \ln \frac{x}{1 - y}.$$

Corollary 2. *Let $N \geq 2$, let $\alpha_{ii} = \alpha$ for all $i = 0, 1, \ldots, N - 1$, and let $\alpha_{ij} = (1 - \alpha)/(N - 1)$, $i \neq j$. Then*

$$M_i \tau \geq \frac{\frac{\alpha N - 1}{N - 1} \ln \frac{\alpha(N - 1)}{1 - \alpha}}{\min_{j \neq i} M_i \ln \frac{p_i(\xi_1)}{p_j(\xi_1)}}. \tag{4.49}$$

4.1.8

To complete the discussion of the conditionally extremal problem of testing two simple hypotheses we need to note that by virtue of Theorem 3

$$\ln \frac{\alpha}{1-\beta} \le \tilde{A}, \qquad \tilde{B} \le \ln \frac{1-\alpha}{\beta}, \tag{4.50}$$

and the mathematical expectations $M_0\tilde{\tau}$ and $M_1\tilde{\tau}$ are estimated by (4.48). It is interesting to note that these inequalities become exact equalities in the problem of testing two simple hypotheses for a Wiener process, which follows.

4.2 Sequential testing of two simple hypotheses on the mean of a Wiener process

4.2.1

For Bayes' formulation, assume that on a probability space $(\Omega, \mathscr{F}, P^\pi)$ we are given a random variable $\theta = \theta(\omega)$ and a standard Wiener process $w = (w_t, t \ge 0)$ mutually independent and such that

$$P^\pi\{\theta = 1\} = \pi, \qquad P^\pi\{\theta = 0\} = 1 - \pi, \tag{4.51}$$

where $0 \le \pi \le 1$ and

$$w_0 \equiv 0,$$

$$M^\pi w_t = 0,$$

$$M^\pi[w_t - w_s]^2 = t - s, \qquad t \ge s \ge 0.$$

We assume that we observe the random process

$$\xi_t = r\theta t + \sigma\omega_t, \qquad \sigma^2 > 0, \qquad r \ne 0. \tag{4.52}$$

By analogy with the case of discrete time considered above (Section 1) we introduce here the concepts of decision rules $\delta = (\tau, d)$, the risk $\rho^\pi(\delta) = M^\pi[c\tau + W(\theta, d)]$, and the π-Bayes decision rule δ_π^*.

We now show that the problem of searching for the rule δ_π^* can be reduced to a problem of optimal stopping.

Let $A \in \mathscr{F}^\xi$ where $\mathscr{F}^\xi = \sigma\{\omega : \xi_s, s \ge 0\}$. Also let $P^0(A) = P^\pi(A \mid \theta = 0)$ and $P^1(A) = P^\pi(A \mid \theta = 1)$. (If $(C, \mathscr{B})$ is a measure space of continuous functions on $[0, \infty)$, then the measures $P_\xi^0(B) = P^0\{\xi \in B\}$ and $P_\xi^1(B) = P^1\{\xi \in B\}$, $B \in \mathscr{B}$, are nothing but Wiener measures whose local drifts are σ and r respectively, and whose diffusion coefficient is σ^2; see [69], chap. 7, §9).

Let $\pi_t^\pi = P^\pi\{\theta = 1 \mid \mathscr{F}_t^\xi\}$, $\mathscr{F}_t^\xi = \sigma\{\omega : \xi_s, s \le t\}$, and let

$$\varphi_t = \frac{dP^1}{dP^0}(\mathscr{F}_t^\xi)(\omega).$$

It is a well-known fact (see, for instance, (7.5) in [69]), that (P^π-a.s.)

$$\varphi_t = \exp\left\{\frac{r}{\sigma^2}\left(\xi_t - \frac{r}{2}t\right)\right\}, \tag{4.53}$$

and since ($\pi < 1$)

$$\pi_t^\pi = \pi \frac{dP^1}{d[\pi P^1 + (1-\pi)P^0]}(\mathscr{F}_t^\xi)(\omega), \tag{4.54}$$

it follows that

$$\pi_t^\pi = \frac{\dfrac{\pi}{1-\pi}\varphi_t}{1 + \dfrac{\pi}{1-\pi}\varphi_t}. \tag{4.55}$$

Taking Ito differentials (see [69], chap. 4, §3) of the right-hand sides in (4.53) and (4.55), we find that

$$d\varphi_t = \frac{r}{\sigma^2}\varphi_t d\xi_t, \qquad \varphi_0 = 1, \tag{4.56}$$

$$d\pi_t^\pi = -\frac{r^2}{\sigma^2}(\pi_t^\pi)^2(1-\pi_t^\pi)dt + \frac{r}{\sigma^2}\pi_t^\pi(1-\pi_t^\pi)d\xi_t, \tag{4.57}$$

$$\pi_0^\pi = \pi.$$

(Let us note that in our case we have $\pi_t^\pi = P^\pi\{\theta = 1 | \mathscr{F}_t^\xi\} = M^\pi[\theta | \mathscr{F}_t^\xi]$).

We consider the process

$$\bar{w}_t = \xi_t - \int_0^t M^\pi[\theta | \mathscr{F}_s^\xi]ds. \tag{4.58}$$

The process (with continuous trajectories)

$$\bar{w} = (\bar{w}_t, \mathscr{F}_t^\xi, P^\pi), \qquad t \geq 0,$$

is a square integrable martingale,

$$M^\pi[\bar{w}_t | \mathscr{F}_s^\xi] = \bar{w}_s \qquad (P^\pi\text{-a.s.}),\ t \geq s,$$

with

$$M^\pi[(\bar{w}_t - \bar{w}_s)^2 | \mathscr{F}_s^\xi] = t - s \qquad (P^\pi\text{-a.s.}),\ t \geq s$$

(see [69], Theorem 7.12).

Therefore (see [69], theor. 4.1), the process $\bar{w} = (\bar{w}_t, \geq 0)$ is a Wiener process.

We find from (4.57) and (4.58) that

$$d\pi_t^\pi = \frac{r}{\sigma}\pi_t^\pi(1-\pi_t^\pi)d\bar{w}_t, \qquad \pi_0^\pi = \pi. \tag{4.59}$$

The results given in [99] imply that the process $\Pi^\pi = (\pi_t^\pi, \mathscr{F}_t^\xi, P^\pi)$ having differential (4.59) is a (strictly) Markov random function. Similarly, as in the case of discrete time (Section 1), the family of Markov random functions $\{\Pi^\pi, 0 \le \pi \le 1\}$, defined on probability spaces $(\Omega, \mathscr{F}, P^\pi)$, $0 \le \pi \le 1$, can naturally be associated with a Markov process $\Pi = (\pi_t, \mathscr{F}_t, P_\pi)$, $t \ge 0$, having zero drift and diffusion coefficient equal to $\sigma^2(\pi) = [(r/\sigma)\pi(1 - \pi)]^2$.

By analogy with Section 1 we can show that the terminal decision d_π^* in the π-Bayes decision rule $\delta_\pi^* = (\tau_\pi^*, d_\pi)$ is to be defined by the formula

$$d_\pi^* = \begin{cases} 1, & a\pi_{\tau_\pi^*}^\pi \ge b(1 - \pi_{\tau_\pi^*}^\pi), \\ 0, & a\pi_{\tau_\pi^*}^\pi < b(1 - \pi_{\tau_\pi^*}^\pi), \end{cases} \tag{4.60}$$

and to find the time τ_π^* we need only to solve the problem of optimal stopping:

$$\rho(\pi) = \inf M_\pi[c\tau + g(\pi_\tau)], \tag{4.61}$$

where inf is taken over the class of stopping times

$$\mathfrak{M}^1 = \{\tau \in \mathfrak{M}[F^\xi] : M_\pi \tau < \infty, \quad 0 \le \pi \le 1\}.$$

To find the function $\rho(\pi)$ and to show that the time

$$\tau_0 = \inf\{t \ge 0 : \rho(\pi_t) = g(\pi_t)\} \tag{4.62}$$

is optimal, we shall make use of the results of Section 3.8 illustrating the ways in which problems of optimal stopping are related to generalized Stefan problems.

Similarly, as in the case of discrete time, we can show here that there exist two numbers A^* and $B^*, 0 \le A^* \le B^* \le 1$, such that the domain of continuing observations is

$$C_0^* = \{\pi : A^* < \pi < B^*\}.$$

Thus, according to Section 3.8, the risk function $\rho(\pi)$ in the domain C_0^* belongs to the domain of definition of the operator $\mathfrak{U}$ and

$$\mathfrak{U}\rho(\pi) = -c, \qquad \pi \in C_0^*. \tag{4.63}$$

In this case

$$\rho(A^*) = g(A^*), \qquad \rho(B^*) = g(B^*). \tag{4.64}$$

If, in addition, the conditions of Theorem 3.16 are satisfied, then

$$\rho'(A^*) = g'(A^*), \qquad \rho'(B^*) = g'(B^*), \tag{4.65}$$

at the points A^* and B^*.

We use the following technique for determining the function $\rho(\pi)$ and the boundary points A^* and B^*.

Let us solve the problem posed by (4.63)–(4.65) in the class of twice continuously differentiable functions $f = f(\pi)$. The operator $\mathfrak{U}$ of the process Π coincides on these functions ([31], theor. 5.7) with the differential operator

$$\mathscr{D} = \frac{r^2}{2\sigma^2}[\pi(1 - \pi)]^2 \frac{d^2}{d\pi^2}, \tag{4.66}$$

and, therefore, the problem posed by (4.63)–(4.65) becomes the problem posed by

$$\frac{r^2}{2\sigma^2}[\pi(1-\pi)]^2\frac{d^2f}{d\pi^2} = -c, \qquad A^* < \pi < B^*, \tag{4.67}$$

$$f(\pi) = g(\pi), \qquad \pi \notin (A^*, B^*), \tag{4.68}$$

$$f'(A^*) = g'(A^*) = a, \qquad f'(B^*) = g'(B^*) = -b. \tag{4.69}$$

It is not clear, *a priori*, that the risk function $\rho(\pi)$ belongs to the class of twice continuously differentiable functions in the interval (A^*, B^*). However, we can show that, first, the solution of the problem posed by (4.67)–(4.69) exists and is unique in this class, and, second, this solution coincides with the risk $\rho(\pi)$. Thus, we shall consider the problem posed by (4.67). Let us fix some number A, $0 \le A \le b/(a+b)$. It can be easily shown that the solution $f(\pi)$ of Equation (4.67) in the domain $\pi \ge A$ satisfying the conditions $f(A) = aA$ and $f'(A) = a$ is given by the formula

$$f(\pi) = aA + (\pi - A)\{a - C\varphi(A)\} + c\{\Psi(\pi) - \Psi(A)\}, \tag{4.70}$$

where

$$C = c\left(\frac{r^2}{2\sigma^2}\right)^{-1}$$

and

$$\Psi(\pi) = (1-2\pi)\ln\frac{\pi}{1-\pi}, \tag{4.71}$$

$$\psi(\pi) = \Psi'(\pi) = \left(\frac{1-\pi}{\pi} - \frac{\pi}{1-\pi}\right) + 2\ln\frac{1-\pi}{\pi}. \tag{4.72}$$

If we use now the boundary conditions at the point B ($f(B) = b(1-B)$, $f'(B) = -b$), we obtain from (4.70) the following system of equations for (unknown) points A and B:

$$b + a = C\{\psi(A) - \psi(B)\}, \tag{4.73}$$

$$b(1-B) = aA + (B-A)\{a - C\psi(A)\} + c\{\Psi(B) - \Psi(A)\}. \tag{4.74}$$

We shall show that unknown constants A and B ($0 \le A \le B \le 1$) are uniquely defined by (4.73) and (4.74).

For this purpose we shall transform the system of equations given by (4.73)–(4.74) to

$$(b-a) + (a + C\psi(B)) = -(a - C\psi(A)), \tag{4.75}$$

$$b + C[B\psi(B) - \Psi(B)] = C[A\psi(A) - \Psi(A)]. \tag{4.76}$$

Let $x = A/(1 - A)$, $y = B/(1 - B)$. Then by taking into account (4.71) and (4.72) we have from (4.75) and (4.76) that

$$(b - a) + \left[a + C\left(\frac{1}{y} - y - 2 \ln y\right)\right] = -\left[a - C\left(\frac{1}{x} - x - 2 \ln x\right)\right], \tag{4.77}$$

$$b - C[y + \ln y] = -C[x + \ln x]. \tag{4.78}$$

It follows from (4.78) that for each x, $0 \leq x < \infty$, there is a unique value $y = y_1(x) \geq 0$, $y_1(0) = 0$, and that for all $x > 0$

$$\frac{dy_1(x)}{dx} = \frac{1 + \dfrac{1}{x}}{1 + \dfrac{1}{y}} > 0.$$

Hence $y = y_1(x)$, $x \geq 0$, is a nondecreasing function.

Similarly, it follows from (4.77) that for each x there is a unique value $y = y_2(x)$, $y_2(0) = \infty$, $y_2(\infty) = 0$, and that for all $x > 0$

$$\frac{dy_2(x)}{dx} = -\frac{\dfrac{1}{x^2} + 1 + \dfrac{2}{x}}{\dfrac{1}{y^2} + 1 + \dfrac{2}{y}} < 0.$$

Therefore, there exists a unique value $x^* > 0$ for which $y_1(x^*) = y_2(x^*)$. It obviously follows from this that the system of equations given by (4.75)–(4.76) has the unique solution (A^*, B^*) with $0 < A^* \leq B^* < 1$. Thus, the solution of the problem proposed by (4.67)–(4.69) in the class of twice continuously differentiable functions and constants $0 \leq A \leq B \leq 1$ exists and is unique. Let us denote by $f^*(\pi)$ the solution of this problem and show that $\rho(\pi) = f^*(\pi)$. Then

$$\begin{aligned} \rho(\pi) &= \inf_{\tau \in \mathfrak{M}^1} M_\pi[c\tau + g(\pi_\tau)] \\ &\geq \inf_{\tau \in \mathfrak{M}^1} M_\pi[c\tau + f^*(\pi_\tau)] + \inf_{\tau \in \mathfrak{M}^1} M_\pi[g(\pi_\tau) - f^*(\pi_\tau)]. \end{aligned} \tag{4.79}$$

Let $\pi \in (A^*, B^*)$. Since $\mathcal{D}f^*(\pi) = -c$, then (see [31], cor. to theor. 5.1) for any Markov time τ with $M_\pi \tau < \infty$

$$M_\pi f^*(\pi_\tau) - f^*(\pi) = -cM_\pi \tau, \tag{4.80}$$

i.e.,

$$M_\pi[c\tau + f^*(\pi_\tau)] = f^*(\pi). \tag{4.81}$$

We shall note that for all $\pi \in [0, 1]$, $g(\pi) \geq f^*(\pi)$. Hence

$$\inf_{\tau \in \mathfrak{M}^1} M_\pi[g(\pi_\tau) - f^*(\pi_\tau)] \geq 0 \tag{4.82}$$

and, therefore,

$$\rho(\pi) \geq \inf_{\tau \in \mathfrak{M}^1} M_\pi[c\tau + f^*(\pi_\tau)]. \tag{4.83}$$

From this and (4.81) we find that for any $\pi \in (A^*, B^*)$, $\rho(\pi) \geq f^*(\pi)$.

It is clear (see [31], theor. 13.16) that for any $\pi \in [0, 1]$ the time

$$\tau^* = \inf\{t \geq 0 : \pi_t \notin (A^*, B^*)\}, \qquad 0 < A^* \leq B^* < 1,$$

has the finite mathematical expectation $M_\pi \tau^*$. Therefore, $\tau^* \in \mathfrak{M}^1$.

Let us note that for each time τ^*

$$M_\pi[c\tau^* + g(\pi_{\tau^*})] = M_\pi[c\tau^* + f^*(\pi_{\tau^*})] = f^*(\pi).$$

From this and (4.83) we find that for $\pi \in (A^*, B^*)$

$$\rho(\pi) \geq f^*(\pi) = M_\pi[c\tau^* + g(\pi_{\tau^*})],$$

and since, on the other hand,

$$\rho(\pi) \leq M_\pi[c\tau^* + g(\pi_{\tau^*})],$$

it follows that

$$\rho(\pi) = f^*(\pi) = M_\pi[c\tau^* + g(\pi_{\tau^*})], \qquad \pi \in (A^*, B^*).$$

Since the function $g(\pi)$ is linear on intervals $[0, A^*]$ and $[B^*, 1]$, the function $\rho(\pi)$ is convex upward on the interval $[0, 1]$, and $\rho(A^*) = g(A^*)$ and $\rho(B^*) = g(B^*)$, it follows that $\rho(\pi) = g(\pi)$ outside of the interval (A^*, B^*) and, therefore, $\rho(\pi) = f^*(\pi)$.

Therefore, the solution $f^*(\pi)$ of the Stefan problem thus obtained coincides with the risk function $\rho(\pi)$.

Similarly, as in the case of discrete time, we can deduce from the above the following result on the structure of the π-Bayes decision rule.

Theorem 5. *In the problem of testing two simple hypotheses $H_0 : \theta = 0$ and $H_1 : \theta = 1$ on the observations of the process given by* (4.52), *the π-Bayes decision rule $\delta_\pi^* = (\tau_\pi^*, d_\pi^*)$ exists and is*

$$\tau_\pi^* = \inf\{t \geq 0 : \pi_t^\pi \notin (A^*, B^*)\},$$
$$d_\pi^* = \begin{cases} 1, & \pi_{\tau_\pi^*}^\pi \geq B^*, \\ 0, & \pi_{\tau_\pi^*}^\pi \leq A^*, \end{cases} \tag{4.84}$$

where the constants A^* and B^* are uniquely defined by the system of transcendental equations

$$b + a = C\{\psi(A^*) - \psi(B^*)\},$$
$$b(1 - B^*) = aA^* + (B - A^*)\{a - C\psi(A^*)\} + c\{\Psi(B^*) - \Psi(A^*)\}; \tag{4.85}$$

here $C = c(r^2/2\sigma^2)^{-1}$ and the functions $\Psi(\pi)$ and $\psi(\pi)$ are defined by (4.71) *and* (4.72).

Remark. In the symmetric case ($a = b$) $B^* = 1 - A^*$ and A^* is defined as the (unique) root of the equation

$$\frac{a}{C} = \frac{1 - A^*}{A^*} - \frac{A^*}{1 - A^*} + 2\ln\frac{1 - A^*}{A^*}.$$

4.2.2

For the fixed error probability formulation let $w = (w_t, t \geq 0)$ be a standard Wiener process given on a probability space $(\Omega, \mathscr{F}, P)$. We shall assume that we observe the process

$$\xi_t = r\theta t + \sigma w_t, \qquad \sigma^2 > 0, r \neq 0,$$

where θ is the unknown parameter taking on one of two values: $\theta = 1$ (hypothesis H_1); or $\theta = 0$ (hypothesis H_0). Let $\mathscr{F}_0^\xi = \{\varnothing, \Omega\}$, $\mathscr{F}_t^\xi = \sigma\{\omega : \xi_s, s \leq t\}$, $\mathscr{F}_\infty^\xi = \sigma(\bigcup_{t\geq 0} \mathscr{F}_t^\xi)$, and P_i be probability measures on $(\Omega, \mathscr{F}_\infty^\xi)$ induced by the process $(\xi_t, t \geq 0)$ for $\theta = i, i = 0, 1$. We shall denote by $\mathfrak{M}^\xi = \{\tau\}$ the class of stopping times (with respect to $F^\xi = \{\mathscr{F}_t^\xi\}, t \geq 0$), and denote by $\mathscr{D}^\xi = \{d\}$ the aggregate of $\mathscr{F}_\tau^\xi$-measurable functions $d = d(\omega)$ taking on values 0 and 1. As in Section 1, let $\Delta^\xi(\alpha, \beta)$ be the class of the decision rules $\delta = (\tau, d)$ with $\tau \in \mathfrak{M}^\xi$, $d \in \mathscr{D}^\xi$, for which $M_0\tau < \infty$, $M_1\tau < \infty$, and the error probabilities of the first and second kinds are (respectively)

$$\alpha(\delta) = P_1\{d(\omega) = 0\} \leq \alpha \quad \text{and} \quad \beta(\delta) = P_0\{d(\omega) = 1\} \leq \beta.$$

The result which follows is similar to Theorem 2 of the previous section.

Theorem 6. *Let positive numbers α and β be such that $\alpha + \beta < 1$. Then in the class $\Delta^\xi(\alpha, \beta)$ there exists an (optimal) decision rule $\tilde{\delta} = (\tilde{\tau}, \tilde{d})$ such that for all $\delta = (\tau, d) \in \Delta^\xi(\alpha, \beta)$*

$$M_0\tilde{\tau} \leq M_0\tau, \qquad M_1\tilde{\tau} \leq M_1\tau. \tag{4.86}$$

In this case

$$\tilde{\tau} = \inf\{t \geq 0 : \lambda_t \notin (\tilde{A}, \tilde{B})\},$$

$$\tilde{d} = \begin{cases} 1, & \lambda_{\tilde{\tau}} \geq \tilde{B}, \\ 0, & \lambda_{\tilde{\tau}} \leq \tilde{A}, \end{cases} \tag{4.87}$$

where

$$\lambda_t = \frac{r}{\sigma^2}\left(\xi_t - \frac{r}{2}t\right), \tag{4.88}$$

$$\tilde{A} = \ln\frac{\alpha}{1 - \beta}, \qquad \tilde{B} = \ln\frac{1 - \alpha}{\beta}. \tag{4.89}$$

The mathematical expectations are

$$M_0\tilde{\tau} = \frac{\omega(\beta, \alpha)}{\rho}, \qquad M_1\tilde{\tau} = \frac{\omega(\alpha, \beta)}{\rho}, \tag{4.90}$$

where

$$\omega(x, y) = (1 - x) \ln \frac{1 - x}{y} + x \ln \frac{x}{1 - y} \tag{4.91}$$

and $\rho = r^2/2\sigma^2$.

To prove this theorem we need some auxiliary results, which follow.

4.2.3

Let

$$\lambda_t^x = x + \frac{r}{\sigma^2}\left\{\xi_t - \frac{r}{2} t\right\},$$

$$\tau_{(A, B)}^x = \inf\{t \geq 0 : \lambda_t^x \notin (A, B)\}$$

and

$$\alpha(x) = P_1\{\lambda_{\tau_{(A, B)}^x}^x = A\}, \qquad \beta(x) = P_0\{\lambda_{\tau_{(A, B)}^x}^x = B\},$$

where $x \in [A, B]$.

Lemma 4. *For* $x \in [A, B]$

$$\alpha(x) = \frac{e^A(e^{B-x} - 1)}{e^B - e^A}, \qquad \beta(x) = \frac{e^x - e^A}{e^B - e^A}. \tag{4.92}$$

PROOF. It is a known fact (see, for instance, [31], theor. 13.16, or [69], lem. 17.8) that $\alpha(x)$ is the solution of the differential equation

$$\alpha''(x) + \alpha'(x) = 0, \qquad A < x < B,$$

satisfying the boundary conditions $\alpha(A) = 1$ and $\alpha(B) = 0$. Similarly, $\beta(x)$ satisfies the equation

$$\beta''(x) - \beta'(x) = 0, \qquad A < x < B,$$

with $\beta(B) = 1$, $\beta(A) = 0$. Solving these equations yields (4.92).

Lemma 5. *Let* $m_i(x) = M_i \tau_{(A, B)}^x$, $x \in [A, B]$. *Then*

$$m_1(x) = \frac{1}{\rho}\left\{\frac{(e^B - e^{A+B-x})(B - A)}{e^B - e^A} + A - x\right\}, \tag{4.93}$$

$$m_0(x) = \frac{1}{\rho}\left\{\frac{(e^B - e^x)(B - A)}{e^B - e^A} - B + x\right\}. \tag{4.94}$$

PROOF. To deduce (4.93) and (4.94) we need only to note that the functions $m_i(x)$, $i = 0, 1$, are solutions of

$$\frac{r^2}{2\sigma^2} m_i''(x) + (-1)^{1-i} \frac{r^2}{2\sigma^2} m_i'(x) = -1,$$

satisfying the boundary conditions $m_i(A) = m_i(B) = 0$ (see [31], theor. 13.16, and [69], lem. 17.9). □

Lemma 6 (Wald's identities for a Wiener process). *Let* $W = (w_t, \mathscr{F}_t)$, $t \geq 0$, *be a standard Wiener process and let* τ *be a Markov time (with respect to* $F = \{\mathscr{F}_t\}$, $t \geq 0$). *If* $M\tau < \infty$, *then*

$$Mw_\tau = 0, \tag{4.95}$$

$$Mw_\tau^2 = M\tau. \tag{4.96}$$

PROOF. Since $\int_0^\infty I_{\{t \leq \tau\}}(\omega)dt = \tau(\omega) < \infty$ with probability 1, the Ito stochastic integral $\int_0^\infty I_{\{t \leq \tau\}}(\omega)dw_t$ is defined, and

$$w_\tau = \int_0^\infty I_{\{t \leq \tau\}}(\omega)dw_t.$$

By the hypotheses of the lemma

$$\int_0^\infty MI_{\{t \leq \tau\}}^2(\omega)dt = \int_0^\infty P\{\tau \geq t\}dt = M\tau < \infty.$$

Hence we find from the familiar properties of stochastic integrals (see (4.48) and (4.49) and also [69], lem. 4.8) that

$$Mw_\tau = M\int_0^\infty I_{\{t \leq \tau\}}(\omega)dw_t = 0,$$

$$Mw_\tau^2 = M\left[\int_0^\infty I_{\{t \leq \tau\}}(\omega)dw_t\right]^2 = \int_0^\infty MI_{\{t \leq \tau\}}^2(\omega)dt = M\tau.$$

This proves the lemma. □

Remark 1. To prove (4.95) it suffices to prove only the condition $M\sqrt{\tau} < \infty$ (see [78]).

Remark 2. It is, in general, impossible to weaken the condition, $M\tau < \infty$, guaranteeing the equality $Mw_\tau^2 = M\tau$; this fact can be illustrated by the following example.

Let $\tau = \inf\{t \geq 0 : w_t = 1\}$. Then $P\{\tau < \infty\} = 1$, $M\tau = \infty$, but $1 = Mw_\tau^2 \neq M\tau = \infty$.

Remark 3. Let $\tau = \inf\{t \geq 0 : |w_t| = A\}$ where $A < \infty$. Then $M\tau = A^2$. In fact, let us assume $\tau_N = \min(\tau, N)$. Then by Lemma 6, $Mw_{\tau_N}^2 = M\tau_N$, from which $M\tau_N \leq A^2$ and, therefore, $M\tau = \lim_N M\tau_N \leq A^2 < \infty$. By applying Lemma 6 we find that $Mw_\tau^2 = M\tau$. Since $P(\tau < \infty) = 1$, $M\tau = Mw_\tau^2 = A^2$.

Remark 4. Let $\tau = \inf\{t \geq 0 : |w_t| = a\sqrt{t + b}\}$ where $0 < b < \infty$, $0 \leq a < 1$. Then

$$M\tau = \frac{a^2 b}{1 - a^2}.$$

To prove this fact we shall assume $\tau_N = \min(\tau, N)$. Then $M\tau_N = Mw_{\tau_N}^2 \leq a^2 M(\tau_N + b)$, i.e., $M\tau_N \leq a^2 b/(1 - a^2)$. Therefore, $M\tau = \lim_N M\tau_N \leq a^2 b/(1 - a^2) < \infty$. Hence

$$M\tau = Mw_\tau^2 = M[w_\tau^2 I_{\{\tau<\infty\}}] = a^2 M[(\tau + b)I_{\{\tau<\infty\}}] = a^2[M\tau + b],$$

which yields the required formula for $M\tau$.

Remark 5. Let a Markov time τ (with respect to the system $F = \{\mathcal{F}_t\}$, $t \geq 0$) be such that $M \exp\{(\lambda^2/2)\tau\} < \infty$ where $-\infty < \lambda < \infty$, and let $w = (w_t, \mathcal{F}_t)$, $t \geq 0$, be a standard Wiener process.

Then

$$M \exp\left\{\lambda w_\tau - \frac{\lambda^2}{2}\tau\right\} = 1. \tag{4.97}$$

(Compare with the fundamental identity from the sequential analysis (4.40).)

For proof see [69], theor. 6.1. In particular, (4.97) holds for any λ, $-\infty < \lambda < \infty$, in the case of bounded Markov times ($P\{\tau \leq N\} = 1$, $N < \infty$).

Lemma 7. *For any decision rule $\delta = (\tau, d) \in \Delta^\xi(\alpha, \beta)$*

$$M_0\tau \geq \frac{\omega(\beta, \alpha)}{\rho}, \qquad M_1\tau \geq \frac{\omega(\alpha, \beta)}{\rho}, \tag{4.98}$$

where the function $\omega(x, y)$ was defined in (4.91) *and* $\rho = r^2/2\sigma^2$.

Proof. By virtue of (4.53) and (4.96), we have

$$\begin{aligned} M_1 \ln \varphi_\tau &= M_1 \ln \frac{dP_1}{dP_0}(\mathcal{F}_\tau^\xi)(\omega) \\ &= M_1\left\{\frac{r}{\sigma^2}\xi_\tau - \frac{r^2}{2\sigma^2}\tau\right\} \\ &= M_1\left\{\frac{r^2}{2\sigma^2}\tau + \frac{r}{\sigma}w_\tau\right\} \\ &= \frac{r^2}{2\sigma^2}M_1\tau = \rho M_1\tau. \end{aligned} \tag{4.99}$$

On the other hand,[6]

$$
\begin{aligned}
M_1 \ln \varphi_\tau &= -M_1 \ln \frac{dP_0}{dP_1}(\mathscr{F}_\tau^\xi)(\omega) \\
&= -\int_{\{d(\omega)=1\}} \ln \frac{dP_0}{dP_1}\, dP_1 - \int_{\{d(\omega)=0\}} \ln \frac{dP_0}{dP_1}\, dP_1 \\
&= -P_1\{d(\omega)=1\} \int_\Omega \ln \frac{dP_0}{dP_1} P_1(d\omega \,|\, d(\omega)=1) \\
&\quad -P_1\{d(\omega)=0\} \int_\Omega \ln \frac{dP_0}{dP_1} P_1(d\omega \,|\, d(\omega)=1) \\
&\geq -P_1\{d(\omega)=1\} \ln \int_\Omega \frac{dP_0}{dP_1} P_1(d\omega \,|\, d(\omega)=1) \\
&\quad -P_1\{d(\omega)=0\} \ln \int_\Omega \frac{dP_0}{dP_1} P_1(d\omega \,|\, d(\omega)=0), \qquad (4.100)
\end{aligned}
$$

where we have taken advantage of the Jensen inequality

$$\ln M\eta \geq M \ln \eta,$$

which holds for any nonnegative random variable η.

By transforming the right-hand side in (4.100) we find that

$$
\begin{aligned}
M_1 \ln \varphi_\tau &= -P_1\{d(\omega)=1\} \ln \left[\frac{1}{P_1\{d(\omega)=1\}} \cdot \int_{\{d(\omega)=1\}} \frac{dP_0}{dP_1} P_1(d\omega) \right] \\
&\quad -P_1\{d(\omega)=0\} \ln \left[\frac{1}{P_1\{d(\omega)=0\}} \cdot \int_{\{d(\omega)=0\}} \frac{dP_0}{dP_1} P_1(d\omega) \right] \\
&= -P_1\{d(\omega)=1\} \ln \frac{P_0\{d(\omega)=1\}}{P_1\{d(\omega)=1\}} \\
&\quad -P_1\{d(\omega)=0\} \ln \frac{P_0\{d(\omega)=0\}}{P_1\{d(\omega)=0\}} \\
&= -(1-\alpha) \ln \frac{\beta}{1-\alpha} \\
&= -\alpha \ln \frac{1-\beta}{\alpha} \\
&= (1-\alpha) \ln \frac{1-\alpha}{\beta} + \alpha \ln \frac{\alpha}{1-\beta} = \omega(\alpha, \beta). \qquad (4.101)
\end{aligned}
$$

[6] If $P_1\{d(\omega) = i\} = 0$, then the product

$$P_1\{d(\omega)=i\} \cdot \int_\Omega \ln \frac{dP_0}{dP_1} P_1(d\omega \,|\, d(\omega)=i)$$

is taken to be zero.

Comparing (4.99) and (4.101) gives us the second formula in (4.98).
The first inequality in (4.98) can be proved in a similar way.

4.2.4

PROOF OF THEOREM 6. Let us consider the decision rule $\tilde{\delta} = (\tilde{\tau}, \tilde{d})$ defined in (4.87).

By virtue of Lemma 4

$$P_1\{\tilde{d}(\omega) = 0\} = \alpha(0) = \frac{e^{\tilde{A}}(e^{\tilde{B}} - 1)}{e^{\tilde{B}} - e^{\tilde{A}}} = \alpha, \tag{4.102}$$

$$P_0\{\tilde{d}(\omega) = 1\} = \beta(0) = \frac{1 - e^{\tilde{A}}}{e^{\tilde{B}} - e^{\tilde{A}}} = \beta. \tag{4.103}$$

Further, by virtue of Lemma 5

$$M_1\tilde{\tau} = \frac{1}{\rho}\left\{\frac{(e^{\tilde{B}} - e^{\tilde{A}+\tilde{B}})(\tilde{B} - \tilde{A})}{e^{\tilde{B}} - e^{\tilde{A}}} + \tilde{A}\right\}$$
$$= \frac{1}{\rho}\frac{\tilde{B}e^{\tilde{B}}(1 - e^{\tilde{A}}) + \tilde{A}e^{\tilde{A}}(e^{\tilde{B}} - 1)}{e^{\tilde{B}} - e^{\tilde{A}}} = \frac{1}{\rho}\,\omega(\alpha, \beta), \tag{4.104}$$

$$M_0\tilde{\tau} = \frac{1}{\rho}\left\{\frac{(e^{\tilde{B}} - 1)(\tilde{B} - \tilde{A})}{e^{\tilde{B}} - e^{\tilde{A}}} - \tilde{B}\right\}$$
$$= \frac{1}{\rho}\frac{\tilde{B}(e^{\tilde{A}} - 1) + \tilde{A}(1 - e^{\tilde{B}})}{e^{\tilde{B}} - e^{\tilde{A}}} = \frac{1}{\rho}\,\omega(\beta, \alpha). \tag{4.105}$$

It follows from (4.102)–(4.105) that the decision rule $\tilde{\delta} = (\tilde{\tau}, \tilde{d}) \in \Delta^{\xi}(\alpha, \beta)$. By comparing (4.98) with (4.104) and (4.105) we have that for any decision rule $\delta = (\tau, d) \in \Delta^{\xi}(\alpha, \beta)$

$$M_0\tau \geq M_0\tilde{\tau}, \qquad M_1\tau \geq M_1\tilde{\tau}.$$

This proves that the decision rule $\tilde{\delta} = (\tilde{\tau}, \tilde{d})$ is optimal in the conditionally extremal case. □

4.2.5

Using the problem of testing two simple hypotheses for a Wiener process as our example, we shall compare the means $M_1\tilde{\tau}$ and $M_0\tilde{\tau}$. These will correspond to the optimal decision rule $\tilde{\delta} = (\tilde{\tau}, \tilde{d}) \in \Delta^{\xi}(\alpha, \beta)$ with fixed observation time $t(\alpha, \beta)$ necessary for testing hypotheses $H_1: 0 = 1$ and $H_0: \theta = 0$, if a most powerful classical rule is used (see [66]) which requires that the error probabilities of the first and second kinds do not exceed α and β (respectively).

Let $\mathscr{F}_0^{\xi} = \{\varnothing, \Omega\}$, $\mathscr{F}_t^{\xi} = \sigma\{\omega: \xi_s(\omega), s \leq t\}$. We shall denote by $\tau_t(\omega)$ an arbitrary $\mathscr{F}_0^{\xi}$-measurable function such that $\tau_t(\omega) \equiv t$, and let $d_t(\omega)$ be any $\mathscr{F}_t^{\xi}$-measurable function assuming the values 1 and 0.

Each pair of functions $\delta_t = (\tau_t(\omega), d_t(\omega))$ determines a certain classical rule with duration of observation time $\tau_t(\omega) \equiv t$ and terminal decision $d_t(\omega)$. If $d_t(\omega) = 1$, then we accept the hypothesis H_1; in the case $d_t(\omega) = 0$ we accept the hypothesis H_0.

Let $\Delta_0^\xi(\alpha, \beta)$ be the aggregate of the classical rules $\delta_t = (\tau_t(\omega), d_t(\omega))$, $t \geq 0$, for which the error probabilities

$$P_1\{d_t(\omega) = 0\} \leq \alpha, \qquad P_0\{d_t(\omega) = 1\} \leq \beta.$$

It is seen that $\Delta_0^\xi(\alpha, \beta) \subseteq \Delta^\xi(\alpha, \beta)$.

By virtue of the Neyman–Pearson fundamental lemma [66], for the most powerful classical rule

$$\delta_{t(\alpha, \beta)} = (t(\alpha, \beta), d_{t(\alpha, \beta)}) \in \Delta_0^\xi(\alpha, \beta)$$

the terminal decision $d_{t(\alpha, \beta)}$ is to be defined by

$$d_{t(\alpha, \beta)} = \begin{cases} 1, & \lambda_{t(\alpha, \beta)} \geq h(\alpha, \beta), \\ 0, & \lambda_{t(\alpha, \beta)} < h(\alpha, \beta), \end{cases} \tag{4.106}$$

where the duration of observation $t(\alpha, \beta)$ and the threshold $h(\alpha, \beta)$ are chosen according to the rule $\delta_{t(\alpha, \beta)} \in \Delta_0^\xi(\alpha, \beta)$.

We shall show that

$$t(\alpha, \beta) = \frac{(C_\alpha + C_\beta)^2}{2\rho}, \tag{4.107}$$

$$h(\alpha, \beta) = \frac{C_\beta^2 - C_\alpha^2}{2}, \tag{4.108}$$

where C_γ is the root of the equation

$$\frac{1}{\sqrt{2\pi}} \int_{C_\gamma}^\infty e^{-x^2/2}\, dx = \gamma, \qquad 0 \leq \gamma \leq 1.$$

In fact, for any rule $\delta_t = (t, d_t(\omega))$ such that

$$d_t(\omega) = \begin{cases} 1, & \lambda_t \geq h, \\ 0, & \lambda_t < h, \end{cases}$$

the probability ($\sigma = +\sqrt{\sigma^2}$)

$$\begin{aligned} P_0\{d_t(\omega) = 1\} &= P_0\{\lambda_t \geq h\} \\ &= P_0\left\{\frac{r}{\sigma^2}\left[\xi_t - \frac{r}{2}t\right] \geq h\right\} \\ &= P\left\{\frac{w_t}{\sqrt{t}} \geq \frac{h + \rho t}{\frac{r}{\sigma}\sqrt{t}}\right\} \\ &= \Phi\,\frac{h + \rho t}{\frac{r}{\sigma}\sqrt{t}} \end{aligned} \tag{4.109}$$

where

$$\Phi(x) = \frac{1}{\sqrt{2\pi}} \int_x^\infty e^{-u^2/2}\, du.$$

Similarly, we have

$$P_1\{d_t(\omega) = 0\} = 1 - \Phi\left(\frac{h - \rho t}{\frac{r}{\sigma}\sqrt{t}}\right). \tag{4.110}$$

By setting the right-hand sides in (4.109) and (4.110) equal to β and α, respectively, we get for $t = t(\alpha, \beta)$ and $h = h(\alpha, \beta)$, the system of two equations

$$\frac{h + \rho t}{\frac{r}{\sigma}\sqrt{t}} = C_\beta, \qquad \frac{h - \rho t}{\frac{r}{\sigma}\sqrt{t}} = -C_\alpha,$$

which yields immediately (4.107) and (4.108).

Therefore, for given α and β, $\alpha + \beta < 1$, we have by virtue of (4.90) and (4.107) that

$$\frac{M_0\tilde{\tau}}{t(\alpha, \beta)} = 2\frac{\omega(\beta, \alpha)}{(C_\alpha + C_\beta)^2}, \tag{4.111}$$

$$\frac{M_1\tilde{\tau}}{t(\alpha, \beta)} = 2\frac{\omega(\alpha, \beta)}{(C_\alpha + C_\beta)^2}. \tag{4.112}$$

The calculation gives that for $\alpha, \beta \leq 0.03$ (see [1])

$$M_0\tilde{\tau} \leq \frac{17}{30}\, t(\alpha, \beta), \qquad M_1\tilde{\tau} \leq \frac{17}{30}\, t(\alpha, \beta).$$

Furthermore, if $\alpha = \beta$, then (see [1])

$$\lim_{\alpha \downarrow 0} \frac{M_0\tilde{\tau}}{t(\alpha, \alpha)} = \lim_{\alpha \downarrow 0} \frac{M_1\tilde{\tau}}{t(\alpha, \alpha)} = \frac{1}{4}.$$

4.3 The problem of disruption (discrete time)

4.3.1

In the problem considered above of testing two simple hypotheses, the one-dimensional probability distribution of random variables $\xi_1, \xi_2, \ldots$ remained unchanged (although unknown) during the whole observation process.

We frequently encounter in the theory of detection and statistical control problems in which the probability characteristics of the observable variables may change at a random instant of time $\theta = \theta(\omega)$ (the time at which *disruption* occurs). We shall state some problems of this type, and suggest methods for solving them using the general theory of optimal stopping rules discussed in Chapters 2 and 3.

4.3.2

For the Bayes formulation, let us assume that on a measure space $(\Omega, \mathscr{F})$ we are given random variables $\theta, \xi_1, \xi_2, \ldots$ and a probability measure P^π such that:

$$P^\pi\{\theta = 0\} = \pi,$$
$$P^\pi\{\theta = n\} = (1 - \pi)(1 - p)^{n-1}p, \qquad n \geq 1, \tag{4.113}$$

where p and π are known constants with $0 < p \leq 1$ and $\pi \in [0, 1]$; for each set $A = \{\omega : \xi_1 \leq x_1, \ldots, \xi_n \leq x_n\}$

$$P^\pi(A) = \pi P^1(A) + (1 - \pi) \sum_{i=0}^{n-1} p(1 - p)^i P^0\{\xi_1 \leq x_1, \ldots, \xi_i \leq x_i\}$$
$$\times P^1\{\xi_{i+1} \leq x_{i+1}, \ldots, \xi_n \leq x_n\} + (1 - \pi)(1 - p)^n P^0(A), \tag{4.114}$$

where P^1 and P^0 are probability measures on $(\Omega, \mathscr{F}^\xi)$, $\mathscr{F}^\xi = \sigma\{\omega : \xi_1, \xi_2, \ldots\}$, independent of π and having the property that

$$P^j\{\xi_1 \leq x_1, \ldots, \xi_n \leq x_n\} = \prod_{k \leq n} P^j\{\xi_k \leq x_k\}. \tag{4.115}$$

We can assume without loss of generality that the distribution functions $F^j(x) = P^j\{\xi_1 \leq x\}$ have the densities $p_j(x)$, $j = 0, 1$ (with respect to a σ-finite measure μ).

The obvioús meaning of the conditions given by (4.113)–(4.115) is as follows. If $\theta = 0$, we observe the sequence of independent identically distributed random variables $\xi_1, \xi_2, \ldots$ with the probability density $p_1(x)$.

If $\theta = i$ the random variables $\xi_1, \ldots, \xi_{i-1}, \xi_i, \ldots$ are mutually independent, $\xi_1, \ldots, \xi_{i-1}$ being identically distributed with the probability density $p_0(x)$; $\xi_i, \xi_{i+1}, \ldots$ are also identically distributed, but with the different probability density $p_1(x)$. The problem of the earliest detection of the time at which the disruption occurs is meaningful for distributions more general than (4.113). For the sake of simplicity, we shall consider the *geometric* distribution for the time at which the disruption occurs.

Let τ be a stopping time (with respect to the system of σ-algebras $F^\xi = \{\mathscr{F}_n^\xi\}, n \geq 0)$ where $\mathscr{F}_0^\xi = \{\varnothing, \Omega\}$ and $\mathscr{F}_n^\xi = \sigma\{\omega : \xi_1, \ldots, \xi_n\}$. For our problem it is convenient to interpret τ as the time at which the "alarm" is sounded to signal the change in distribution based on an observed process. It is clearly desirable to choose τ, the time at which the "alarm signal" is given, as close as possible to the time θ. As the variable characterizing the risk associated with τ we consider $(c > 0)$

$$\rho^\pi(\tau) = P^\pi\{\tau < \theta\} + cM^\pi \max\{\tau - \theta, 0\}, \tag{4.116}$$

where $P^\pi\{\tau < \theta)$ is naturally interpreted as probability of false alarm and $M^\pi \max\{\tau - \theta, 0\}$ as the average delay of detecting the occurence of disruption correctly, i.e., when $\tau \geq \theta$.

Definition 1. For a given $\pi \in [0, 1]$ we call the stopping time τ_π^* *a* π*-Bayes time* if

$$\rho^\pi(\tau_\pi^*) = \inf \rho^\pi(\tau), \tag{4.117}$$

where inf is taken over the class of all stopping times $\tau \in \mathfrak{M}[F^\xi]$ (with respect to the system F^ξ).

Theorem 7. *Let* $c > 0$, $p > 0$, *and let*

$$\pi_n^\pi = P^\pi\{\theta \leq n \mid F_n^\xi\} \tag{4.118}$$

be the a posteriori probability of disruption occuring before time n; $\pi_0^\pi = \pi$. *Then the time*

$$\tau_\pi^* = \inf\{n \geq 0 : \pi_n^\pi \geq A^*\}, \tag{4.119}$$

where A^*, *a constant, is a* π*-Bayes time.*

Proof. As in Section 1, we shall show first that the problem of determining the π-Bayes stopping time can be reduced to an optimal stopping problem for a Markov sequence.

By Bayes' formula, (P^π-a.s.)

$$\pi_{n+1}^\pi = \frac{\pi_n^\pi p_1(\xi_{n+1}) + (1 - \pi_n^\pi) p p_1(\xi_{n+1})}{\pi_n^\pi p_1(\xi_{n+1}) + (1 - \pi_n^\pi) p p_1(\xi_{n+1}) + (1 - \pi_n^\pi)(1 - p) p_0(\xi_{n+1})}, \tag{4.120}$$

from which it follows (compare with Section 2.15) that the system $\Pi^\pi = (\pi_n^\pi, \mathscr{F}_n^\xi, P^\pi)$, $n \geq 0$, forms a Markov random function (for a given π).

We shall transform the risk given by (4.116) into a more convenient form. Let us note first that for each $\tau \in \mathfrak{M}[F^\xi]$

$$P^\pi\{\tau < \theta\} = M^\pi[1 - \pi_\tau^\pi]. \tag{4.121}$$

Further, for each $n \geq 0$

$$\begin{aligned}
M^\pi[\max(n - \theta, 0) \mid \mathscr{F}_n^\xi] &= \sum_{k=0}^{n} (n - k) P^\pi\{\theta = k \mid \mathscr{F}_n^\xi\} \\
&= \sum_{k=0}^{n-1} P^\pi\{\theta \leq k \mid \mathscr{F}_n^\xi\} \\
&= \sum_{k=0}^{n-1} [P^\pi\{\theta \leq k \mid \mathscr{F}_n^\xi\} \\
&\quad - P^\pi\{\theta \leq k \mid \mathscr{F}_k^\xi\}] + \sum_{k=0}^{n-1} P^\pi\{\theta \leq k \mid \mathscr{F}_k^\xi\} \\
&= \sum_{k=0}^{n-1} [P^\pi\{\theta \leq k \mid \mathscr{F}_n^\xi\} - P^\pi\{\theta \leq k \mid \mathscr{F}_k^\xi\}] + \sum_{k=0}^{n-1} \pi_k^\pi;
\end{aligned}$$

let

$$\psi_n^\pi = \sum_{k=0}^{n} [P^\pi\{\theta \le k | \mathscr{F}_n^\xi\} - P^\pi\{\theta \le k | \mathscr{F}_k^\xi\}]$$

$$= -\sum_{k=0}^{n} [P^\pi\{\theta > k | \mathscr{F}_n^\xi\} - P^\pi\{\theta > k | \mathscr{F}_k^\xi\}].$$

The sequence $(\psi_n^\pi, \mathscr{F}_n^\xi, P^\pi)$, $n \ge 0$, forms a martingale for each $\pi \in [0, 1]$. In fact, it is obvious that $M^\pi|\psi_n^\pi| < \infty$ and

$$\psi_{n+1}^\pi = \sum_{k=0}^{n} [P^\pi\{\theta \le k | \mathscr{F}_{n+1}^\xi\} - P^\pi\{\theta \le k | \mathscr{F}_k^\xi\}],$$

from which it follows that (P^π-a.s.)

$$M^\pi\{\psi_{n+1}^\pi | \mathscr{F}_n^\xi] = \psi_n^\pi.$$

Since

$$|\psi_n^\pi| \le \sum_{k=0}^{\infty} P^\pi\{\theta > k | \mathscr{F}_n^\xi\} + \sum_{k=0}^{\infty} P^\pi\{\theta > k | \mathscr{F}_k^\xi\},$$

where (due to the assumption $p > 0$)

$$M^\pi \sum_{k=0}^{\infty} P^\pi\{\theta > k | \mathscr{F}_n^\xi\} = \sum_{k=0}^{\infty} P^\pi\{\theta > k\} = M^\pi\theta < \infty$$

and

$$M^\pi \sum_{k=0}^{\infty} P^\pi\{\theta > k | \mathscr{F}_k^\xi\} = M^\pi\theta < \infty,$$

for $\tau \in \mathfrak{M}[F^\xi]$ we have

$$\lim_n \int_{\{\tau > n\}} |\psi_n^\pi| dP^\pi = 0. \tag{4.122}$$

It can be also seen that

$$M^\pi|\psi_\tau^\pi| < \infty, \qquad \tau \in \mathfrak{M}[F^\xi]. \tag{4.123}$$

It follows from (4.122) and (4.123) (see Theorem 1.12) that for any $\tau \in \mathfrak{M}[F^\xi]$

$$M^\pi\psi_\tau^\pi = M^\pi\psi_0^\pi = 0.$$

Hence if $\tau \in \mathfrak{M}[F^\xi]$, then

$$\rho^\pi(\tau) = P^\pi\{\tau < \theta\} + cM^\pi \max\{\tau - \theta, 0\}$$

$$= M^\pi\left\{(1 - \pi_\tau^\pi) + c\sum_{k=0}^{\tau-1} \pi_k^\pi + c\psi_\tau^\pi\right\}$$

$$= M^\pi\left\{(1 - \pi_\tau^\pi) + c\sum_{k=0}^{\tau-1} \pi_k^\pi\right\},$$

and, therefore, for

$$\rho^\pi = \inf_{\tau \in \mathfrak{M}[F^\xi]} \rho^\pi(\tau),$$

we find that

$$\rho^\pi = \inf_{\tau \in \mathfrak{M}[F^\xi]} M^\pi \left\{ (1 - \pi_\tau^\pi) + c \sum_{k=0}^{\tau-1} \pi_k^\pi \right\}. \tag{4.124}$$

The process $\Pi^\pi = (\pi_n^\pi, \mathscr{F}_n^\xi, P^\pi)$, $n \geq 0$, forms a submartingale

$$M^\pi[\pi_{n+1}^\pi | \mathscr{F}_n^\xi] \geq \pi_n^\pi \qquad (P^\pi\text{-a.s.}).$$

Hence (see Theorem 1.9) $\lim_n \pi_n^\pi$ exists with P^π-probability 1. It is clear that $\lim_n \pi_n^\pi \leq 1$, $\lim_n M^\pi \pi_n^\pi = 1$. By Fatou's lemma $1 = \lim_n M^\pi \pi_n^\pi \leq M^\pi \lim_n \pi_n^\pi$ and, therefore, $\lim_n \pi_n^\pi = 1$ with P^π-probability 1 for any $\pi \in [0, 1]$. It follows from the above that

$$\lim_n \sum_{k=0}^{n} \pi_k^\pi = \infty \qquad (P^\pi\text{-a.s.}),\ \pi \in [0, 1]. \tag{4.125}$$

As in Section 1, the family of Markov random functions $\{\Pi^\pi, 0 \leq \pi \leq 1\}$ can be associated with a Markov process with discrete time $\Pi = (\pi_n, \mathscr{F}_n, P_\pi)$, $n \geq 0$, having the same transition probabilities as each Markov random function Π^π, $\pi \in [0, 1]$.

According to the results of Section 2.15, to find the π-Bayes time τ_π^* we need only to find the optimal stopping time in the problem

$$\rho(\pi) = \inf M_\pi \left[(1 - \pi_\tau) + c \sum_{k=0}^{\tau-1} \pi_k \right], \tag{4.126}$$

where inf is taken over the class of stopping times

$$\mathfrak{M}^1 = \left\{ \tau \in \mathfrak{M} : M_\pi \sum_{k=0}^{\tau-1} \pi_k < \infty, \qquad \pi \in [0, 1] \right\}. \tag{4.127}$$

Let $g(\pi) = 1 - \pi$, and let

$$Qg(\pi) = \min\{g(\pi), c\pi + Tg(\pi)\}, \tag{4.128}$$

where $Tg(\pi) = M_\pi g(\pi_1)$.

It follows from Theorem 2.23 that the risk function

$$\rho(\pi) = \lim_n Q^n g(\pi),$$

$$\rho(\pi) = \min\{(1 - \pi), c\pi + T\rho(\pi)\},$$

and the time

$$\tau_0 = \inf\{n \geq 0 : \rho(\pi_n) = 1 - \pi_n\}$$

is an optimal stopping time.

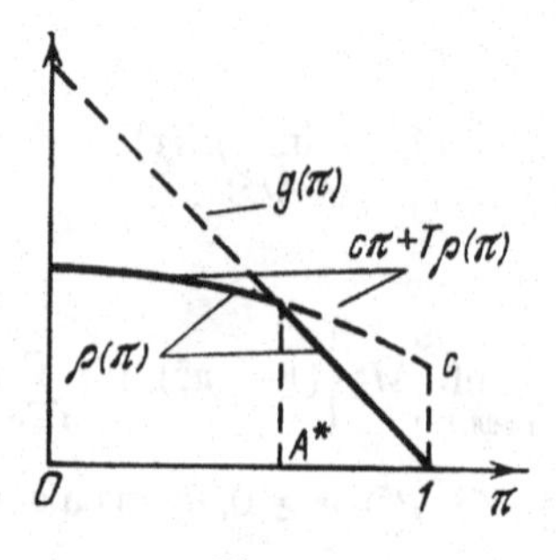

Figure 7

A simple check shows that each function $Q^n g(\pi)$ is convex upward. Hence the risk function $\rho(\pi)$ is also convex upward and, therefore, there is (see Figure 7), a constant A^* such that

$$\begin{aligned}\tau_0 &= \inf\{n \geq 0 : \rho(\pi_n) = 1 - \pi_n\} \\ &= \inf\{n \geq 0 : \pi_n \geq A^*\}.\end{aligned} \tag{4.129}$$

Thus, we have found the optimal stopping time in the problem posed by (4.126). It follows from the above by Section 2.15 that the time $\tau_\pi^* = \inf\{n \geq 0 : \pi_n^\pi \geq A^*\}$ is π-Bayes for any $\pi \in [0, 1]$ (in this case the threshold A^* is independent of π), so that the theorem is proved. □

4.3.3

For the fixed false alarm formulation, let $\pi \in [0, 1)$, $p \in (0, 1]$. We shall denote by $\mathfrak{M}^\xi(\alpha; \pi)$ the class of stopping times $\tau \in \mathfrak{M}[F^\xi]$ for which

$$P^\pi\{\tau < \theta\} \leq \alpha, \tag{4.130}$$

where α is a given constant, $\alpha \in [0, 1)$.

Definition 2. The time $\tilde{\tau} \in \mathfrak{M}^\xi(\alpha; \pi)$ is called *optimal* (*in the conditionally extremal version*) if for all $\tau \in \mathfrak{M}^\xi(\alpha; \pi)$

$$M^\pi \max\{\tilde{\tau} - \theta, 0\} \leq M^\pi \max\{\tau - \theta, 0\}. \tag{4.131}$$

Before describing the structure of optimal times $\tilde{\tau}$, we shall note that only the values $\alpha < 1 - \pi$ are of primary interest for a given π.

In fact, if $\alpha \geq 1 - \pi$, then, assuming $\tilde{\tau} \equiv 0$, we shall have

$$P^\pi\{\tilde{\tau} < \theta\} = P^\pi\{\theta > 0\} = 1 - \pi \leq \alpha$$

and $M^\pi \max\{\tilde{\tau} - \theta, 0\} = 0$. It follows from this fact that the time $\tilde{\tau} \equiv 0$ is optimal in the class $\mathfrak{M}^\xi(\alpha; \pi)$ for $\alpha \geq 1 - \pi$.

Therefore, we assume that $\alpha < 1 - \pi$.

Let

$$\tau_A^\pi = \inf\{n \geq 0 : \pi_n^\pi \geq A\}$$

and

$$\alpha_A^\pi = P^\pi\{\tau_A^\pi < \theta\} = M^\pi[1 - \pi_{\tau_A^\pi}^\pi]. \tag{4.132}$$

It is seen that $\alpha_0^\pi = 1 - \pi$, $\alpha_1^\pi = 0$, and the function α_A^π does not decrease as A increases, $0 \leq A \leq 1$. We shall consider only the case where α_A^π is a continuous function of A.

Let $\alpha < 1 - \pi$ and let $A(\alpha)$ be the smallest A for which $\alpha_A^\pi = \alpha$.

We shall consider the risk

$$\rho_c^\pi = \inf_{\tau \in \mathfrak{M}^\xi(\alpha;\, \pi)} [P^\pi(\tau < \theta) + cM^\pi \max\{\tau - \theta, 0\}], \tag{4.133}$$

where index c in ρ_c^π is introduced to emphasize the fact that the risk is dependent of c.

We shall denote by $A^* = A^*(c)$ the value of the threshold A^* (as a function of c) included in the definition of the π-Bayes time given by (4.119). The function ρ_c^π is convex upward for fixed π (this fact follows immediately from (4.133)), and, therefore, ρ_c^π is continuous with respect to c on $(0, \infty)$.

This implies that the function $A^*(c)$ is a continuous nonincreasing function of c, $A^*(0) = 1$, and $\lim_{c \to \infty} A^*(c) = 0$. We denote by $c^*(A^*)$ the minimum c for which $A^*(c) = A^*$.

Let $0 < \alpha < 1 - \pi$ and let also $c_\alpha = c^*(A(\alpha))$. We consider the risk

$$\rho_{c_\alpha}^\pi = \inf_{\tau \in \mathfrak{M}^\xi(\alpha;\, \pi)} [P^\pi\{\tau < \theta\} + c_\alpha M^\pi \max\{\tau - \theta, 0\}]. \tag{4.134}$$

By virtue of Theorem 7 the π-Bayes time in the problem posed by (4.134) is

$$\tau_\pi^* = \inf\{n \geq 0 : \pi_n^\pi \geq A^*(c_\alpha)\}. \tag{4.135}$$

By the definition of c_α

$$P^\pi\{\tau_\pi^* < \theta\} = \alpha.$$

Next let τ be a time from the class $\mathfrak{M}^\xi(\alpha, \pi)$. We have from (4.134) and (4.135) that

$$\begin{aligned} P^\pi\{\tau < \theta\} &+ c_\alpha M^\pi \max\{\tau - \theta, 0\} \\ &\geq P^\pi\{\tau_\pi^* < \theta\} + c_\alpha M^\pi \max\{\tau_\pi^* - \theta, 0\} \\ &= \alpha + c_\alpha M^\pi \max\{\tau_\pi^* - \theta, 0\}. \end{aligned} \tag{4.136}$$

By the definition of the class $\mathfrak{M}^\xi(\alpha; \pi)$, $P^\pi\{\tau < \theta\} \leq \alpha$. Hence we find from (4.136) that

$$c_\alpha M^\pi \max\{\tau - \theta, 0\} \geq c_\alpha M^\pi \max\{\tau_\pi^* - \theta, 0\}. \tag{4.137}$$

Let us note that if $0 < \alpha < 1 - \pi$, then $c_\alpha \neq 0$. In fact, if $c_\alpha = 0$, then $A^*(0) = 1$ and

$$P^\pi\{\tau_\pi^* < \theta\} = M[1 - \pi_{\tau_\pi^*}^\pi] = 0.$$

Thus, $c_\alpha \neq 0$ and we have from (4.137) that

$$M^\pi \max\{\tau - \theta, 0\} \geq M^\pi \max\{\tau_\pi^* - \theta, 0\}. \tag{4.138}$$

Therefore, we have proved:

Theorem 8. *Let* $0 < \alpha < 1, 0 \leq \pi < 1, p > 0$, *and let* $\mathfrak{M}^\xi(\alpha; \pi)$ *be the aggregate of the stopping times* $\tau \in \mathfrak{M}[F^\xi]$ *for which* $P^\pi\{\tau < \theta\} \leq \alpha$. *Then, if the function* α_A^π *is continuous with respect to A, the Markov time*

$$\tilde{\tau} = \inf\{n \geq 0 : \pi_n^\pi \geq \tilde{A}_\alpha\},$$

where $\tilde{A}_\alpha = A^*(c_\alpha)$, *is optimal in the sense that for any* $\tau \in \mathfrak{M}^\xi(\alpha; \pi)$

$$M^\pi \max\{\tau - \theta, 0\} \geq M^\pi \max\{\tilde{\tau} - \theta, 0\}. \tag{4.139}$$

Remark 1. It is difficult to find the exact value of the threshold $\tilde{A}_\alpha = A^*(c_\alpha)$ for each α, $0 < \alpha < 1$. Hence the following estimate for $\tilde{A}_\alpha$ is useful:

$$\tilde{A}_\alpha \leq 1 - \alpha. \tag{4.140}$$

To prove (4.140) we need only to note that for each A, $0 \leq A \leq 1$, and $\tau_A^\pi = \inf\{n \geq 0 : \pi_n^\pi \geq A\}$ we have

$$M^\pi[1 - \pi_{\tau_A^\pi}^\pi] \leq 1 - A$$

and, therefore,

$$\alpha = M^\pi[1 - \pi_{\tilde{\tau}}^\pi] \leq 1 - \tilde{A}_\alpha.$$

Remark 2. The theorem holds if we replace the requirement that the function α_A^π be continuous with respect to A with the assumption that for a given $\alpha \in (0, 1)$ there is a threshold A_α such that the value of

$$M^\pi[1 - \pi_{\tau_{A_\alpha}^\pi}^\pi]$$

is equal to α (compare with the statement of Theorem 2).

Remark 3. The function α_A^π is continuous with respect to A if the distribution function $F_n(x) = P^\pi\{\pi_n^\pi \leq x\}$ is continuous with respect to x for each n. This condition will, in turn, be satisfied if, for instance, the densities $p_0(x)$ and $p_1(x)$ (with respect to a Lebesgue measure) are Gaussian.

4.4 The problem of disruption for a Wiener process

4.4.1

We shall assume that on a probability space $(\Omega, \mathscr{F}, P^\pi)$ a random variable θ with values in $[0, \infty)$ and a standard Wiener process $w = (w_t, t \geq 0)$ mutually independent are given such that

$$P^\pi\{\theta = 0\} = \pi,$$

$$P^\pi\{\theta \geq t \mid \theta > 0\} = e^{-\lambda t}, \qquad t > 0, \tag{4.141}$$

where λ is the known constant, $0 < \lambda < \infty$, $0 \leq \pi \leq 1$, and

$$w_0 \equiv 0, \qquad M^\pi w_t = 0, \qquad M^\pi[w_t - w_s]^2 = t - s,$$
$$t \geq s \geq 0. \tag{4.142}$$

We also assume that we observe the random process $\xi = (\xi_t, t \geq 0)$ with the stochastic differential

$$d\xi_t = r\chi(t - \theta)dt + \sigma\, dw_t, \qquad \xi_0 = 0, \tag{4.143}$$

where

$$\sigma^2 > 0, \qquad r \neq 0, \qquad \chi(t) = \begin{cases} 1, & t \geq 0, \\ 0, & t < 0. \end{cases}$$

In other words, the structure of the process observed is such that

$$\xi_t = \begin{cases} \sigma w_t, & t < \theta, \\ r(t - \theta) + \sigma w_t, & t \geq \theta. \end{cases} \tag{4.144}$$

By analogy with the case of discrete time, we shall consider the problem of the earliest detection of θ in the Bayes and fixed error probability formulations.

4.4.2

For the Bayes formulation, let

$$\rho^\pi = \inf\{P^\pi\{\tau < \theta\} + cM^\pi \max\{\tau - \theta, 0\}\}, \tag{4.145}$$

where inf is taken over the class of all stopping times $\tau \in \mathfrak{M}[F^\xi]$. As in Section 3, we say that the time τ_π^* is *π-Bayes* if its risk function

$$\rho^\pi(\tau_\pi^*) = P^\pi\{\tau_\pi^* < \theta\} + cM^\pi \max\{\tau_\pi^* - \theta, 0\} \tag{4.146}$$

coincides with ρ^π.

Theorem 9. *The π-Bayes time*

$$\tau_\pi^* = \inf\{t \geq 0 : \pi_t^\pi \geq A^*\}$$

where $\pi_t^\pi = P^\pi\{\theta \leq t | \mathscr{F}_t^\xi\}$, and the threshold A^* is the (unique) root of the equation

$$\int_0^{A^*} e^{-\Lambda[H(A^*) - H(x)]} \frac{dx}{x(1 - x)^2} = C^{-1}, \tag{4.147}$$

where $C = c(r^2/2\sigma^2)^{-1}$, $\Lambda = \lambda(r^2/2\sigma^2)^{-1}$, and

$$H(x) = \ln \frac{x}{1 - x} - \frac{1}{x}.$$

Furthermore,

$$\rho^\pi = \begin{cases} (1 - A^*) + C \displaystyle\int_{1/A^*}^{1/\pi} \frac{e^{\Lambda x}(x-1)^\Lambda}{x^2} \left[\int_x^\infty \frac{e^{-\Lambda u} u}{(u-1)^{2+\Lambda}} du \right] dx & \pi \in [0, A^*], \\ 1 - \pi, & \pi \in [A^*, 1]. \end{cases} \tag{4.148}$$

PROOF. Let us consider the random function $\Pi^\pi = (\pi_t^\pi, \mathscr{F}_t^\xi, P^\pi)$, $t \geq 0$, for a given $\pi \in [0, 1]$. It follows from the results of [69] (see chap. 9, equat. (9.84)) that the process $(\pi_t^\pi, t \geq 0)$ permits the stochastic differential

$$d\pi_t^\pi = \lambda(1 - \pi_t^\pi)dt + \frac{r}{\sigma^2} \pi_t^\pi(1 - \pi_t^\pi)(d\xi_t - r\pi_t^\pi dt), \tag{4.149}$$

where $\pi_0^\pi = \pi$.

The process $\bar{w} = (\bar{w}_t, \mathscr{F}_t^\xi, P^\pi)$, $t \geq 0$, with

$$\bar{w}_t = \frac{1}{\sigma}\left[\xi_t - r \int_0^t \pi_s^\pi \, ds\right]$$

is a Wiener process ([69], theor, 7.12) and hence

$$d\pi_t^\pi = \lambda(1 - \pi_t^\pi)dt + \frac{r}{\sigma} \pi_t^\pi(1 - \pi_t^\pi)d\bar{w}_t, \ \pi_0^\pi = \pi. \tag{4.150}$$

As in Section 2 (see (4.59)), it follows from (4.150) that the process $\Pi^\pi = (\pi_t^\pi, \mathscr{F}_t^\xi, P^\pi)$, $t \geq 0$, is a (strictly) Markov random function.

We can construct from the family of Markov random functions $\{\Pi^\pi, 0 \leq \pi \leq 1\}$ defined on probability spaces $(\Omega, \mathscr{F}, P^\pi)$, $0 \leq \pi \leq 1$, the corresponding Markov process $\Pi = (\pi_t, \mathscr{F}_t, P_\pi)$, $\pi \in [0, 1]$, which has the drift coefficient

$$a(\pi) = \lambda(1 - \pi) \tag{4.151}$$

and the diffusion coefficient

$$\sigma^2(\pi) = \left[\frac{r}{\sigma} \pi(1 - \pi)\right]^2. \tag{4.152}$$

By analogy with the proof in Theorem 7, we can easily show here that for $\lambda > 0$ and $\tau \in \mathfrak{M}[F^\xi]$

$$\begin{aligned} \rho^\pi(\tau) &= P^\pi\{\tau < \theta\} + cM^\pi \max\{\tau - \theta, 0\} \\ &= M^\pi\left\{(1 - \pi_\tau^\pi) + c \int_0^\tau \pi_s^\pi \, ds\right\}. \end{aligned}$$

It follows from this (compare with Section 2.15) that to find the π-Bayes time τ_π^* we need only to find the optimal stopping time in the problem

$$\rho(\pi) = \inf M_\pi\left\{(1 - \pi_\tau) + c \int_0^\tau \pi_s \, ds\right\}, \tag{4.153}$$

where $\Pi = (\pi_t, \mathscr{F}_t, P_\pi)$, $\pi \in [0, 1]$, is a diffusion-type Markov process with local characteristics $a(\pi)$ and $\sigma^2(\pi)$ given by (4.151) and (4.152). It is sufficient to take the inf in (4.153) over the class

$$\mathfrak{M}^1 = \left\{\tau \in \mathfrak{M}[F^\xi] : M_\pi \int_0^\tau \pi_s \, ds < \infty, \pi \in [0, 1]\right\}. \tag{4.154}$$

As in our solution to the problem of testing two simple hypotheses on the mean of a Wiener process (Section 2), in order to find the risk function $\rho(\pi)$ and prove the fact that the time

$$\tau_0 = \inf\{t \geq 0 : \rho(\pi_t) = 1 - \pi_t\}$$

is optimal we shall consider the Stefan problem:

$$\begin{aligned} \mathscr{D}f(\pi) &= -c\pi, & 0 \leq \pi < A, \\ f(\pi) &= 1 - \pi, & A \leq \pi \leq 1, \\ f''(\pi)|_{\pi = A} &= -1, \end{aligned} \tag{4.155}$$

where $\mathscr{D}$ is the differential operator

$$\lambda(1 - \pi)\frac{d}{d\pi} + \frac{r^2}{2\sigma^2}[\pi(1 - \pi)]^2 \frac{d^2}{d\pi^2},$$

A is an unknown constant in $[0, 1]$, and $f(\pi)$ is the unknown function from the class $\mathbb{F}$ of nonnegative convex upward, twice continuously differentiable functions.

The general solution of the equation $\mathscr{D}f(\pi) = -c\pi$ contains two undetermined constants. In addition, we have the unknown point A which defines the domain $[0, A)$ in which the equation $\mathscr{D}f(\pi) = -c\pi$ holds. Therefore the two conditions ($f(\pi) = 1 - \pi$, $\pi \in [A, 1]$, and $f'(A) = -1$) do not determine a unique solution of the Stefan problem (4.155). It turns out, however, that in the class $\mathbb{F}$ there is a solution which is unique and has the property that $f'(0) = 0$.

Let $C = c/p$, $\Lambda = \lambda/\rho$ and $\psi(\pi) = f'(\pi)$. We find from the equation $\mathscr{D}f(\pi) = -c\pi$ that

$$\psi''(\pi) = -\frac{C\pi + \Lambda(1 - \pi)\psi(\pi)}{[\pi(1 - \pi)]^2}. \tag{4.156}$$

This equation contains a singular point $\pi = 0$ and a separatrix $\psi^*(\pi)$ in coming to this point ($\psi^*(0) = 0$). It is not difficult to see that

$$\psi^*(\pi) = -C \int_0^\pi e^{-\Lambda[H(\pi) - H(y)]} \frac{dy}{y(1 - y)^2}, \tag{4.157}$$

where

$$H(y) = \ln \frac{y}{1-y} - \frac{1}{y}.$$

Let A^* be the root of the equation

$$\psi^*(A^*) = -1 \tag{4.158}$$

and let

$$f^*(\pi) = \begin{cases} (1 - A^*) - \int_\pi^{A^*} \psi^*(x)dx, & \pi \in [0, A^*), \\ 1 - \pi, & \pi \in [A^*, 1]. \end{cases} \tag{4.159}$$

The function $f^*(\pi)$ is nonnegative, convex upward, and is also a solution of problem (4.155). Let us show that this solution is unique in the class $\mathbb{F}$.

To this end we shall consider the family of integral curves in Equation (4.156).

Let the point $A > A^*$ and let $\psi_A(\pi)$ be a solution of this equation satisfying the condition $\psi_A(A) = -1$. Then $\psi_A(0) = +\infty$ and, therefore, the solution of the system of equations given by (4.155) with $f(A) = 1 - A$ and $f'(A) = \psi_A(A) = -1$, is not a convex upward function.

Let the point $A < A^*$ and let $\psi_A(\pi)$ be a solution of Equation (4.156) with $\psi_A(A) = -1$. Then $\psi_A(0) = -\infty$ and the solution of the system of equations given by (4.155) with $f(A) = 1 - A$ and $f'(A) = \psi_A(A) = -1$ is such that $f(0) < 0$.

Thus, the pair $(A^*, f^*(\pi))$ is the unique solution of the problem posed by (4.155) (in the class of functions $f(\pi) \in \mathbb{F}$).

Next, we shall show that the function $f^*(\pi)$ thus found coincides with the risk function $\rho(\pi)$. To this end we shall use the same method as the one that was used in proving the similar assertion in Theorem 5.

It is seen from (4.151) and (4.152) (see also (4.150)) that for any stopping time τ (with respect to $F = \{\mathscr{F}_t\}, t \geq 0$)

$$M_\pi \pi_{\tau \wedge N} = \pi + \lambda M_\pi(\tau \wedge N) - \lambda M_\pi \int_0^{\tau \wedge N} \pi_s\, ds,$$

and, therefore,

$$M_\pi(\tau \wedge N) = M_\pi \int_0^{\tau \wedge N} \pi_s\, ds + \frac{1}{\lambda}[M_\pi \pi_{\tau \wedge N} - \pi], \qquad \lambda > 0.$$

This fact implies that if $M_\pi \int_0^\tau \pi_s\, ds < \infty$, then $M_\pi \tau < \infty$ (the converse is obvious). Hence, if the time τ is such that $M_\pi \int_0^\tau \pi_s\, ds < \infty$, then according to Dynkin ([31], cor. to theor. 5.1)

$$M_\pi f^*(\pi_\tau) - f^*(\pi) = -cM_\pi \int_0^\tau \pi_s\, ds, \qquad \pi \in [0, A^*).$$

Therefore,

$$\rho(\pi) = \inf_{\tau \in \mathfrak{M}^1} M_\pi\left\{(1 - \pi_\tau) + c \int_0^\tau \pi_s \, ds\right\}$$

$$\geq \inf_{\tau \in \mathfrak{M}^1} M_\pi\left\{f^*(\pi_\tau) + c \int_0^\tau \pi_s \, ds\right\} + \inf_{\tau \in \mathfrak{M}^1} M_\pi\{(1 - \pi_\tau) - f^*(\pi_\tau)\}$$

$$= f^*(\pi) + \inf_{\tau \in \mathfrak{M}^1} M_\pi\{(1 - \pi_\tau) - f^*(\pi_\tau)\}.$$

But $1 - \pi \geq f^*(\pi)$ for all $\pi \in [0, 1]$; hence $\rho(\pi) \geq f^*(\pi)$ for $\pi \in [0, A^*)$.

Further, the time $\tau^* = \inf\{t \geq 0 : \pi_t \in A^*\}$ belongs to the class $\mathfrak{M}^1$ (by virtue of [31], theor. 3.16), and the equation

$$M_\pi\left\{(1 - \pi_{\tau^*}) + c \int_0^{\tau^*} \pi_s \, ds\right\} = M_\pi\left\{f^*(\pi_{\tau^*}) + c \int_0^{\tau^*} \pi_s \, ds\right\} = f^*(\pi).$$

Hence for all $\pi \in [0, A^*)$ the risk function $\rho(\pi)$ coincides with $f^*(\pi)$. It follows from the fact that the function $\rho(\pi)$ is convex upward that $\rho(\pi) = f^*(\pi) = 1 - \pi$ for $\pi \geq A^*$ as well.

Thus, we have found the structure of the optimal stopping times in the problem posed by (4.153) and the function $\rho(\pi)$. As noted above, this fact implies Theorem 9.

4.4.3

For the fixed false alarm formulation, let $\mathfrak{M}(\alpha; \pi)$ be the class of stopping times for which

$$P^\pi\{\tau < \theta\} \leq \alpha.$$

Theorem 10. *Let* $0 < \alpha < 1, 0 \leq \pi < 1, 0 < \lambda < \infty$. *Then the stopping time*

$$\tilde{\tau}_\alpha = \inf\{t \geq 0 : \pi_t^\pi \geq \tilde{A}_\alpha\}, \tag{4.160}$$

where $\tilde{A}_\alpha = 1 - \alpha$, *is optimal in the sense that for any* $\tau \in \mathfrak{M}(\alpha; \pi)$ $M_\pi \max\{\tilde{\tau}_\alpha - \theta, 0\} \leq M^\pi \max\{\tau - \theta, 0\}$.

Proof. The proof of this theorem is similar to that of Theorem 8. We note only that the equality $\tilde{A}_\alpha = 1 - \alpha$ follows from the fact that for all $\pi \leq \tilde{A}_\alpha$

$$M^\pi[1 - \pi_{\tilde{\tau}_\alpha}^\pi] = 1 - \tilde{A}_\alpha,$$

and that if $\pi > \tilde{A}_\alpha$, then

$$M^\pi[1 - \pi_{\tilde{\tau}_\alpha}^\pi] = 1 - \pi. \qquad \square$$

4.4.4

We shall discuss, in addition, the question of the delay time for a given probability α of the false alarm signal:

$$\mathbb{R}(\alpha; \lambda) = M^0\{\tilde{\tau}_\alpha - \theta \mid \tilde{\tau}_\alpha \geq \theta\} \tag{4.161}$$

(we restrict ourselves to the case $\pi = 0$ for the sake of simplicity).

Let c_α be the constant c contained in (4.145) for which the 0-Bayes time τ_0^* coincides with the time $\tilde{\tau}_\alpha$ defined in (4.160). (The existence of c_α follows from considerations similar to those used in proving Theorem 8.)

Then, by virtue of (4.148),

$$\rho^0 = \alpha + \frac{c_\alpha}{\rho}\int_{1/A_\alpha}^{\infty} \frac{e^{\Lambda x}(x-1)^\Lambda}{x^2}\left[\int_x^\infty \frac{e^{-\Lambda u}u\,du}{(u-1)^{2+\Lambda}}\right]dx. \tag{4.162}$$

On the other hand,

$$\begin{aligned}\rho^0 &= \inf_{\tau\in\mathfrak{M}^1}[P^0\{\tau<\theta\} + c_\alpha P^0\{\tau\geq\theta\}M^0\{\tau-\theta\,|\,\tau\geq\theta\}]\\ &= P^0\{\tilde{\tau}_\alpha<\theta\} + c_\alpha P^0\{\tilde{\tau}_\alpha\geq\theta\}\mathbb{R}(\alpha;\lambda)\\ &= \alpha + c_\alpha(1-\alpha)\mathbb{R}(\alpha;\lambda).\end{aligned} \tag{4.163}$$

Comparing (4.162) with (4.163), we find that

$$\mathbb{R}(\alpha;\lambda) = \frac{\displaystyle\int_{1/(1-\alpha)}^{\infty}\frac{e^{\Lambda x}(x-1)^\Lambda}{x^2}\left[\int_x^\infty\frac{e^{-\Lambda u}u}{(u-1)^{2+\Lambda}}\,du\right]dx}{(1-\alpha)\rho}. \tag{4.164}$$

We shall investigate this formula in the case $\lambda\to 0$, which is more interesting from the engineering point of view. It is natural that for $\lambda\to 0$, i.e., when the mean time at which the disruption occurs $M^0\theta = \lambda^{-1}$ tends to infinity, it is wise to assume that $\alpha\to 1$. Let $\lambda\to 0$, $\alpha\to 1$, but such that $(1-\alpha)/\lambda = \mathbb{T}$ where $\mathbb{T}$ is fixed. Then we find from (4.164) for $\alpha\to 1$, $\lambda\to 0$ and fixed $(1-\alpha)/\lambda = \mathbb{T}$ that

$$\begin{aligned}\mathbb{R}(\mathbb{T}) &= \lim_{\alpha\to 1,\lambda\to 0}\mathbb{R}(\alpha;\lambda)\\ &= \lim_{\alpha\to 1,\lambda\to 0}\frac{\Lambda\displaystyle\int_{\Lambda/(1-\alpha)}^{\infty}\frac{e^y}{y^2}\left(\int_y^\infty\frac{e^{-z}}{z}\,dz\right)dy}{\rho(1-\alpha)}\\ &= \frac{1}{\rho}\lim_{\alpha\to 1,\lambda\to 0}\Lambda/1-\alpha\int_{\Lambda/1-\alpha}^{\infty}\frac{e^y}{y^2}\left(\int_y^\infty\frac{e^{-z}}{z}\,dz\right)dy\\ &= \frac{b}{\rho}\int_b^\infty\frac{e^y}{y^2}[-\mathrm{Ei}(-y)]dy,\end{aligned}$$

where $b = (\rho\mathbb{T})^{-1}$ and

$$-\mathrm{Ei}(-y) = \int_y^\infty\frac{e^{-z}}{z}\,dz$$

is an integral exponential.

Making simple transformations (see [86]) we obtain

$$\begin{aligned}\mathbb{R}(\mathbb{T}) &= \frac{b}{\rho}\int_b^\infty e^y \operatorname{Ei}(-y)d\left(\frac{1}{y}\right)\\ &= \frac{b}{\rho}\left\{-\frac{1}{b}e^b \operatorname{Ei}(-b) - \int_b^\infty \frac{e^z}{z}\operatorname{Ei}(-z)dz - \int_b^\infty \frac{dz}{z^2}\right\}\\ &= \frac{1}{\rho}\left\{e^b\left[-\operatorname{Ei}(-b) - 1 + b\int_b^\infty \frac{e^z}{z}[-\operatorname{Ei}(-z)\right]dz\right\}.\end{aligned}$$

But

$$-\operatorname{Ei}(-z) = e^{-z}\int_0^\infty \frac{e^{-y}}{y+z}dy,$$

from which

$$\begin{aligned}\int_b^\infty \frac{e^z}{z}[-\operatorname{Ei}(-z)]dz &= \int_0^\infty e^{-y}\left(\int_b^\infty \frac{dz}{z(z+y)}\right)dy\\ &= \int_0^\infty \frac{e^{-z}\ln\left(1+\dfrac{z}{b}\right)}{z}dz.\end{aligned}$$

Therefore,

$$\mathbb{R}(\mathbb{T}) = \frac{1}{\rho}\left\{e^b[-\operatorname{Ei}(-b)] - 1 + b\int_0^\infty e^{-bz}\frac{\ln(1+z)}{z}dz\right\}, \tag{4.165}$$

where $b = (\rho\mathbb{T})^{-1}$.

In the case of large $\mathbb{T}$, from (4.165) we have (see [86])

$$\mathbb{R}(\mathbb{T}) = \frac{1}{\rho}\left\{\ln(\rho\mathbb{T}) - 1 - C + O\left(\frac{1}{\rho\mathbb{T}}\right)\right\}, \tag{4.166}$$

where $C = 0.577\ldots$ is the Euler constant.

Notes to Chapter 4

4.1. The Bayesian and variational formulations of the problem of sequential testing of two simple hypotheses are due to Wald [106]. The proof given of Theorem 1 follows Chow and Robbins [18] and Shiryayev [92]. Theorem 2 was proved by Wald and and Wolfowitz [103]. For the proof of this theorem see also Lehmann [66] and Zaks [111]. Estimates (4.29) in Theorem 3 were obtained by Wald [106]. Lemma 2 is due to Stein (see Wald [106]). Wald's identities (Lemmas 3 and 6) have been studied by Wald [106], Blackwell [10], Doob [28], Chow, Robbins, and Teicher [20], Shepp [82], and Brown and Eagleson [14]. Theorem 4 for the case $N = 2$ is due to Wald [106], and for the general case to V. Hoeffding who conveyed it to the author in 1965. See also Simons [96], and Bechhofer, Kiefer, and Sobel [8] (theor. 3.5.1).

4.2. Equations (4.85) in Theorem 5 were obtained by Mikhalevich [74]. A different proof of Theorem 5 can be found in Shiryayev [92]. Theorem 6 and Lemma 7 are due to Wald [106]. The proof of Lemma 6 follows Shepp [82]. The optimality properties of the Neymann–Pearson method were compared with the Wald sequential probability ratio test by Aivazjan [1].

4.3. The disruption (discontinuity) problem was discussed for the first time by Kolmogorov and Shiryayev at the Sixth All-Union Symposium on Probability Theory and Mathematical Statistics (Vilnius, 1960, USSR). The results given in this section can be found in Shiryayev [84], [86], [88].

4.4. The disruption problem for a Wiener process, as well as other statements of the problems of the earliest detection of disruption, was studied by Shiryayev in [85], [86], [89], [92]. The problem was examined also by Stratonovich [98] and by Bather [7]. (4.165)–(4.166) are due to Shiryayev [86].

Bibliography

[1] Aivazjan, S. A., A comparison of the optimal properties of the Neyman–Pearson and the Wald sequential probability ratio tests, *Teoria Verojatn. i Primenen.* 4(1) (1959), 86–93. (Russian.)

[2] Aleksandrov, P. S., *Theory of Sets and Functions* (New York: Chelsea, 1974).

[3] Arrow, K. I., D. Blackwell, and M. A. Girshick, Bayes and minimax solutions of sequential decision problems, *Econometrica* 17 (1949), 213–214.

[4] Bahadur, R. R., Sufficiency and statistical decision functions, *Ann. Math. Statist.* 25(3) (1954), 423–462.

[5] Basharinov, A. E., and B. S. Fleishman, Methods of statistical sequential analysis and its application to radio engineering, *Izdat.* "*Sovetskoje Radio*" (Moscow: 1962). (Russian.)

[6] Bather, I. A., Bayes procedures for deciding the sign of a normal mean, *Proc. Cambr. Phil. Soc.* 58(4) (1962), 226–229.

[7] Bather, I. A., On a quickest detection problem, *Ann. Math. Statist.* 38(3) (1967), 711–724.

[8] Bechhofer, R. E., I. Kiefer, and M. Sobel, *Sequential Identification and Ranking Procedures* (Chicago: U. of Chicago Press, 1968).

[9] Bellman, R., *Dynamic Programming* (Princeton: Princeton U. Press, 1957).

[10] Blackwell, D. H., On an equation of Wald, *Ann. Math. Statist.* 17(1) (1946), 84–87.

[11] Blackwell, D. H., and M. A. Girshick, *Theory of Games and Statistical Decisions* (Moscow: IL, 1958).

[12] Blumenthal, R. M., and R. K. Getoor, *Markov Processes and Potential Theory* (New York and London: Academic Press, 1968).

[13] Brieman, L., Problems of stopping rules, *Sbornik* "*Prikladnaja Kombinatornaja Matematika*" (Moscow: "Mir', 1968), pp. 159–202.

[14] Brown, B. M., and G. K. Eagleson, Simpler conditions for Wald equations, *J. Appl. Prob.* 10 (1973), 451–455.

[15] Burkholder, D. L., and R. A. Wijsman, Optimum properties and admissibility of sequential tests, *Ann. Math. Statist.* 34(1) (1963), 1–17.

[16] Chernoff, H., Sequential tests for the mean of a normal distribution, *Proc. Fourth Berkeley Symp. Math. Statist. Prob.* (Berkeley: Cal. Press, 1961), pp. 79–92.

[17] Chow, Y. S., and H. Robbins, A martingale system theorem and applications, *Proc. Fourth Berkeley Symp. Math. Statist. Prob.* (Berkeley: Cal. Press, 1961), pp. 93–104.

[18] Chow, Y. S., and H. Robbins, On optimal stopping rules, *Z. Wahrscheinlichkeitstheorie und verw. Gebiete* 2 (1963), 33–49.

[19] Chow, Y. S., S. Moriguti, H. Robbins, and S. M. Samuels, Optimal selection based on relative rank (the "secretary problem"), *Israel J. Math.* 2(2) (1964), 81–90.

[20] Chow, Y. S., H. Robbins, and H. Teicher, Moments of randomly stopped sums, *Ann. Math. Statist.* 36(4) (1965), 789–799.

[21] Chow, Y. S., and H. Robbins, On values associated with a stochastic sequence, *Proc. Fifth Berkeley Symp. Math. Statist. Prob.* (Berkeley: U. Cal. Press, 1967), pp. 427–440.

[22] Chow, Y. S., H. Robbins, and D. Siegmund, *Great Expectations: The Theory of Optimal Stopping* (Boston: Houghton Mifflin, 1971).

[23] Courrège, P., and P. Priouret, Temps d'arrât d'une fonction aléatoire: relations d'équivalence associées et propriétes de décomposition, *Publ. Inst. Statist. U. Paris* 14 (1965), 245–274.

[24] Davis, M. H. A., A note on the Poisson disorder problem, *Proc. of International Conference on Control Theory, Zakopane, Poland* (Zakopane: 1974).

[25] De Groot, M. H., *Optimal Statistical Decisions* (New York: McGraw-Hill, 1970).

[26] Dieudonné, Y., *Foundations of Modern Analysis* (New York: Academic Press, 1969).

[27] Dochviri, V., On the supermartingale characterization of the payoff in the problem of optimal stopping of Markov processes, *Soobstchenia Academii Nauk Gruz. SSR* 59(1) (1970), 29–31.

[28] Doob, J. L., *Stochastic Processes* (New York: John Wiley, 1953).

[29] Dvoretzky, A., J. Kiefer, and J. Wolfowitz, Sequential decision processes with continuous time parameter: testing hypotheses, *Ann. Math. Statist.* 24(2) (1953), 254–264.

[30] Dynkin, E. B., *Foundations of the Theory of Markov Processes* (Moscow: Fizmatgiz, 1959).

[31] Dynkin, E. B., *Markov Processes* (Berlin, Heidelberg, New York: Springer–Verlag, 1965).

[32] Dynkin, E. B., The optimum choice of the instant for stopping a Markov process, *Doklady Akademii Nauk SSSR* 150(2) (1963), 238–240.

[33] Dynkin, E. B., and A. A. Yushkevich, *Theorems and Problems in Markov Processes* (New York: Plenum Press, 1967).

[34] Dynkin, E. B., Sufficient statistics for the optimal stopping problem, *Teoria Verojatn. i Primenen.* 13(1) (1968), 150–151.

[35] Engelbert, A., On ε-optimality of Markov times in the problem of stopping Markov processes with continuous time, *Mathematische Nachrichten* 70 (1975), 251–257.

[36] Engelbert, A., *Über die Konstruktion des "Wertes" $s(x)$ beim optimalen Stoppen von Standard-Markow-Prozessen* (Jena: Preprint, 1975).

[37] Engelbert, A., *Optimal Stopping Problems in a Standard Markov Process* (Jena: Preprint, 1975).

[38] Engelbert, G. Yu., On the theory of optimal stopping rules of Markov processes, *Teoria Verojatn. i Primenen.* 18(2) (1973), 312–320.

[39] Engelbert, G. Yu., On optimal stopping rules of Markov random processes with continuous time, *Teoria Verojatn. i Primenen.* 19(2) (1974), 289–307.

[40] Engelbert, G. Yu., On the construction of the payoff $s(x)$ in the problem of optimal stopping of a Markov sequence, *Mathematische Operations Forschung und Statistik* 3(6) (1975), 493–498.

[41] Fakejev, A. G., On optimal stopping of random processes with continuous time, *Teoria Verojatn. i Primenen.* 15(2) (1970), 336–344.

[42] Gardner, M., Mathematical games, *Sci. Amer.* 202(1) (1960), 150–156; 202(3) (1960), 173–182.

[43] Gikhman, I. I., and A. V. Skorokhod, *Introduction to the Theory of Random Processes* (Moscow: Nauka, 1965).

[44] Gikhman, I. I., and A. V. Skorokhod, *Stochastic Differential Equations* (Berlin, Heidelberg, New York: Springer–Verlag, 1972).

[45] Gikhman, I. I., and A. V. Skorokhod, *The Theory of Random Processes*, Vol. I–II (Berlin, Heidelberg, New York: Springer–Verlag, 1974–1975); Vol. III (Moscow: Nauka, 1975).

[46] Gilbert, Y. P., and F. Mosteller, Recognizing the maximum of a sequence, *J. Amer Statist. Assoc.* 61(313) (1966), 35–73.

[47] Graves, L. M., *The Theory of Functions of Real Variables* (New York and London: McGraw-Hill, 1946).

[48] Grigelionis, B. I., and A. N. Shiryayev, The "truncation" criteria for the optimal stopping time in the sequential analysis, *Teoria Verojatn. i Primenen.* 10(4) (1965), 601–613.

[49] Grigelionis, B. I., and A. N. Shiryayev, On the Stefan problem and optimal stopping rules for Markov processes, *Teoria Verojatn. i Primenen.* 11(4) (1966), 612–631.

[50] Grigelionis, B. I., The optimal stopping of Markov processes. *Litovsk. Matemat. Sbornik* 7(2) (1967), 265–279.

[51] Grigelionis, B. I., Conditions for the uniqueness of the solution of Bellman's equations, *Litovsk. Matemat. Sbornik* 8(1) (1968), 47–52.

[52] Grigelionis, B. I., Sufficiency in optimal stopping problems, *Litovsk. Matemat. Sbornik* 9(3) (1969), 471–480.

[53] Gusein-Zade, S. M., The problem of choice and the optimal stopping rule for a sequence of independent tests, *Teoria Verojatn. i Primenen.* 11(3) (1966), 534–537.

[54] Haggström, G. W., Optimal stopping and experimental design, *Ann. Math. Statist.* 37(1) (1966), 7–29.

[55] Hunt G. A., *Markov Processes and Potentials* (Moscow: IL, 1962).

[56] Ito, K., On a formula concerning stochastic differentials, *Matematika, Sbornik Perevodov Inostr. Statej* 3(5) (1959), 131–141.

[57] Ito, K., Probability processes, Vol. I–II (Moscow: IL, 1960–1963).

[58] Kolmogorov, A. N., *Foundations of the Theory of Probability* (New York: Chelsea, 2nd ed. 1974).

[59] Kolmogorov, A. N., and S. V. Fomin, *Elements of the Theory of Functions and Functional Analysis* (Baltimore: Graylock, 1961).

[60] Krylov, N. V., On optimal stopping of a control circuit, *Sbornik "Optimal Control and Information Theory"* (*Abstracts of Reports at the Seventh All-Union Conf. Theory of Probability and Math. Statist., Tbilisi, 1963*) (Kiev: Izdat. Inst. Mat. Akad. Nauk Ukrain. SSR, 1963), pp. 11–15.

[61] Krylov, N. V., A free boundary problem for an elleptic equation and optimal stopping of a Markov process, *Doklady Akademii Nauk SSSR* 194(6) (1970), 1263–1265.

[62] Krylov, N. V., On control of the solution of a degenerate stochastic integral equation, *Izv. Akademii Nauk SSSR* 36(1) (1972), 248–261.

[63] Kudzhma, R., Optimal stopping of semi-stable Markov processes, *Litovsk. Matem. Sbornik* 13(3) (1973), 113–117.

[64] Kullback, S., *Information Theory and Statistics* (Moscow: Nauka, 1967).

[65] Lazrijeva, N. L., On solutions of the Wald–Bellman equations, *Litovsk. Matem. Sbornik* 19(2) (1974), 79–88.

[66] Lehmann, E., *Testing Statistical Hypotheses* (New York: John Wiley, 1959).

[67] Lindley, D. V., Dynamic programming and decision theory, *Appl. Statist.* 10 (1961), 39–51.

[68] Liptser, R. Sh., and A. N. Shiryayev, Nonlinear filtering of Markov diffusion processes, *Trudy Matem. Inst. Imeni Steklova* 104 (1968), 135–180.

[69] Liptser, R. Sh., and A. N. Shiryayev, *Statistics of Random Processes* (New York, Heidelberg, Berlin: Springer–Verlag, 1977).

[70] Loève, M., *Probability Theory* (New York, Heidelberg, Berlin: Springer–Verlag, 6th ed. 1977).

[71] Matskjavichus, V., On optimal stopping of a Markov chain with reestimation, *Litovsk. Matem. Sbornik* 11(1) (1971), 153–157.

[72] Meyer, P. A., *Probability and Potentials* (New York: Blaisdell, 1966).

[73] Meyer, P. A., *Processus de Markov* (Berlin, Heidelberg, New York: Springer–Verlag, 1967).

[74] Mikhalevich, V. S., Bayesian choice between two hypotheses for the mean value of a normal process, *Visnik Kiiv. Univ.* 1(1) (1958), 101–104. (Ukranian.)

[75] Miroshnichenko, T. P., Optimal stopping of the integral of a Wiener process, *Teoria Verojatn. i Primenen.* 20(2) (1975), 397–401.

[76] Natanson, I. P., *Theory of Functions of a Real Variable* (Moscow: Gostekhizdat, 1957).

[77] J. Neveu, *Mathematical Foundations of the Calculus of Probability* (New York: Holden–Day, 1965).

[78] Novikov, A. N., On moment inequalities for stochastic integrals, *Teoria Verojatn. i Primenen.* 16(3) (1971), 548–550.

[79] Presman, E. L., and I. M. Sonin, The problem of the best choice with a random number of objects, *Teoria Verojatn. i Primenen.* 17(4) (1972), 695–706.

[80] Ray, S. N., Bounds on the maximum sample size of a Bayes sequential procedure, *Ann. Math. Statist.* 36(3) (1965), 859–878.

[81] Rubinshtein, L. I., *The Stefan Problem* (Riga: Izdat. Zvajgzne, 1967).

[82] Shepp, L. A., A first passage problem for the Wiener process, *Ann. Math. Statist.* 38(6) (1967), 1912–1914.

[83] Shepp, L. A., Explicit solutions of some problems of optimal stopping, *Ann. Math. Statist.* 40(3) (1969), 993–1010.

[84] Shiryayev, A. N., The detection of spontaneous effects, *Doklady Akademii Nauk SSSR* 138(4) (1961), 794–801.

[85] Shiryayev, A. N., The problem of the earliest detection of a disturbance in a stationary process, *Doklady Akademii Nauk SSSR* 138(5) (1961), 1039–1042.

[86] Shiryayev, A. N., On optimal methods in earliest detection problems, *Teoria Verojatn. i Primenen.* 8(1) (1963), 26–51.

[87] Shiryayev, A. N., On the theory of decision functions and control of a process of observation based on incomplete information, *Trans. Third Prague Conference on Information Theory, Statistical Decision Functions, Random processes, Prague, 1964* (Prague: 1964), pp. 557–581.

[88] Shiryayev, A. N., On Markov sufficient statistics in non-additive Bayes problems of sequential analysis, *Teoria Verojatn. i Primenen.* 9(4) (1964), 670–686.

[89] Shiryayev, A. N., Some explicit formulas for a problem of disruption, *Teoria Verojatn. i Primenen.* 10(2) (1965), 380–385.

[90] Shiryayev, A. N., Stochastic equations of non-linear filtering of jump Markov processes, *Problemy Peredachi Informatsii* 2(3) (1966), 3–22.

[91] Shiryayev, A. N., Some new results in the theory of controlled random processes, *Trans. Fourth Prague Conference on Information Theory, Statistical Decision Functions, Random Processes, Prague, 1967* (Prague: 1967), pp. 131–203.

[92] Shiryayev, A. N., On two problems of sequential analysis, *Kibernetika* 2 (1967), 79–80.

[93] Shiryayev, A. N., Studies in the statistical sequential analysis. *Matemat. Zametki* 3(6) (1968), 739–754.

[94] Shiryayev, A. N., *Statistical Sequential Analysis* (Moscow: Nauka, 1969).

[95] Siegmund, D. O., Some problems in the theory of optimal stopping rules, *Ann. Math. Statist.* 38(6) (1967), 1627–1640.

[96] Simons, G., Lower bounds for average sample number of sequential multi-hypothesis tests, *Ann. Math. Statist.* 38(5) (1967), 1343–1364.

[97] Snell, I. L., Applications of martingale system theorems, *Trans. Amer. Math. Soc.* 73 (1953), 293–312.

[98] Stratonovich, R. L., *Conditional Markov Processes and their Application to the Theory of Optimal Control* (Moscow: Izdat. MGU, 1966).

[99] Stroock, D. W., and S. R. S. Varadhan, Diffusion processes with continuous coefficients, I, II, *Comm. Pure Appl. Math.* 12 (1969), 345–400, 479–530.

[100] Taylor, H. M., Optimal stopping in Markov processes, *Ann. Math. Statist.* 39(4) (1968), 1333–1344.

[101] Thompson, M. E., Continuous parameter optimal stopping problems, *Z. Wahrscheinlichkeitstheorie und verw. Gebiete* 19 (1971), 302–318.

[102] Tobias, T., Optimal stopping of diffusion processes and parabolic variation inequations, *Differentsialnye Uravnenia* 9(4) (1973), 702–708.

[103] Wald, A., and J. Wolfowitz, Optimum character of the sequential probability ratio test, *Ann. Math. Statist.* 19(3) (1948), 326–339.

[104] Wald, A., and J. Wolfowitz, Bayes solutions of sequential decision problems, *Ann. Math. Statist.* 21(1) (1950), 82–99.

[105] Wald, A., *Statistical Decision Functions* (New York: John Wiley, 1950).

[106] Wald, A., *Sequential Analysis* (Moscow: Fizmatgiz, 1960).

[107] Walker, L. H., Optimal stopping variables for stochastic processes with independent increments, *Ann. Probability* 2(2) (1974), 309–316.

[108] Walker, L. H., Optimal stopping variables for Brownian motion, *Ann. Probability*, 2(2) (1974), 317–320.

[109] Wetherill, G. B., *Sequential Methods in Statistics* (London: 1966).

[110] Whittle, P., Some general results in sequential design, *J. Royal Statist. Soc.*, Ser. B 27(3) (1965), 371–387.

[111] Zaks, Sh., *The Theory of Statistical Deductions* (Moscow: Mir, 1975).

[112] Zvonkin, A. K., On sequentially controlled Markov processes, *Matemat. Sbornik* 86(108) (1971), 611–621.

Index

A

B

C

D

F

Applications of Mathematics

Volume1
W. H. Fleming and R. W. Rishel
Deterministic and Stochastic Optimal Control
1975. ix, 222p. 4 illus. cloth

Volume 2
G. I. Marchuk
Methods of Numerical Mathematics
1975. xii, 316p. 10 illus. cloth

Volume 3
A. V. Balakrishnan
Applied Functional Analysis
1976. x, 309p. cloth

Volume 4
A. A. Borovkov
Stochastic Processes in Queueing Theory
1976. xi, 280p. 14 illus. cloth

Volume 5
R. S. Liptser and A. N. Shiryayev
Statistics of Random Processes I
General Theory
1977. x, 394p. cloth

Volume 6
R. S. Liptser and A. N. Shiryayev
Statistics of Random Processes II
Applications
1977. x, 339p. cloth

Volume 7
N. N. Vorob'ev
Game Theory
Lectures for Economists
and Systems Scientists
1977. xi, 178p. 60 illus. cloth